T0260491

TOWARD
A UNIFIED ECOLOGY

COMPLEXITY IN ECOLOGICAL SYSTEMS

COMPLEXITY IN ECOLOGICAL SYSTEMS SERIES

Timothy F. H. Allen and David W. Roberts, Editors

Robert V. O'Neill, Adviser

TOWARD
A UNIFIED ECOLOGY

TIMOTHY F. H. ALLEN
AND THOMAS W. HOEKSTRA

WITH ILLUSTRATIONS
BY JOYCE V. VANDEWATER

SECOND EDITION

COLUMBIA UNIVERSITY PRESS   NEW YORK

Columbia University Press
Publishers Since 1893
New York   Chichester, West Sussex

cup.columbia.edu
Copyright © 2015 Timothy F. H. Allen and Thomas W. Hoekstra
Illustrations © Joyce V. VanDeWater

All rights reserved

Library of Congress Cataloging-in-Publication Data

Allen, T. F. H.
Toward a unified ecology / Timothy F.H. Allen and Thomas W. Hoekstra. — Second edition.
pages cm. — (Complexity in ecological systems series)
Includes bibliographical references and index.
ISBN 978-0-231-16888-5 (cloth : alk. paper)—
ISBN 978-0-231-16889-2 (pbk. : alk. paper)—ISBN 978-0-231-53846-6 (ebook)
1. Ecology—Philosophy.   I. Hoekstra, T. W.   II. Title.
III. Series: Complexity in ecological systems series.

QH540.5.A55 2015
577.01—dc23
2014033202

Columbia University Press books are printed on permanent
and durable acid-free paper.
This book is printed on paper with recycled content.
Printed in the United States of America
c 10 9 8 7 6 5 4 3 2 1
p 10 9 8 7 6 5 4 3 2 1

COVER DESIGN: Noah Arlow

References to websites (URLs) were accurate at the time of writing.
Neither the author nor Columbia University Press is responsible for URLs
that may have expired or changed since the manuscript was prepared.

CONTENTS

PREFACE

THIS SECOND edition of *Toward a Unified Ecology* comes some twenty years after the first. Since the first edition, notions of complexity theory have become current. The first edition was ahead of its time, but this new edition is able to take advantage of more recent sources that were not available in the early 1990s. We have since been involved with two books that pressed ideas of complexity and hierarchy theory. Ahl and Allen presented a slim book in 1996, *Hierarchy Theory: A Vision, Vocabulary and Epistemology*. That was the condensed version of Allen and Starr's 1982 *Hierarchy: Perspectives for Ecological Complexity*. In 2003, with Joseph Tainter, we published *Supply-Side Sustainability*, one of the few books to integrate social and ecological science into a workable whole. We have taken advantage of that integration in this new edition. There was a set of principles there, and we have folded them into the management sections of the new edition.

Another important idea in our work in the last decade has been notions of high gain and low gain. Ecologists think about using resources, but focus in particular on running out of resource. By contrast, economists know we do not run out of anything; things just get more expensive. As a result, economists watch their systems adapt, while ecologists cry doom, gloom, and extinction. The idea of profit in using resource is alien to ecology, but is well embedded in notions of high and low gain. Gain is profit. High gain and low gain relate to r versus K selection in ecology, but there is more generality in gain. For instance, high gain predicts system behavior from flux driving the system. Low gain looks at constraints as predictors. The high/low distinction pits rate dependence against rate independence, and notes that the two classes of explanation cannot be readily applied at once. If you can see the river flowing, you cannot see the effect of the dam; if you can see the dam creating a lake, you cannot see the river (a wider purview would see them both, but the rate dependence of the river would be lost in the ribbon). To an extent, high versus low gain is a way of cleaving apart levels of analysis. Gain is the complexity science version of r and K selection, where K gives structure and r is the dynamics that links structure from different levels.

We were committed to narratives in the first edition, but could not be explicit then for fear of out-and-out rejection. Narrative is now more acceptable in science. Storytelling was always there, but was not recognized; it was used as a cheat in the realist world of modernist science. The power of narrative is now accepted, but the persistent overt realism in the mainstream of ecology has been unhappy that narratives are not about truth. Stories are statements of points of view. Focus on verity and truth is associated with entrenched modernism, which insists that science approaches truth. It may or may not, we can never know. There are subtleties here, such as true and false versus not true and not false.

Reflexive realism will be frustrated at our position because we are denying some of its premises. But we ask for tolerance; the reason is that we only want to rein in realist excesses, not deny realism. In the end, we are realists too, but ours is a realism that is so measured as not to be an easy target in more sophisticated logical discourse. Our realism does not invite methods of investigation that get muddled. We say to realists, we are on your side. But we do want to modify your claims about reality so they are as fully valid as they can be in the final analysis. The device of science from its inception is to investigate what appears before us, and indeed what appears to be true. But scientific investigation always aims at showing our first impressions to be untrue. Investigation never proves anything; it just shows that such and such cannot be true. Science moves by disproving things, not by proving anything. And science shows appearance at first glance to be wrong and illogical time and time again, such that it would be unwise to expect anything else. Asserting truth would seem reckless, at least until the investigative process is seen as concluded.

To press the point home, we note that science is even detecting at a metalevel that important reliable findings are incorrect across a range of fields. Jonathan Schooler has reported the decline effect.[1] That work shows significance and reliability to decline linearly as work is repeated by other scientists. Perhaps feeling that parapsychology had it coming, many will be delighted that the effect was first noted in the 1930s, where statistically reliable results in parapsychology were shown to decline in attempts to validate early work. But not so fast; the decline effect has been reported in a number of fields, tellingly in the effectiveness of drugs for mental health. Drugs such as aripiprazole (e.g., Abilify) were at first shown to be effective with robust statistical significance, but attempts to reproduce those effects decline some 30 percent in significance each time. Decline effect has been shown in ecology with regard to symmetry of tail feathers and mating success in birds.[2] Reliability of results declines faster in popular fields, where repetition is most likely to occur.[3] If well-performed experiments, accepted in peer review, achieving the highest professional standards cannot be validated and reproduced, it would seem that a certain caution should be applied to truth and reality with regard to science.

Science is designed to be a naysayer. So assertions at the outset that such and such is true enough to be used as a benchmark are unwise. Such insistence inserts a point of reference that exactly blunts science as a device. Science is as good a device as we have, so keeping it honed is important. The danger with hasty realism in science is that it interferes with doing scientific measurement and testing. In the end, the authors here are hard bitten, driven by data and experience; there is nothing airy-fairy about our posture, despite our proclivity for abstraction. We were both empiricists in our training and early careers. All we are saying is that using truth as a reference at the base of measurement and provisional notions is unwise. We can never know ultimate truth, so it is not a reliable reference; science itself shows us that fact time and again. As scientists take positions, we insist that they take proper responsibility for those decisions; otherwise, they will fool themselves. But we must hasten to add that once the decision making and testing is over, and we agree the result is good, then saying, "We are now closer to reality" does no harm and has intuitive appeal. So we are not bottom-line antirealist. Reality as a conceptual device has its uses. The feeling of reality in stories is one of the appeals of narrative, even if stories are neither true nor false. We do not need to be tethered to reality as a device, but that does not stop it from being a powerful, workable context.

The organization in this new edition follows that of the first edition. Indeed, our philosophy has not changed. The introduction and chapter 1 lay out the general premise that cleaves scale from type in investigating complexity. Clarity on scale and type allows organized movement around a labyrinth that is complicated by being tiered. The tools we introduce are ways of dealing with complexity in remarkably practical ways.

The middle section of this edition works its way through types of ecology, investigating scaling issues within each type. We use the same types in the same order as in the first edition. It worked then, and we are happy to use the same framework again. Reference to new papers and recent research amplifies what we said two decades ago. So there is old and new here.

New to this edition is an explicit chapter on the use of narratives and models in ecology in particular, but also in science in general. That precision in philosophy of science informs the last two chapters, which, as in the first edition, are about management and basic science ecology. We feel the whole is better integrated in this edition.

Allen did use the first edition to teach his general ecology course. He made it work well. The book has the advantage of covering the whole discipline with an intellectual challenge. We invite others to do the same. But this is not a normal textbook. Maybe it is time to rethink how textbooks might work. There is more thinking here and fewer facts. It is sad that students like facts because facts appear concrete for memorization for the test. But one does not have to use this volume as a textbook for teaching ecology; a large part of our audience is likely to be seminars in ecology. As teachers ourselves, we are aware of something like fifteen working weeks in a semester. Here we have eleven units of text, so the number is not quite right for one chapter a week. There are several ways to extend the readings to get the fit with the weeks in a semester course. First, chapter 1 introduces material that is likely to be very new and distinctive. It might make sense to give chapter 1 two weeks of discussion: the first week for the conceptual setup, and the second week to do justice to our solution, the cone diagram. We are also aware that some of the chapters on types of ecology are large. Landscapes are given a long treatment. Ecosystems and communities are behemoths. Each could be treated first as theory and second with examples. In fact, even used as just one chapter for a whole discussion, it makes sense to emphasize the theory/example tension in the big chapters.

But everyone in ecology needs this work. It anchors us in the world when modernist mechanism is breaking down. Our book is not so much a security blanket while the old intellectual framework fidgets itself apart as it is something to hold on to as ecology melts into economics and the humanities come to stay as collaborators.

Timothy F. H. Allen
Thomas W. Hoekstra

ACKNOWLEDGMENTS

THE CRITICAL role of Clyde Fasick in supporting our work was acknowledged in the first edition dedication and described in the acknowledgments of that edition, all of which are still well deserved and appropriate for the publication of the second edition. This edition relies heavily on the first, and so acknowledgments there still apply.

There were many groups who played critical roles in developing Allen's ideas in the new edition. There are so many people that they are ordered alphabetically in their respective groups. All are significant. At the outset of the focused effort to create the second edition, Allen's graduate students met weekly for a semester to discuss what should be done: Peter Allen, Marc Brakken, Julie Collins, Keith Doyon, Nissa Enos, Brian Garcia, Cassandra Garcia, Megan Pease, Steve Thomforde, and Devin Wixon. Peter Allen there resigned to the fact, saying, "Tim will write what Tim will write." That does tend to be Allen's method, but the influence of the graduate students is not to be underestimated. Another group that has been crucial in Allen's development met every Tuesday during the semester, year after year until 2012. It was called "Sandbox," and was a testing ground for many ideas. David Hall was the longest inhabitant, and attended and contributed handsomely for almost two decades. Preston Austin and Jerry Federspiel came from off campus to make wonderful Sandbox contributions. Undergraduates figured large in Sandbox, notably but not exclusively: Mike Chang, Ed Engler, David Evans, Amelia Krug, Noel Lawrence, and Anna Marie Vascan. Some Sandbox undergraduates have been wonderful coauthors with Allen on some big ideas on high and low gain and narrative: Elizabeth Blenner, John Flynn, Michael Flynn, Amy Malek, Kristina Nielsen, and Rachel Steller. Kirsten Kesseboehmer and Amanda Zellmer, as undergraduates, made huge contributions to the notions of modeling and narrative and, as authors, on some of the most important papers. Another crucial group met after Sandbox, Allen's "My engineers": Gregori Kanatzidis, Nathan Miller, Samantha Paulsen, and Edmond Ramly. They all played a role in developing the ideas for chapter 8 here. Julie Collins, Kirsten Kesseboehmer, and Noel Lawrence have backgrounds in the humanities, which they presented with unusual confidence to the scientists. Bruce Milne and his graduate student Mike Chang were generous with their new ideas on foodsheds.

Various senior scientists have been helpful as colleagues for Allen to discuss aspects of the work between editions: Thomas Brandner, Martin Burd, Steve Carpenter, Charles Curtin, Billy Dawson, Roydon Fraser, Robert Gardner, Mario Giampietro, Philip Grime, Alan Johnson, James Kay, Ronald McCormick, John Norman, Robert O'Neill, David Roberts, Edward Rykiel, John Sharpless, Duncan Shaw, Hank Shugart, Joseph Tainter, and David Waltner-Toews. Henry Horn and Tony Ives provided the anecdotes of their childhood memories. Judith Rosen was most helpful in giving the inside story on her father, Robert Rosen, and for chasing down several of his crucial quotations.

Hoekstra was fortunate to have the support of the U.S. Forest Service (USFS) in integrating his work assignments there with work on and application of *Toward a Unified Ecology*. That support allowed Hoekstra and his colleagues to test the application of concepts in the first edition. The applied research studies were carried out within the mandate for resource management by the agency and with the narrative modeling scheme described in chapter 9 of the first edition, "A Unified Approach to Basic Research." They were able to develop and apply the concepts in *Toward a Unified Ecology* from a national forest to the international scale. The results of these applied efforts support and confirm our confidence in the value of these concepts, which are expanded in this edition.

The geographic and ecological scope of these applied research studies required the involvement of many people. Unfortunately, it is not possible to individually recognize everyone involved, but only to mention the key participants. In addition, it should be mentioned that many of these studies on *Toward a Unified Ecology* concepts were also created within the context of *Supply-Side Sustainability* concepts found in our companion book with our colleague Joseph Tainter.

The most significant of *Toward a Unified Ecology* tests of concept are briefly described here. First, a 1994 workshop in Israel and its publication involved several hundred scientists and managers from more than thirty nations focused on the topic of Hoekstra and Moshe Shachak's book *Arid Land Management: Toward Ecological Sustainability*. Steve Archer, Linda Joyce, Joe Landsberg, David Saltz, No'am Seligman, and Moshe Shachak led the technical sessions of the workshop. Second, a North American applied research study in 1999 used the concepts in

Toward a Unified Ecology for developing ecological protocols to monitor forest management unit scale sustainability, the Local Unit Criteria and Indicators Development (LUCID) test. Gregory Alward, Brent Tegler, Matt Turner, and Pamela Wright worked with Hoekstra on the test that involved more than one hundred individuals in six USFS national forests and universities. In addition, there was one site in both Canada and Mexico. An applied research study using a hierarchy of ecological, administrative, and political units in 2005–2006 integrated the framework of *Toward a Unified Ecology* with social and administrative requirements. The objective was to develop local scale inventory and monitoring criteria and indicators necessary to work within the regional and national information and reporting requirements of the Forest and Rangeland Renewable Resources Planning Act. Joseph Tainter and Cedric Tyler, working with Hoekstra, led various phases of the study. There were several other applications implemented on a smaller scale.

We want to especially acknowledge the following individuals for making a special effort to provide photographs and illustrations that are used and duly credited in the book: Martin Burd, Mike Clayton, Kelly Elder, Robert Gardner, Mario Giampietro, Philip Grime, William Hereford, Alan Johnson, Ronald McCormick, Bruce Milne, Itshack Moshe, John Norman, Richard T. Reynolds, Charles Rhoades, and Monica Turner. We also deeply appreciate Matt Turner's editorial suggestions on national forest policies and protocols.

Hoekstra is doubly indebted to his wife, Joyce VanDeWater—first, for having her personal life seriously impacted by his unpredictable and challenging priorities associated with work on this book; and second, as the illustrator and partner, patiently developing and revising high-quality illustrations with us while we produced multiple versions of the book and equally as many versions of the illustrations. It is a small token of our appreciation to have Joyce identified on the title page as our partner in what is published here.

Allen remains eternally grateful to his wife, Valerie Ahl, for her huge intellectual contributions, pulled together in their 1996 book and with ongoing influence right up to the second edition. He greatly appreciates her patience and support as the work dragged on and cluttered their lives since his retirement in 2010.

TOWARD
A UNIFIED ECOLOGY

INTRODUCTION

I T MIGHT seem an odd place to start, but the Beaufort scale is a combination of science and poetry. This is apt for a book on our subject because ecology is also about capturing nature and overcoming the difficulty of unambiguously communicating an overwhelming amount of information. Scott Huler wrote a splendid book, *Defining the Wind*, reporting the life of Rear Admiral Beaufort.[1] In 1806, Sir Francis Beaufort (1774–1857) was a captain in the British Admiralty. For his own use, he wrote down a set of systematized observations for wind speed.[2] One captain might write in the log that the wind was a moderate breeze, while another might call it a fresh breeze. Beaufort's words were accounts of the sails that were up on the ship: At force 0, the ship would be in full sail, but the sailors would be unable to steer because there was no flow of water past the rudder. To steer a big ship would take a force 2 wind. At force 6, a strong breeze, half the sails would be taken down in an orderly manner, with some sails only taken in halfway as opposed to stowed. Any captain would be aware of the exact conditions. At force 6, it would be "the wind to which a well-conditioned man-of-war could just carry in chase, full and by single-reefed topsails and top-gale sail." Force 7 is a moderate gale, "to which a well-conditioned man-of-war could just carry in chase, full and by double reefed topsails, jib, &c." At force 12, all sails would be down. There have been studies on past weather conditions from times long ago, based on ships logs. With the coming of steam power, sails were absent and so the description became the condition of the sea, for instance, foam streaking on the waves.

Beaufort did not pen the words that are so poetic for the wind on land; that was done in a most unlikely way. In 1906, a committee of engineers, of all things, created a lovely set of descriptions! A recitation of the Beaufort scale sounds like descriptions of a collection of Turner paintings, from idyllic to dramatic (figures 1A and 1B). Huler (2004) notes that at force 5, "Small trees in leaf begin to sway." It is in iambic form. The second half goes on in trochaic pentameter, "Wavelets form on inland waters." Table 1 provides a complete listing in the poetic wording of the 1906 British engineers; with two parts to most entries, it rings of a set of Japanese haiku.[3]

Beaufort and the 1906 engineers used their senses to capture a richly textured fabric of experience and standardize it so that it could be reliably communicated. Scott Huler appeared on a National Public Radio (NPR) interview, where he said he bought a small anemometer so he could calibrate his experience of the wind.[4] On a visit to a convenience store, there were some interesting wind conditions and he was frustrated when he realized that he did not have his machine. But then Huler made exactly the point we are making. He remembered the reason for his whole study: to observe conditions so that he could calibrate them by making comparisons. So he made those observations, one of which was that it was difficult to open the car door. Not as poetic as the engineers of 1906, but fully functional.

FIGURE 1A.   Turner Painting, "Calm, from the Liber Studiorum," 1812. Joseph Mallord William Turner, 1775–1851, Collection Tate, 178 × 267 mm.

FIGURE 1*B*.   Turner Painting, "The Shipwreck," 1802. Joseph Mallord William Turner, 1775–1851, Bridgeman Art Library, Fitzwilliam Museum, University of Cambridge, United Kingdom, 1,705 × 2,416 mm.

TABLE 1   The Beaufort Scale

0. Smoke rises vertically.
1. Direction of wind shown by smoke, but not by wind vanes.
2. Wind felt on face; leaves rustle; ordinary vane moved by wind.
3. Leaves and small twigs in constant motion; wind extends light flag.
4. Raises dust and loose paper; small branches are moved.
5. Small trees in leaf begin to sway; wavelets form on inland waters.
6. Large branches in motion; whistling heard in telegraph wires; umbrellas used with difficulty.
7. Whole trees in motion; inconvenience felt when walking against the wind; umbrellas discarded in exposed places.
8. Breaks twigs off trees; generally impedes progress.
9. Slight structural damage occurs (chimney pots and slates removed).
10. Seldom experienced inland; trees uprooted; considerable structural damage occurs.
11. Very rarely experienced; accompanied by widespread damage.
12. Countryside is devastated.

Huler's interview was a call-in show, and one of the callers reported that he had standardized his experience of the wind at Candlestick Park, the baseball stadium by San Francisco Bay, a windy place. The caller's observation that clinches his level 10 is a wind so strong it blows the foam off his beer as he watches the game—a spirited scale indeed! There has been an official update to Beaufort beyond the 1906 poetry. As of the late twentieth century, bureaucrats without much soul now reference plastic trash cans blown over, and other godless things.

The scale is used in places scattered all over the world. Canada uses it, but not the United States. Allen lived in a trailer during graduate school in North Wales. It was a rickety old thing, made of painted hardboard, not metal. He had it up on a hill with nothing substantial between him and Brazil. It was next to a hedge on the leeward side, and on the windward side it had a milk churn full of rocks with a rope over the roof that was tied to the leeward axle. The British Broadcasting Corporation (BBC) radio closed its broadcast at night with a report of sea conditions around the coast. Allen fell asleep at night listening. The places seemed romantic: "Faroe Islands: westerly, gale force 8. Scilly Isles: southwesterly, strong breeze, force 7." Allen knew he was in for a rough night if he heard "Irish Sea: westerly, storm force 10." That was as bad as it got, and the milk churn did its job.

Ecology is one of a handful of disciplines whose material study is part of everyday encounters: birds, bees, trees, and rivers. But it is a mistake to imagine that this familiarity makes ecology an easy pursuit. In fact, that is the main issue in this introduction: ease of observation makes the study harder. Being outdoors, simply enjoying nature is one thing, but a formal study is a very different matter. From the outset, we state that the very familiarity of ecological objects presents difficulties, some of the same difficulties Beaufort addressed. The familiar slips through human portals with such ease that memory is quickly overwhelmed by the sheer quantity.

To deal with the excess of human experience, ecologists must have a plan, a method of choosing significance. In any scientific study, there are the obvious steps that compress the rich experience down to a set of formal models.[5] Formal models can be equations or graphs, and may be word models. Scientists are familiar with that sort of compression. Less obvious is a prior compression down to the general area of discourse in which the model sits. This compression decides what the science is going to be about. The prior compression is less obvious than erecting the model because it is quickly forgotten as context to the second compression to the model. The first compression is taken for granted, so while it is still there, it disappears.

The first compression is to a paradigm. Thomas Kuhn identified a paradigm as a shared vocabulary, a shared methodology, and a shared view of what matters.[6] The last thing a scientific fish would discover is water, for much the same reason that devotees of a paradigm are not conscious that they have a paradigm, let alone know what theirs is. The interaction between models and paradigms is a continuing theme throughout this book. Paradigms name the things to be studied. Scientists then go on to build the narrative into which the named things fit. Paradigms are narratives; one has the story framed out so the next compression down to the model is directed at part of the narrative paradigm, filling in the gaps. Models are used to improve the quality of the narrative and press the paradigm forward. A less positive description is that a paradigm is a tacit agreement not to ask certain questions. Defense of old paradigms can be a nasty business, full of jealousy and deceit. Ecology has its fair share of paradigm fights, and we do not shrink from including the politics of them in this volume.

In natural history, there is less attention to compression, so natural historical accounts are often an accumulation of natural historical experience. A natural historical account is very narrative, but the process of improving the tale with the second compression down to models is largely neglected. In ecology, the scientist is supposed to do more than that. Failure in that responsibility since our last edition might call a lot of what passes for ecology as "natural history with numbers." We insist on theory over quantification.

Experiencing something is an act separate from deciding what that something might be called and how it might be measured. Beaufort gave nature names like "fresh breeze." The name comes from the group of equivalent wind conditions to which the experience is assigned. One hopes that when you have seen one fresh breeze, you have seen them all. The name comes from the class to which the entity is said to belong. Thus, a tree could be an organism, a plant, or a member of a particular species like maple. By being conscious of the steps toward model building, we can keep tabs on what we are doing. Beaufort makes a nice parallel exposition.

Measurement comes quite late in compressing experience into things to be investigated. You cannot measure something until a boundary for it has been decided. Before measurement, you have to have a thing to measure. That thing for Beaufort is the complex of observables about the wind. Note here that we emphasize it is *decided*; that way, we can take responsibility for our decisions. There is no excuse for abdicating our decisions to nature, as if nature could make decisions for ecological observers. No! Definitions and names do not come from nature; they come from us, just like the Beaufort scale came from Beaufort. Things may exist in the world, but they do not exist in the world as things. We create "fresh breezes," although nature makes the wind. The thing behind the names comes from human decisions that carve out a piece of the continuous experience stream so as to freeze some of nature into a thing. All sciences do this, but in ecology, the familiarity with the nature that we see might lull us into thinking that we are looking at the true nature of nature. No, we only have experience, which only later is given significance and formalized into observation.

Ecology, as with all science, is a matter of tunneling forward with decisions. Ecologists always could have tunneled their way to somewhere else, so investigators will constantly need a good record of where they have been. Formal accounts force ecologists to remember which are decisions, as opposed to which are aspects of nature. Ecologists must be aware that over time, as they find out more, the situation may change, as when Beaufort's excellent descriptions of sails became moot when sails disappeared with steamships.

Constructivist ideas about learning assert that there is no blank slate on which the world writes to human experience.[7] Remembering prior experience opens us up to recognizing certain things. Human experience is a product of prior experience. The past experience makes the present. We all see what we expect to see. With Beaufort's experience and ideas impressed on them, the 1906 engineers assisting the British Meteorological Office could easily erect new criteria for inland conditions. Beaufort opened their eyes to a fresh breeze force 5: "Small trees in leaf begin to sway; wavelets form on inland waters." Notice in constructivism, it is not the physical fresh breeze that is constructed; it is the construction of the observers' understanding and capacity for recognition (figure 2).

Constructivism

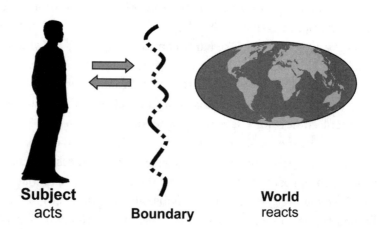

Subject
acts

Boundary

World
reacts

Interaction=Source of order

Development=Progressive change
in subject as function of
interaction of the world

FIGURE 2.   Subjects cannot experience the world itself, only sensory inputs through the filter of their senses, or through some contrivance like a pH meter. To deal with the world, the observer acts, and in some mysterious way gets an apparent reaction from the presumed world. Change occurs (sometimes imagined as progress) as the observer's understanding is constructed by interaction.

Deciding on what an ecological thing is assigns it to an equivalence class, a set of things that have something critical in common, like wavelets forming on inland waters. Classes are the organizing principle that helps human observers deal with the tidal wave of different things flooding their senses. In a step further, classes can themselves be organized. We do this intuitively much of the time, but intuition may not be an adequate organizer. At the outset, the class is simply a set of things that have something in common. But the classes that are recognized are often related one to another, whereupon the classes may become levels.

The body of theory here is hierarchy theory, and it has been particularly useful in ecology. As Eric Knox told taxonomists, since they use levels all the time (species, genera, families, etc.), it might be smart to have a theory of levels and hierarchies (figure 3).[8] Level of organization is only one sort of level, which is why we need a body of theory to keep things straight. Other sorts of levels may be based on size, that is, scalar levels. But other levels may be control levels, such as the governor on a steam engine belonging to a higher level than the machine it controls. Scale-based levels reflect the size of the things assigned to the level. Notice that in contrast, levels defined by control are not size ordered; the Watt governor is smaller than the steam engine it controls. Whatever the steam engine does, the governor

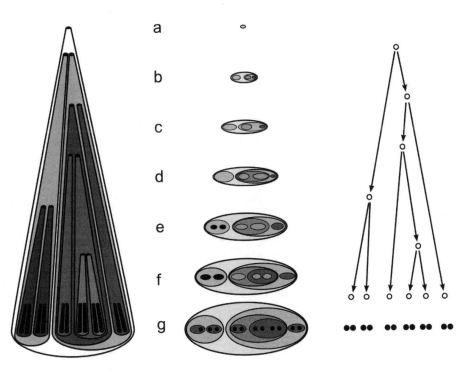

FIGURE 3. Knox (1998) shows different sorts of hierarchies that speak for themselves, but only after Knox pointed them out. The systematists he addressed were mired in what are the real hierarchies, not understanding that each hierarchy is a point of view.

has an answer; it fills the control space above the engine. Classes of larger, more inclusive things are seen as being at a higher scalar level. Scalar levels invite, but do not insist on, nesting of levels.

We have scaling tools that let us observe at different scales. To see big things at a distance, a telescope might be useful. To see small things, one might employ a microscope. Small things that move fast, like birds, might require binoculars to establish a context for the observation. Embedded in the use of a given tool are two scaling considerations: grain and extent or scope. Grain is the smallest distinction we make in a set of observations.[9] Extent is the size of the widest thing that can fit into a set of measurements. Ecology has been particularly helped by new technology that allows wide extent while retaining fine grain. The study of landscape ecology of a modern sort was not possible until remote sensing, sometimes from hundreds of miles up in the sky. There is a security issue with oceanic satellite images with pixels smaller than 30 meters on a side because at 10 meters resolution, you can distinguish water from water with a submarine in it. The extent is the scope of all the measurements that were made or would have been pertinent had they been made. You cannot see something in an image if its signal is wider than the scope of the data.

The scaling tools are part of deciding levels of observation.[10] These are in contrast to levels of organization, which are explicitly not scalar. The latter are defined not by their size but by organizational characteristics of the things at those levels. Levels of organization are the familiar levels that define the different sorts of ecological subdisciplines like organismal ecology. Scalar levels and levels of organization are likely to be at odds.

And all of this is going on with an act as simple as seeing and recognizing a bee. It may be smaller and in the context of the flower it pollinates. But with a different sort of level, it can be seen as an organism at the organism level that may be in a population. Population is also a level of organization, but a level that is not necessarily bigger than the organism level. The reason for that apparent contradiction is that population is not just a collection of organisms; it is a collection of organisms that are required to be in some way equivalent. You, the reader, as an organism, are not in the population of mites that feed on flakes that come off your skin. They are called dust mites and are everywhere in most houses, and are responsible for some people having allergies to dust. Thus, you as an organism are physically bigger than the whole population of mites. Population is not necessarily higher than organism, unless the organisms are equivalent. The two organisms in figure 7 (a flea and an elephant) are so different that they cannot be aggregated in a population no matter how proximate they are. Populations have a distinctive relationship to energetics. Populations do not so much get bigger through birth.[11] In a population births only present vessels that may or may not be filled, depending on the resources available.

So it is important to distinguish level of observation from level of organization because the same thing can be in two entirely different sorts of level without any contradiction. Scientists have to know what they are talking about and how they are talking about it. It is only through the formalism described earlier that casual natural history can become the science of ecology. It was only after Beaufort that impressions of the wind at sea became formalized. So what seems like the proverbial walk in the woods can become scientific once formality is introduced. In this way, scientists can handle the cacophony of ecological players. There is almost always a rich and textured set of things and happenings, even in a simple ecological study.

The joy of models is that they are internally consistent. The burden of models is that they have to be consistent. Rosen notices that formal models are scale independent.[12] They use scaling equations as laws to identify relationships that apply across scales. Beaufort addressed an expression of the laws of aerodynamics. Notice that the same laws of fluid motion apply to large and small flying objects linked through an exponential equation. An example would be the equation for drag. Drag depends on the fluid medium through which the object passes. It also depends on the size, shape, and speed of the object. One way to express these relationships is by means of the equations for drag. There is no need to spell them out here, but drag relates to: (1) density of the fluid, (2) speed relative to the fluid, (3) cross-sectional area, and (4) drag coefficient, a number without units.

Intuitively, we know that as something moves faster through air, the air applies increased resistance. On a graph with air speed on the horizontal axis (the abscissa) and the drag plotted on the other axis (the ordinate), drag increases on an ever-steepening curve (figure 4A). But some objects, like flags, deform to catch the wind more, so drag increases faster than expectations for a rigid form. Other objects, like trees, deform so as to streamline. At first, leaves flutter to catch the wind like a flag, but then the tree bends and so streamlines to avoid the worst of the drag at times when the tree is in danger of toppling (figure 4B). Drag still goes up, but not as much as it would were the tree rigid.[13]

The power of the exponent in the drag equation is significant, which is why a gale of 40 miles per hour buffets you, but a hurricane of 70 miles per hour throws you about, and may even turn over vehicles. Here is Beaufort again. First, there is a cottage with smoke going straight up. Then leaves rustle and branches bend. In no time, force 8 for a gale is tearing off branches. By hurricane force 12, trees are uprooted and buildings are damaged. And all of this is because of the exponent on the drag equation. Double the speed of the wind and you get much more than simply twice the wind damage. The laws of aerodynamics are captured in a whole set of equations that give the rules for relating things of different size in different wind. The laws of aerodynamics are therefore scale independent. A paper airplane and a DC-10 both yield to those laws (figure 5).

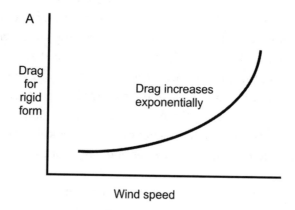

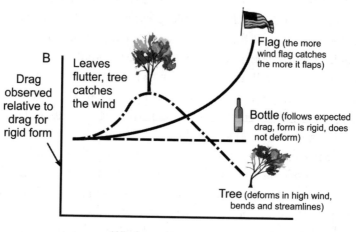

FIGURE 4.   *A.* The exponential increase on wind speed against drag for a given rigid form. Graph (*B*) shows a dimensionless number, which is drag as expected on *A* for a rigid form divided by observed drag on the structure. Drag (observed) divided by drag (expected) cancels out the units to give a dimensionless number. Rigid structures like bottles do not deform and so measured drag over expected drag is a unit value. But flags flutter and keep catching the wind more as they flap harder. Trees have fluttering leaves and so in a light breeze they catch the wind more, like a flag. But then in high winds, trees bend away from the wind and so streamline, giving drags ever lower than expected, even as raw drag increases.

You do not want size to be specified explicitly in a model because then other sizes will not fit the model, even if the same principles prevail. Small and big things both fly, so we need a model that works for both big and small flying things. Ecology often makes the mistake of including size in models. We do not wish to anticipate the biome chapter in advance, but suffice it to say that the notion of biome, which is a model, is conventionally seen to apply only to big things, like the prairie biome that covers the Great Plains. In fact, the biome way of looking at

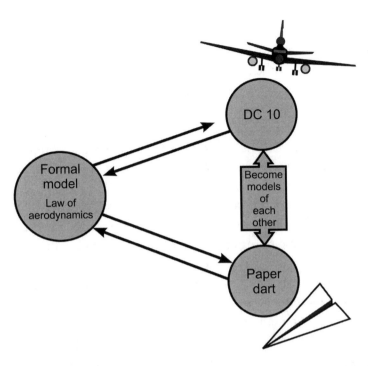

FIGURE 5.   Rosen's (1991) modeling relation showing how a formal model of the equations of aerodynamics can be decoded into different flying objects. In turn, the material systems may be encoded back into the formal model. When that can be done, the two material systems become analog models of each other by compressing them down to only what the two have in common. The model for the child's toy is the DC-10 itself. The toy is also a model for the DC-10, so the analogy goes both ways. The analogy is not coded into words or equations; it is simply taken as a material equivalence. But the formal model *is* coded in symbols.

things also applies profitably to small things, like frost pockets. Frost pockets are an acre or so, but they are recognized in the same way we recognize big biomes. Conventional definitions of biomes would unnecessarily exclude frost pockets, even though the same principles apply.

The thing itself is not coded like the formal model, it just is. Translating material observables into the model is called encoding; you check to see how a material thing fits the coded specifications in the model. Yes, paper darts do greatly increase their drag if you throw them hard. Now it gets exciting! If you can encode and decode two material observables with a single formal model, something wonderful happens. The two material systems become equivalent so that you can experiment on one of them and use the results to predict behavior of the other. The principle is analogy, and all experimentation depends on that relationship. The formal model is a metaphor for the material system; it is a description, a representation. The relationship between the two material systems in an analogy is a compression down to only what the two things have in common. You do not refer to paper in the analogy between a paper airplane and the DC-10 because the DC-10 is not made of paper. Following the characterization of common ecological criteria in chapters 1 to 7, we devote chapter 8 to an in-depth explanation of the narrative and model relationship as a basis for a metatheory of ecology and the unification of ecology for use in management and research.

In ecology, there is a need for a framework that the scientist can use to organize experience; that is the challenge of ecology. We have just gone through the narrative description and basis for this book. Now, let us stand back to see the model and its rules that we propose for a unification of ecology.

This book erects that framework. Given the central role of scale in using formal models, the framework turns on scale. The concept is rich, requiring this whole book for a complete accounting of scale in ecology. However, at this early stage, we need to introduce briefly what we mean by scale, so that the word can pass from jargon to working vocabulary. Scale pertains to size in both time and space; size is a matter of measurement, so scale does not exist independent of the scientists' measuring scheme. Something is large-scale if perceiving it requires observations over relatively long periods of time, across large parcels of space, or both. With all else equal, the more heterogeneous something is, the larger its scale. An example here might be in comparing vegetated tracts of the same size. The vegetation that has more types of plants in more varied microhabitats more evenly represented is larger scale. It is more inclusive. Not only do things that behave slowly generally occupy larger spaces but they are also more inclusive and therefore heterogeneous. There are nuances of size all over the place.

In several topic areas of the ecological literature there is confusion because of opposite meanings between vernacular and technical terms. Evenly spread means low variation between locales. Ecologists use quadrats, square areas that are laid down for purposes of sampling vegetation. We often count plants in quadrats. Evenly dispersed vegetation leaves no sample area without hits because all the space is evenly covered. There will be no high values with lots of hits because a hit on one plant comes with a local surrounding space with no individuals in it; there are no clumps with which to score big. A frequency distribution is a statistical device where we rank order the values in samples. We note how frequently we encounter a given score. So with quadrat samples of evenly spaced vegetation where plants are counted in sample areas, there are no values of zero and no high values either. The variance of a frequency distribution is a measure of statistical dispersion that is a general measure of how wide the span of the numbers is in the samples. So, evenly spaced plants show low variance in the sample. That is whence the "underdispersion" comes from between samples in vegetation that is evenly spread out. With clumped plants, the sampling quadrats tend to hit clumps or miss them. So there are lots of zero samples and many full samples too. The

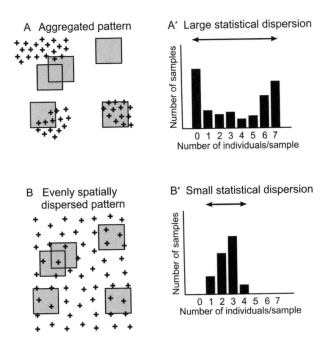

FIGURE 6.   Two patterns of spatial distribution and their corresponding frequency distributions. A. Sampling clumped individuals with a quadrat gives mostly either empty or full samples,
but with few in between. The variability of such a collection of samples is large, giving an overdispersed frequency distribution. B. With populations that are evenly distributed, all samples
contain about the same number of individuals with little variation; the frequency distribution is
accordingly underdispersed. Quadrats A and B are the patterns on the ground, while A′ and B′
are the respective statistical distributions.

frequency distribution shows a wide span of frequencies of zero to many, and so
clumped vegetation is counterintuitively called overdispersed. The difference is
that vernacular meanings refer directly to the thing or behavior, whereas the technical meaning refers to how one views the situation from some standard device or
unit (figure 6).

So it is with "scale." Something that is big, we call large-scale because it is large
in and of itself. That is the way we couple the words *large* and *scale* throughout this
book. Accordingly, we say that small things are small-scale. However, cartogra-
phers reading this book, and anyone else using their terminology, will be tearing
their hair out. Our choice is either to ignore the sensitivities of a group of specialists or use *scale* in their counterintuitive way. We choose the former, but for clarity
need to present the geographer's point of view. A small-scale map in geography
indicates that a unit measure like a mile will be very small on the map, so that
a large area is represented. Conversely, a large-scale map shows small things on
the ground with clarity, because a large-scale map makes them large. Therefore, a
large-scale map must be of a relatively small area. A map of the entire globe would

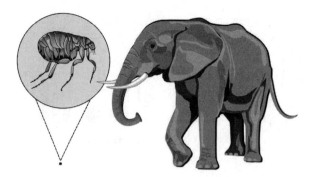

FIGURE 7.   Both fleas and elephants are organisms, but their different sizes demand observation from very different distances. Note that the organismal form is very different with change of size, even though there is a head, a body, and legs in both cases. Those differences relate to scale.

be on the order of 1:50,000,000, a geographer's small-scale map of what we in this book call a large-scale structure. The technical meaning to a cartographer refers to the smallness of the one in relation to the fifty million. The vernacular meaning, the one we use throughout this book, indicates that the fifty million is a big number and the whole world is a big place; it is large-scale.

Ecology includes material and processes ranging from the physiology and genetics of small organisms to carbon balance in the entire biosphere (figure 7). At all scales, there are many ways to study the material systems of ecology. Let us emphasize that the physical size of the system in time and space does not prescribe the pertinent conceptual devices associated with different ways of studying ecology, although it may indicate the pertinent tools for observing, like remote sensing for big landscapes. Each set of devices or points of view embodies a different set of relationships. One ecologist might choose to emphasize physiological considerations while another might look at relationships that make an organism part of a population. But a physiological point of view is not necessarily small-scale. Elephant physiology includes more matter than does a whole population of nematode worms, and it is much bigger than the entire community or ecosystem in a small tidal pool or pothole (figure 8). The physiological differences in photosynthetic mechanisms between grasses define entire biomes in the dry western United States, so physiologists can think as big as almost any sort of ecologist. Brian Chabot and Harold Mooney have published an entire book on the physiological ecology of communities and biomes.[14]

The levels of organization refer to types of ecology. The types of ecological system are often ranked in textbooks: biosphere, biome, landscape, ecosystem, community, population, organism, and cell. We do not find that ranking useful, and call it the conventional biological or ecological hierarchy; each level therein we call

FIGURE 8. Pockets of water much smaller than some large organisms represent fully functional, self-contained ecosystems. *A.* The leaves of pitcher plants that trap insects. *B.* The ponds in the middle of epiphytes (photo courtesy of C. Lipke). *C* and *D.* Pothole ecosystems, from inches to meters across, made by boulders trapped in eddies in glacial outwash of the St. Croix River, Minnesota (photos courtesy of T. Allen).

a conventional level of organization. When seeking mechanisms, it is certainly a mistake to assume that explanatory subsystems must come from lower down the conventional ranking of levels of organization (figure 9).

In the literature of biological levels of organization, there are some branched variants of the conventional scheme, but the simple hierarchy captures the prevailing paradigm for grand, unifying designs for biology. Although many ecologists view themselves as working at a level of organization in the grand hierarchy, the levels defining the different types of ecologists do not strictly depend on the scale used by the respective scientists. The ordered relationships between the conventional levels of biological organization in fact offer relatively few explanations for the configurations that we seem to find in nature.

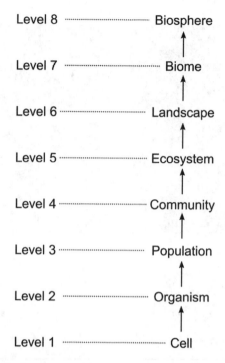

FIGURE 9.   The conventional hierarchy of levels of organization from cell to biosphere.

Hoekstra personally experienced the logical inconsistency within the hierarchy of conventional levels of organization. The experience occurred on a warm, sunny fall afternoon in a beautiful remnant old-growth hardwood forest community, a factor that may well have contributed to the situation being still imprinted in his mind. It was a class field trip to collect plant community data when Hoekstra was an undergraduate student in Alton A. Lindsay's plant ecology class at Purdue University in the early 1960s. The conventional hierarchy puts ecosystems higher than communities because ecosystems are more inclusive and fold in the physical environment. Hoekstra was unable to get resolution to the inconsistency of how an ecosystem could be defined by a log on the forest floor that was at a smaller size than the forest community within which it existed. The conundrum was finally resolved through his work with Allen in the 1980s, and the application of hierarchy theory to provide the effective framework needed to differentiate between scale-defined levels of ecological criteria and system type. Their work on levels of organization distinct from levels of observation in the context of hierarchy theory was able to provide the explanation. This experience was part of Hoekstra's motivation for the research that led to the eventual development of this book.

If the ordering of conventional levels is often unhelpful, where can ecologists find powerful explanations for what they observe? The ordered sequence from cell to biosphere receives lip service as a grand scheme, but it is not the driver of ecological research activity. The conceptual devices that ecologists actually use in practice invoke explicitly scaled structures, not the generalized entities from the conventional hierarchy. The reason is that the conventional hierarchy is not scale-based, although its users think it is. We emphasize that conceptions invoked by conventional levels of organization are very important for ecological understanding, particularly when each is given autonomy separate from the grand conventional scheme.

We call the levels of the conventional biological hierarchy "criteria for observation," or just "criteria," to distinguish them from scale-defined levels. Criteria, as we use them, are not scalar but rather announce how one plans to study a slice of ecological material cut out for research. Criteria are the basis upon which one makes a decision as to what relationships are important in an ecological observation. The principal criteria in this book are: organism, population, community, landscape, ecosystem, biome, and biosphere. However, we do not use them as ordered levels per se, except to explain aptly certain relationships. We do not order the conventional levels by scale. When we do order by scale, it is within each one criterion. Not to anticipate the population criterion as it is considered in the population chapter under the population criterion, but suffice it to say that we look at pronghorn antelope as small populations of a doe and her fawns, as herd groups (which are not held together by relatedness), as a population in a favorable region, and finally as a large population that is the whole species.

We prefer a scheme that explains what ecologists find in practice, even if it lacks the feigned intellectual tidiness of the conventional hierarchy. Rather than a grand ordering scheme, we see the criteria as the prevailing means whereby ecologists categorize themselves. Almost all ecologists would put themselves in one or another of the criteria of the conventional hierarchy. Allen calls himself mostly a plant community ecologist. Hoekstra would answer to the appellation of ecosystem and landscape ecologist, with an applied bent of a wildlife ecologist. In a conversation about lakes Jim Kitchell, the distinguished limnologist, always says, "The zooplankton did it." Kitchell noted that when faced with the self-same situation, Allen always starts looking for big patterns with the statistical tools of the community ecologist. "Tim, you always do that. I've seen you." And it is true. Each type of ecology appears to involve its own style of investigation that follows naturally from the critical characteristics of the preferred conception. The contrast between those stylistic differences gives the relationships between the parts of this book. We dissect out the effect of scale within the type of ecology at hand.

Many general texts on ecology start with either organisms or physiology and build up to larger systems, extending to large ecosystems or the biosphere. Other texts start with the biggest ecological systems and work down through a process of disaggregation, looking for mechanisms and explanatory principles going down the hierarchy. They both make a fundamental mistake of scale ordering unscaled entities, but beyond that, they both invite a misplaced emphasis even if the scaling was adept. By choosing either order, up to biosphere or down to organisms and cells, the ecologist can easily be led away from considering interlevel relationships in the other respective direction. For an adequate understanding leading to robust prediction, you have to consider at least three levels at once: (1) the level in question; (2) the level below that gives mechanisms; and (3) the level above that gives context, role, or significance.[15] The conventional scheme also does not invite linking across to several levels away in the conventional scheme. It does not invite linking pathways at a subcellular level to differences between biomes or even the functioning of the whole biosphere, where indeed reasonable causality can be asserted.

In chapter 8, we give a formal account of linking levels in principle. But at this point, suffice it to say that there is subtlety and richness not found in the conventional scheme. Given our reservations for either a top-down or bottom-up approach separately, we feel that a different organization is appropriate, one that works with types of systems as alternative conceptual devices with equal status. We do not feel compelled to deal with a sequence of levels ordered by definitions that emphasize degrees of inclusiveness, although we use inclusivity when it is pertinent. The conventional ecological hierarchy is not wrong, but it is far too particular to serve as a framework for an undertaking as broad as we have in mind, a unified ecology. In unifying ecology we need to juxtapose levels and structures that come from distant places in the conventional order.

This book offers a cohesive intellectual framework for ecology. We show how to link the various parts of ecology into a natural whole. The prevailing lack of unity that we address comes from ecologists resorting to telling stories about special cases instead of rigorously defining the general condition. We like stories so much that we devote chapter 8 to them, but wish to go further than natural historical stories. There are too many tangibly different cases in ecological subject matter for us to retreat into a description of all the differences that come to mind. We offer an emphasis on what is similar across ecology, so order can emerge from a wide-ranging pattern. We take care not to be blinkered by the conventional hierarchy. When it fits, we use part of it temporarily.

The body of ideas we use has been gaining credence in ecology for more than the last quarter of a century under the rubrics of hierarchy theory, patch dynamics, scale questions, general systems theory, multiple stable points, surprise, chaos, catastrophe, complexity, and self-organizing systems. Although computers are not always used in the application of all these ideas, the mind-set that they have in common came from computer-based modeling of complexity. These collective conceptions are sufficiently mature for us to pull them together, with some new material, into a cohesive theory for ecological systems in general. The first edition of this book is still up to date, but the places where ecology has moved forward are now included.

The principles we use are those of hierarchy: a formal approach to the relationship between upper-level control that limits the outcomes from lower-level possibilities. Hierarchies of hegemony in unfair ranked societies are such a special case that we are not bothered by the political correctness sometimes stirred up by the word. Hierarchy is too important and generally useful for it to be sidelined for political reasons. The observer always has a scale of perception and a level of analysis that deals with the system as a complex observable. Hierarchy theory is not a set of esoteric speculations about ontological reality. It provides a hard-nosed protocol for observing complexity without confusion; it is an epistemology.

There are two separate aspects to observation. In hierarchies, content and context together generate significant behavior at each scale-defined level. We find the level by using a certain scale of observation. However, at a given scale, it is possible to recognize many different types of things. Which types are recognized and which are ignored come from the observer's decisions about what is to be considered important. "Criteria for observation" is the name we give to whatever it is that makes something important enough to be recognized in an observation or set of observations. Our hierarchical framework thus focuses on scale, on the one hand, and criteria for observation on the other. Criteria for observation identify what connections make the whole that appears in the foreground. Different things can be in the foreground at a given scale according to decisions about what to recognize. We get a much clearer view of both scaling effects and the consequences of defining a criterion in a particular way if we emphasize the difference between scale and criterion. The conventional hierarchy is not so careful, and therefore often leads to a muddle.

In ecology, the criteria for observation give rich perceptions, above and beyond the fact that many scales of perception are necessary to do justice to ecological material. For example, the organism can be conceived in many ways: the consequences of a genome and a vessel for housing the genome; a collection of internal regulated processes; an input/output system showing irritability; a system in a loop of action on the world, and a response to it; a structural mechanical system with scaling problems; and so on. This book investigates the richness of character of the objects of study that define the principal subdisciplines in ecology. We unite ecologists by the common strategies for observation that each group has developed parallel to, but separate from, their colleagues in other ecological subdisciplines.

As we apply hierarchy theory to ecology, we pursue relationships of a functional sort. In and around the material that ecology studies there is a tangle of flows of material and energy. We organize our treatment of ecology around those fluxes and the connections they embody. It is thus the unexpected limits on material flows that make ecology more than complicated physics and chemistry. This book looks at the material system, identifying fresh conceptions to account for the limits that appear to be ecological in nature. We attend to physical limits only occasionally and as necessary.

FIGURE 10. Three ways to realize the essence of a bull elephant. Nothing is scaled until it is realized. Notice that the realization in a photograph does not tell which of the other realizations is in the picture. Realizations are generally anchored to a purpose. *A*. The size of an individual free ranging bull elephant in East Africa (photo courtesy of T. Hoekstra). *B*. The size of a bull elephant sculpture. *C*. The size of a postcard about wildlife in Kenya.

We are careful to identify purpose in both the investigator and in the things we model. Organisms model purposefully; they have a model and tell stories. Purpose is distinctly not a physical thing. Behind the observables we see in biology are essences. The things we see are realizations of essences. A scale can be imposed only after there is a realization. Essence as a notion is unpacked fully in chapter 8. Here, we suggest essences of a particular bull elephant in East Africa (figure 10). Realizations are for a purpose. There are several realizations of the bull elephant essence: one offers experience during wildlife fieldwork; another is a sculpture for an office shelf. Yet another realization is a postcard of the bull elephant. The bull elephant essence can be realized as an object weighing several tons or only a few ounces. If you take the picture of the realization in the right way, you cannot tell which realization it is. Species and demes have coded purpose derived from evolution by natural selection. The purpose is transferred to the realizations, which then know and signify things beyond their experience. The biologist and social scientist needs to keep careful tabs on information.

A minimal set of premises underlies the organization of this book. They are:

1. Ecological processes and types of ecological structures are multiscaled. Each particular structure relates to a particular scale used to observe it such that, at that scale of perception, the entity appears cohesive, explicable, and predictable relative to some question the scientist asks. The scale of a process becomes fixed only once the associated scaled structures are prescribed and set in their scaled context. Scaling is done by the observer; it is not a matter of nature independent of observation.

2. The structures that match human scales of unaided perception are the most well known and are the most frequently discussed. Ecological processes are usually couched in terms of those familiar structures. The scale of those processes is prescribed by (a) those tangible structures and (b) a context that is also scaled so as to be readily discernable. A common error is to leave out discussion of the context. Meaning comes from the context, so meaningful discussion must include it. The principles derivable from observing tangibles deserve to be applied to the unfamiliar and the intangible.

3. At some scales of perception, phenomena are simpler than at others; the tangible system thus indicates powerful scales of perception. Predictability is improved if the scale suggested by immediate observables of the system anchors the investigation. The attributes of the particular type of ecological system distinguish foreground from background or the whole from its context. The criteria that distinguish foreground from background are usually independent of scale. It is therefore sensible to determine the appropriate scale of perception separately from choosing the type of system.

There is an observer in the system; only by knowing the location and activities of the observer can we avoid self-deception and start to make ecology a predictive science. In the end, ecologists are addressing the same challenges as did Beaufort. A complicated set of experiences need to be systematized and calibrated so that different observers have some frame of reference for prediction and communication. In this introduction, we have spelled out what Beaufort did so powerfully and eloquently, and have moved it over into the science of ecology. So the unlikely starting point of Beaufort and his scale now looks to be a more suitable place to start this book.

1

THE PRINCIPLES OF ECOLOGICAL INTEGRATION

Some scientific disciplines study objects distant from commonplace human experience. The stars are literally far away and the quarks might as well be, for all the direct experience humans have with them. By contrast, ecology studies a bundle of rich and direct human encounters. Most ecologists have fond memories of some childhood place or activity that not only stimulated a first interest in field biology but now determines what in particular they study. While limnologists might remember a pond behind the house and their first microscope, oceanographers may sit in their offices smelling imaginary sea air and wandering along sea cliffs of summers long ago. Henry Horn has communicated one such story about himself:

> Several blocks from my childhood home in Augusta, Georgia, the street lost its pavement and continued as a dirt track through an abandoned field that was growing to woodland. I was not supposed to go there, especially alone, but I did. What drew me were birds, trees, and butterflies. The butterflies collected in prodigious numbers at the edges of evaporating mud puddles. In my memory they were mostly Tiger Swallowtails (*Papilioglaucus*) and Zebra Swallowtails (*Eurytides marcellus*) and the large Cloudless Sulphur (*Phoebis sennae*), and they sat in little groups according to their kind, all pointing in the same direction. At the time, I wanted to learn more about birds and trees, but I just wanted to enjoy the butterflies, creeping close and lying on my belly very carefully to avoid the stains of red clay that would betray where I had been.[1]

Horn says that as a graduate student, he camped by the Potholes Reservoir of eastern Washington. He was washing camp dishes by a stream between North and South Teal Lakes, with water striders all over. Ever the kid, he sheepishly admits that his curiosity led him to put some detergent on the surface of the pond, which of course wrecked the surface tension on which the striders were striding, and they sank. Surface tension now held the skaters in the water. Filled with remorse, he managed to rescue all but one, patting them off, rinsing them, and air drying them. On release, they strode away. Horn is a particularly curious ecologist. Tony Ives claims, "I spent huge amounts of time when I was a kid outdoors—my Dad was the director of the Mountain Research Station at CU [University of Colorado]. But I was a pretty lousy boy naturalist. Maybe that's why I'm a theoretical ecologist!"[2] It takes all sorts. Ecology is a very "hands-on" study, where the scientist often goes to natural places and looks at other living things in their own habitats.

The things we study in ecology seem very real. Nevertheless, ecology is a science and is therefore about observation and measurement more than about nature independent of observation. It is easy for a physicist tracking subatomic particles in a bubble chamber to remember that science must work through observation and has no direct access to ontological reality. Science in its practice is not about truth and reality; it is about organizing experience and predictive power. In the end, it is about developing compelling narratives. Compelling narratives lead to commensurate experience. Prediction does not so much tell us that we are right about what is happening as it makes our narratives more convincing, so the listener starts to see things the same way as the narrator. Then they can talk and deepen each other's understanding. Almost everything a physicist measures comes through the filter of some gadget. For the ecologist, it is harder to remember that measurement is not reality, for out in the field where birds sing and flowers bloom, all the human senses are flooded with experiences that have an ecological basis, if we would only do the ecology. Even so, ecology is a matter of organizing and challenging perception, with reality always at least one step behind the screen.

Our bottom line is that there may be a real external world out there with things in it, so we are far from committed to antirealism. But there is much that lies between our explanations of our experiences and external physical reality. After the science is all over, many will prefer to believe that what science finds gets ever closer to external reality. Nobody can ever prove that, but such a belief is just fine. It is part of the myth, a belief of modernist science. Since we put narratives front and center in our discussion, we have to respect myths. The myth of science is what keeps some scientists dealing with the tedium of collecting endless data. And in ecology, the data collecting might even be uncomfortable, as when it is raining. Peter Greig-Smith taught Allen ecology, and Greig-Smith had never been known to cancel a field trip because of rain. And it rained a lot in North Wales. Yes, the myth keeps some of us going; it raises science to a metaphysical discourse. So what is our objection to realism in ecology if we are not committed to there being no external reality?

Realism is just fine when the work is over. In retrospect, the consistency of results and the predictions of experiments can reasonably encourage a belief that science has progressed closer to reality. What exactly that means is less clear. As George Box said, "Essentially all models are wrong, but some are useful."[3] Box is emphasizing that in the act of representation, experience passes through a set of symbols that bear an arbitrary relationship to that which is represented. There is no true representation in the way that there can be better or worse approximations. Representation passes through arbitrary symbols, and so is never in terms of that which is represented. Approximation is always in the same terms as that which is approximated. Hourglass, water clock, grandfather clock, and quartz crystal atomic clocks successively keep better time. Critically, it is time all the way, so degrees of being correct are easily understandable. Analog recording devices, such as a long-playing record, have physical connection from sound to groove, and groove back to speaker. In a sense, the record approximates the sound in reproduction through the speaker. But a CD or DVD is different. Digital media introduce arbitrary symbols, for instance a certain number for middle C. In that way there is the critical step to representation that is never in the same terms as the original. Representation can be better or worse for certain stated purposes, but then purpose is not a material thing so is not something that can be approximated. Representations are not approximations and that throws a monkey wrench into the notion of scientific models closer to reality. Better models that serve a purpose is something else, which is captured in Box's "but some models are useful."

The difference between analog and digital is the same difference as between a sign and a signal or symbol. When Sherlock Holmes sees footprints on the ground outside a window, he is seeing marks, signs that someone was there. The sign can be elaborate, as in the pattern of the detail of the sole and its wear. There is no signal in the footprint, it is just the mark left by someone standing. When Holmes develops an account of what happened, and ties the sole of the footprint to the shoe of a suspect, something is added. It has interpretation from the observer. The interpretation is a model, and all models have symbols with meaning, and that is more than just a sign. When Abraham Wald looked at planes returning from air raids in World War II, he saw the places where the planes had been hit.[4] The orthodox view of the remedy was to look at the pattern and reinforce the planes where they had been hit most often. That model suggests that planes tend not to be hit in the places where there were no bullet holes. Wald said to reinforce the planes where there were no bullet holes. His logic was that planes that get hit there never come back. Orthodoxy mistook bullet holes for the reality of planes getting hit. Wald suspended judgment for holes until he had worked out that the holes he saw were places that planes could survive. Data are only data, and are always in context.

Meaning is in the context, and that is a separate part of modeling that involves abstraction. There is also abstraction in the orthodox interpretation, but in taking the data as reality, interpretation is blunted. One can make assertions as to what was really going on with the hits, but they are best suspended until the context is interpreted.

Realism during the conduct of investigation gets badly underfoot while the science is being done. Notice how Holmes is flexible in his interpretation. He is open to the signs implying any of a large number of scenarios. If the terminology is expected to reflect reality, it is not surprising that terms are held with an unreasonable stubbornness that blinds insight. Realism encourages inflexibility. Thus, realism leads to semantic arguments (squabbles over terminology) and so wastes a lot of time. Just because species seem real, there is no reason to suppose that they are real where other taxonomic levels such as genus or family are taken as only abstraction. In the end it is all abstraction, so asserting the real is an unfaithful anchor. Ernst Mayr noted a remarkable coincidence between the numbers of recognized scientific species of birds in New Guinea with the number of bird types recognized by locals not trained in ornithology. "But I discovered that the very same aggregations or groupings of individuals that the trained zoologist called separate species were called species by the New Guinea natives," Dr. Mayr said. "I collected 137 species of birds. The natives had 136 names for these birds—they confused only two of them. The coincidence of what Western scientists called species and what the natives called species was so total that I realized the species was a very real thing in nature."[5] Had he chosen some other group, such as lizards, the coincidence would not have been there. The issue is that birds to humans are like computer clones to the IBM personal computer. Birds are different on the inside, but they share our favored input and output devices, as do computer clones. Humans and birds address the world orally and perceive the world visually. So of course we recognize differences between birds that birds do. Their breeding patterns depend on things we readily recognize. Lizards live in a world where they taste the atmosphere. We are no good at that. Species are not only an abstraction, they are an anthropomorphic abstraction. They represent the degree of variation that humans can readily grasp. For a taxonomist, species have a gestalt. Genera and taxonomic families are simply useful abstractions that are based on criteria different from those taxonomists use on species. All are abstractions. There may be something in reality going on with regard to a particular species in a particular study, but it is still wise to delay asserting reality as long as possible so as to avoid jumping to conclusions. The authors are antirealist only to that point.

If ecology had avoided realism, as Arthur G. Tansley argued before his time,[6] we might not have wasted the twentieth century trying to find out what is a community, really. The Clements/Gleason debate (discussed in chapter 4) is largely semantics driven by different levels of analysis connected by one word, "community." Tansley

met with many thoughtful social scientists, and even Sigmund Freud. When everyone else was arguing whether climax vegetation is a real state, Tansley went into a
very sophisticated caveat:

> One last point, we must always be aware of hypostasizing [reifying] abstractions,
> that is, giving them an unreal substance, for it is one of the most dangerous
> and widespread vices through the whole range of philosophical and scientific
> thought. I mean we must always remain alive to the fact that our scientific con-
> cepts are obtained by "abstracting from the continuum of sense experience," to
> use philosophical jargon, that is by *selecting* certain sets of phenomena from the
> continuum and putting them together to form a concept which we use as an ap-
> paratus to formulate and synthesize thought. This we must continually do, for
> it is the only way in which we can think, in which science can proceed. What
> we should not do is treat the concepts so formed as if they represented entities
> which we could deal with as we should deal, for example, with persons, instead
> of being, as they are, mere thought apparatuses of strictly limited, though of es-
> sential value.* Thus a plant community is an essential concept for purposes of
> the study of vegetation, but is, on the other hand, an aggregation of individual
> plants which we choose to consider an entity, because we are able to recognize
> certain uniformities of vegetational structure and behavior within the aggrega-
> tion by doing so. A climax community is a particular aggregation which lasts, in
> its main features, and is not replaced by another, for a certain length of time; it
> is indispensable as a conception, but viewed from another standpoint it is a mere
> aggregation of plants on some of whose qualities as an aggregation we find it
> useful to insist . . . But we must never deceive ourselves into believing that they
> are anything but abstractions which we make for our own use, partial synthesis
> of partial validity, never covering *all* the phenomena, but always capable of im-
> provement and modification, preeminently useful because they direct our atten-
> tion to the means of discovering connections we should otherwise have missed,
> and thus enable us to penetrate more deeply in the web of natural causation.
>
> *Footnote [of Tansley]: A good example of the hypostatization of an abstrac-
> tion, exceedingly common 40 years ago, but now happily rare, is the treatment
> of the process of natural selection as if it were an active sort of *deux ex machi-
> na* which always and everywhere modified species and created new ones, as a
> breeder might do with conscious design. (Tansley 1926, 685–86, emphasis and
> quotation marks in original)

One manifestation is the attempt by some organismal ecologists to justify assertions that concrete things like organisms are real and material, whereas communities and ecosystems are mere abstractions. Allen, once challenged in this way, was
expected to assert: "No, communities are real." He did not, but rather preferred to

FIGURE 1.1.   Portuguese man-of-war.

argue that organisms are abstractions. Some "organisms" stretch the definition of organism, as when a Portuguese man-of-war (figure 1.1) possesses not one but three genetic lines in its body, one for the sail, one for the body, and a third for the tentacles.

Some argue that a fungal strain found all over the Upper Peninsula of Michigan is a single organism because all the disconnected threads have the same genome

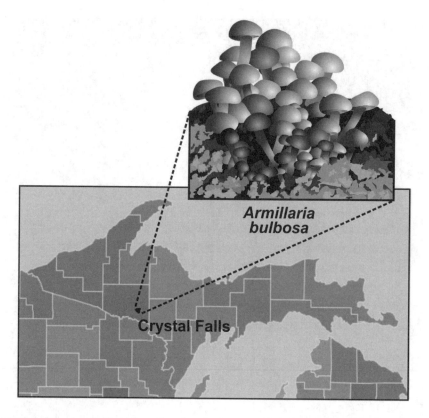

FIGURE 1.2.   *Armillaria bulbosa* found in the vicinity of Crystal Falls in the Upper Peninsula of Michigan.

(figure 1.2).[7] Organisms are a meaningful, significant conception in ecology, but significance is a human assertion, not a material issue.[8]

We do not want data for their own sake; we want them for the more general condition that lies behind the data that we got on a given day. Data for their own sake are a scientific version of stamp collecting. Let us drive that point home by introducing neural nets.[9] They are used, among other things, to calculate whether someone applying for a loan will actually pay it off. The data consist of aspects of the loan seeker reported in the loan application. Some things are obviously connected to the reliability of the prospective borrower, such as income and number of dependents.

Other things are less obviously relevant, like whether the borrower owns a house and how many years the applicant has lived in it. Even numbers like postal zip code might be included. All those bits of information about the applicant are put in a row of information that is fed into the top of the net.

Often, the top row is connected to the net, with each item connected to all the nodes in one level below the row of data. The input to the nodes is determined by a weight as to importance of the input. That is randomly assigned in the beginning, but the connections in the net are reweighted in response to success in a known training set of data. The training set is fed in repeatedly in random order. The inputs to a node are summed in some way, and the sum is fed into a function that says in the end whether the node should fire (like a nerve fires) a signal down to the next lower level of nodes. In the case of a prospective borrower, the net fires on down to give a value for a single node at the bottom of the net. If that value is high, the loan would be granted, but if it is low, the loan is denied. There is a data set coming from people who applied for a loan where the banker knows that the borrower did or did not pay it back. If the net gets it right, the net reinforces the previous adjustment in weights. If the net gets it wrong—say, denying a loan to someone who actually did pay it back—the previous adjustment in weights is reversed. After a lot of training, the net gets good at telling reliable applicants from others in the training set (figure 1.3).

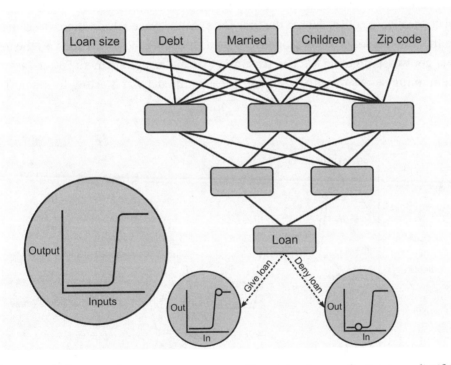

FIGURE 1.3.   Neural nets compress a vector of inputs in a loan application to predict if the applicant will pay off the loan. The training input data sets include who paid on past loans. Output of each node (box) is fed into the node below, with all inputs summed in the node below. The sum is fed into a function. If the input sum is high, the function tells the node to fire. If the input sum is low, the node does not fire an input to the node below. The connecting lines are weighted according to the ultimate success in predicting who paid. Success = adjust weights in same direction. Failure = reverse last time's adjustment. Weights change over time. When weights are found to be accurate predictors, the net can be used to predict the credit rating of new applicants. The usual conversion function is inset here, but others are allowed. Incomplete connections are also allowed in the net, and some may even jump levels to give privileged information to lower nodes.

When the net is calibrated to be good at predicting, it is then used on people who are applying for a loan, to determine whether they will pay it back. If you have ever applied for a mortgage, your information went through this process. In the training process, there are two data sets presented to the net. One is the training set described earlier, to which the net responds; the other is the test data set. The test data are real too; they include people for whom it is also known whether they paid or did not. Critically, the net is not allowed to respond to the test set, only the training set, and so the net forgets any experience of the test set. The net is not allowed to remember what it was told in the test, and it is not allowed to adjust the weights on the nodes in response to the test set. Memory of the training set is embodied in the adjusting weights. As a result, it is possible to track the improving accuracy on the training set, as well as the accuracy on the test set, which the net cannot remember it has seen. Early on in the training process, if prediction on the training set improves, so do the predictions on the test set.

And now we get to the reason why we have discussed neural nets, how they relate to why we collect data in the first place, and why we rely on them in science. As the training set predictions get very good, suddenly the predictions on the test data get worse (figure 1.4). The reason is that the last adjustment in the face of the training set was an improvement because the net was starting to memorize

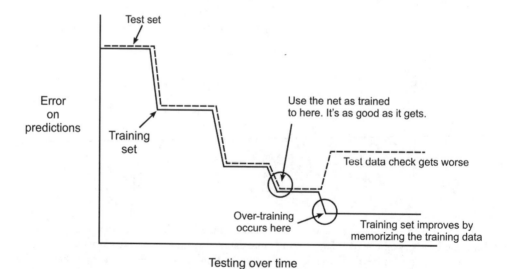

FIGURE 1.4.   Results of training neural nets over time improves prediction in both the training set and in the test set, to which the net is not allowed to respond and so cannot remember. The test set is independent. Improvements occur in both sets over time with training; until, in the end, the training set improves while the test set results get worse. At that point, the net has been overtrained and is memorizing the particulars of the training data, not the generality the data represent. Use the results that the net achieved just before overtraining. It is the best you can do.

the details of the training data. It was coming to the conclusion that this person is a reliable borrower because the net registered that person not as a representative data point, but as a data point it had in particular seen before. In data there are two signals. One is the general pattern that the data represent; that is the signal you want to detect and calibrate. The other signal is the data that you collected in particular on the day you sampled. If you had collected data on some other occasion in exactly the same way you collected the first data set, the new data would be a bit different. You will have sampled different if equivalent people. And the new data would continue to be different on every new occasion you collected more data, even though all those new data sets would still reflect the general phenomenon. Those differences between successive samples of the same universe are incidental and reflect that any data set has its particulars that do not matter. Science does not want data for their own sake; investigators do want data as they reflect the general condition the scientist is investigating.

Ecology may be dealing with a fair reflection of what is behind the veil of our observations, but we have no way to know. As scientists, we deal only with observations, observables, and their implications. We do not rely on assertions that any ecological entity is real in an ultimate sense. We try not to be biased in favor of observables in tune with unaided human perception. However, we do acknowledge that understanding in ecology, as in all science, involves an accommodation between measurements and models that are couched in distinctly human terms.

THE OBSERVER IN THE SYSTEM

If a physicist studying quantum mechanics chooses to suppress the role of the observer in the system, the consequence is wrong predictions. This is called the observation problem, the dilemma of reliance on observation to gain insight into the world that is above and beyond the specifics of the observation. Biologists also have their observation problem, but they prefer, for the most part, to postpone dealing with it. In this volume, we tackle the observation problem head-on. We discuss ecology driven by observation, even if we appear at times nuanced and abstract. The subtleties on which we insist actually take using data more seriously than does the mainstream realism. Data are based on the observer's paradigm as much as externalities. At this juncture, ecology needs a generally acceptable body of theory, for which we propose a theory of observation that ecologists can use to acquire data. If ecology is to become more predictive, it will have to be more careful in recognizing the implications of its observation protocols. Without the ecological observer, there can be no study of ecology. Even at the grossest level of decision making, when the ecologist chooses what to study, that act influences the outcome of the investigation. When one chooses to study shrews, there is an implicit decision not to study everything else. In that implicit decision, most other things ecological, such as trees, rivers, or ants, are excluded from the data.

ECOLOGICAL PHENOMENA AND DEFINITIONS

Phenomena in ecology, as in science in general, are manifestations of change; there can be no phenomenon if everything is constant. Robert Rosen said in *Anticipatory Systems*:

> The observing procedure is the very essence of abstraction. Indeed, no theory, and no understanding, is possible when only a single mode of observation (i.e. a single meter) is available. With only a single meter, there cannot be any science at all; science can only begin when there are at least two meters available, which give rise to two descriptions that may be compared with one another. (2012:214)

Recognizing those changes that constitute a phenomenon must be preceded by observer decisions about what constitutes structure. To accommodate change, there must be some defined structure, a thing to change state. Our observations involve arbitrary structural decisions, many of which revolve around making or choosing definitions. Definitions are not right or wrong, but some give us more leverage against nature's secrets than others. For example, scientific species names or their vernacular counterparts seem to be powerful ways to categorize living things. Nevertheless, the things included under those names are as arbitrary as any other named set. This manifests itself in disagreement between taxonomists about distinctions between species. A misidentified plant involves using the wrong definition; while the species mistakenly used may be a good species, it was the wrong one on that occasion. Right and wrong is a nuanced distinction.

One might argue that species are abstractions and that their very abstractness is the source of their arbitrariness. However, even something as tangible as a tree is, in fact, arbitrary. The entire army of Alexander the Great camped under a single banyan tree. The army was large, but the tree was old and had grown, as is the nature of the species, by sending roots from its limbs down to the ground. At first the roots are threadlike, but after a long time they thicken to become tree trunks in their own right. Thus, one tree can become a forest. Contrast this with a clone of aspen. As the stem establishing the clone becomes large enough to spare reserves, it spreads out its roots. These long roots periodically send a branch above the ground that then grows into an apparently separate new plant. The organic connections between the trees in a clone are quite as strong as the limbs connecting the trunks of an old banyan tree. The only difference between a clone of aspen and a banyan tree is our perception. We do not see underground. An earthworm's eye view might see the

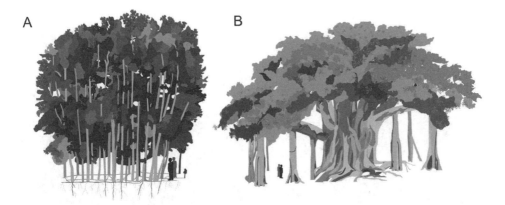

FIGURE 1.5.   *A.* Aspens spread their roots, and new above-ground stems come from those roots. But we cannot see the connection and so call an aspen clone a set of separate trees, even though the clone is as connected as for the single banyan. *B.* Banyan trees come from one tree stock, but roots drop down to make new tree trunks. We can see the connections and realize that it is one tree. Allen and Hoekstra have been introduced into both figures as small people for scale.

banyan as a grove of separate trees but the aspens as all one organism. Thus, even what constitutes a single tree is a matter of arbitrary human judgment (figure 1.5).

Definitions are generally based on discontinuities that have been experienced or are at least conjectured. In the case of the banyan tree and the aspen clone, there are two critical discontinuities. One is the separation between tree trunks; the other is the separation of whole banyan groves or whole aspen clones. Note that the separation between the tree trunks in both banyans and aspens is only a matter of degree. That is why a formal definition is so important, because without it there is ambiguity as to whether the tree trunks or the whole interconnected collection of tree trunks constitute the organism. Note that neither the separate trunks nor the collection of them all are truly the proper level of aggregation to assign to the class "organism." However, a given discourse about aspens or banyan trees has to be consistent in the meaning of the words it employs. Simply make a decision and take responsibility for it.

A definition is a formal description of a discontinuity that makes it easy to assign subsequent experience to the definition. How steep a gradient of change will have to be to be assigned the status of a "discontinuity" is a matter of decision. The observer experiences the world and decides whether the experience fits the definition. With a new definition, some experiences that were once within the defined class are now excluded, while others may be added. By some definitions, a banana plant is a tree; by other definitions, even a banana plant ten meters high would not be a tree. If "tree" is defined as plant material above a certain height, bananas can be trees. The "trunk" of a banana plant is mostly fleshy, sheathing leaf bases, not woody stem (figure 1.6). The stem arises later and grows up the middle of the

A   B

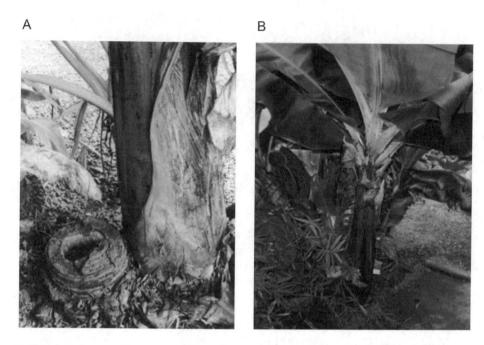

FIGURE 1.6.   A cross-section of a banana tree identifies that the "trunk" is almost entirely fleshy sheathing, not woody stem. (Photos courtesy of C. Lipke.)

sheathing leaf bases when it is time to make bananas. Therefore, a definition of tree that insisted on woody stems could leave even the tallest nonflowering banana out of the class "tree." In a sense, a banana is a tall herb, but the distinction has a human origin; nature does not care what we call it. Behind all acts of naming are implicit definitions. For all we know, nature itself is continuous, but to describe change, we must use definitions to slice the world into sectors. The world either fits into our definitions or not. Either way, all definitions are human devices, not parts of nature independent of human activity.

Definitions, naming, and identifying critical change are not the only arbitrariness in scientific observation. The observer uses a filter to engage the world. The filter chosen by the observer is as much a matter of human decision as is the definition of structures. Because we cannot measure anything in infinite detail, differences too small for the instrument to detect (fine grain) are filtered out of the data. In this case, the smallest differences fall below the level of resolution of the study. Furthermore, any difference that takes too long a time or is too large spatially to fit into the entire sweep of the data (extent) will not survive the filter of the data-collecting protocol. Beyond this, any signal that the instrument cannot detect will be missed. For example, a light meter will not measure pH. Sometimes one hopes for a surrogate signal, as occurred when Carol Wessman used remotely sensed radiation as a measure of lignin concentration in forest canopy, which in turn is a correlate of soil nitrogen.[10] All these aspects of the scientist's input filters are arbitrarily chosen, and they all influence what the ecologist experiences.

Reduction has been a useful tool in that explanations often derive from taking things apart so as to address the parts separately. Reductionism takes reduction one step further into a philosophical realm. It asserts that the whole can be known from its parts. Reductionism takes the question at hand to be self-evidently interesting. It also says that the scientist knows that the chosen level of reduction is somehow the right one. For instance, cancer can be explained and investigated to the point of fighting cancer if one reduces the issue to cell division and failure of moderating signals across cell membranes. Reductionism gets the scientist quickly into data collection. Actually, cancer hormones controlling blood supply appear to be an equally useful explanatory device. Critically holistic approaches also use reduction, but the question asked is not taken as self-evidently interesting. The holist does not assert the level of reduction at the outset. The holistic question says, "I do not know what I need, but it would probably have this and that property. Has anyone seen one?" Ecology needs that level of openness and flexibility.

It is a mistake to presume that studies of small things, like biochemistry, are reductionist. It is just that the holism had been done earlier and well, so the assertion of the level of explanation is well founded. For instance, Hans Krebs took advantage of the many details worked out by other researchers in his discovery of the Krebs cycle.[11] His breakthrough was less in his effort put into the details of *What* and was more in his bigger holistic question of *Why*. Why did the tissue continue longer in its capacity for oxidation when he added citrate? Ah, it closes the loop, an answer that applies to a higher level of analysis, not a lower level of detail. Significance is always a higher-level property. We now know that biochemical cycles like the Krebs cycle are everywhere, and why we should expect them (see chapter 3, figure 3.5A).

EMERGENCE IN BIOSOCIAL SYSTEMS

The difficulty with reduction is that the emergent properties of the whole cannot be derived from the parts or predicted unless they have been seen before. It is not possible to go into water with a device small enough to sort out individual water molecules, and then pull out a wet one. Wetness is a property of water in aggregate. In a universe only the width of a proton, there can be no chemistry. That only occurs in a larger universe. In recent years, much has been made of emergence in complexity science, but at least some of it is far too grand. Emergence need not be a big deal.

Emergence is always some version of finding something new. The trivial case is when you have looked at something in a new way, perhaps more precisely, and found something you did not know about. Sometimes emergence arises because of a wider universe of discourse, as in the proton example, where a wider universe gives chemistry. Less trivial are aspects of emergence that are seen to arise out of positive feedbacks. Positive feedbacks require some sort of gradient to drive the expansion of system components. At the top of the gradient, matter or energy exists concentrated in an unlikely arrangement. In biosocial systems, the gradient starts in the environment. As matter energy is taken in from the top of the environmental gradient, perhaps as food, a gradient is set up inside the biosocial system. The bottom of that gradient is manifested in effluent arising from degrading the inputs. The material or energy at the bottom of the gradient is in a much more likely arrangement, having arrived there by randomizing processes. The randomization follows from the second law of thermodynamics under which things run down. The total gradient from environment back to effluent runs down. But inside the biosocial system, the gradient is fueled by inputs and therefore remains stoked up, far from equilibrium. The internal gradient does not run down unless death occurs. Gradients are needed to drive the positive feedbacks underlying emergence (some positive feedbacks show inexorable decrease, but that is a matter of how the system is specified—a gradient is still needed to drive values down). *A* and *B* increasing under positive feedback do not do so free; it takes an expenditure of energy to drive the increase or decrease. That energy comes from using the external gradient. Physics talks of the second law in closed systems running down, but living systems run up, not down. Quite the opposite of denying the second law of thermodynamics by becoming more organized, life uses the second law for organization.[12] Emergence has been described as some sort of explosion, but that misses a critical part of it. The emergent property is a constant that appears when a racing positive feedback encounters some limit that can usually be cast as a self-correcting negative feedback. An emergent property does not pertain until the explosion encounters some sort of limit.

As we come to the end of the industrial age, there remain many holdovers from the industrial posture. Industrial problems at the cutting edge in Victorian Britain were met by building larger structures and applying more power. Solutions like that have always cost more with increase in size. Meanwhile, problem solving in the postindustrial age involves smaller things that offer immediately cheaper solutions. Telephones and computers are only slightly reduced in price, but each generation comes with greater speed and effectiveness. The price of engines with moving parts generally goes up as fast as inflation because they use industrial principles. A big holdover from the industrial age is ambitious space travel, which should be seen as a bigger engine going further. The idea of colonies on any other celestial body is simply an outdated Industrial Revolution posture at best. Putting up human-made satellites around our own world is a different matter because the issue there is not moving things a long way, but is rather about compressing information about our whole planet, a postindustrial notion. Rockets taking humans to Mars are really overgrown steam engines whose time has passed.

There are general principles here. The difficulties with industrial age ambitions are that they invite emergence. Emergence involves catastrophic failure of the system manifested before. Applying huge force to a larger engine offers a steep gradient that will set off a positive feedback. The crash of the space shuttle *Challenger* started with the failure of O-rings, which were a small part of the system. With that much power applied to takeoff, the gradient was plenty steep enough to cascade dysfunction upscale, which it did. Disaster frequently comes from a process of emergence that had never been seen before. The expensive correction is a matter of prediction, but by then the engineers have seen it before. We cannot predict emergence until we have seen it at least once. Big power and large systems will continue to rise in cost faster than humanity can pay for them. The problem is that emergence in principle is connected to steep gradients. The industrial age is slipping away because it is not a viable strategy in the face of emergent failure and its cost. The postindustrial problems are also a matter of emergence. They have manifested themselves in deep financial global retrenchment caused by steep gradients of information, not steepening gradients of energy and matter, as they were in the industrial age. Any steep gradient becomes an accident waiting to happen, albeit a gradient of information or matter/energy. A cluttered high shelf is an invitation for something to fall.

THE MINIMAL MODEL

When there is danger of over-reduction, the concept of the minimal model is helpful. The minimal model gives predictions from the smallest number of explanatory principles. Thus, the small number of explanatory principles in a minimal model comes from using the highest level that can be contained inside the system to be explained. Occam's razor invokes the principle of parsimony, where one should pursue the simplest explanation. Kevin Kelly has been working toward a deeper justification for Occam's razor.[13] He cites statistics that show those using Occam's razor retract results less often and sooner than those who do not use it. Recently, he has shown why we would expect simpler models to find something with the properties of truth more often. His analysis uses proposed truths, a troublesome notion, but we know what he means. He finds that Occam's razor will find these truths faster, with fewer steps, and more often than more complicated models and procedures.[14] One wants to find the highest logical type that is pertinent. Einstein urged us to make things as simple as possible, but no simpler. The razor cuts away everything else as superfluous.

The relationship between prediction and ultimate reality is not at all clear, so justifying cumbersome modeling by reference to ultimate reality should not be accepted as an excuse for ignoring minimal reduction of Occam's razor. Howard

Pattee suggests that reducing below the level of meaning lets the issue at hand slip through the investigator's fingers like sand.[15] The chemistry of the ink will not enlighten the reader as to the meaning of this text.

THE THEORETICAL BASIS FOR SCALING AND INTEGRATING ECOLOGY

STRUCTURE AND PROCESS

Up to this point we have emphasized structure and entities. However, another facet of levels and scale uses a process-oriented conception of entities and patterns. Biological systems are very much a matter of process. In fact, what appears to be distinctly structural in biology can often be seen as part of a process. For example, the human body consists of material that flushes through in about a seven-year cycle (even the bones). There is very little left of the "you" of seven years ago. In biology, often one process reinforces one or several others; in turn, the first process is reinforced by those it has influenced. That self-reinforcement leads to persistent configurations of processes. That persistence of process clusters explains how biological structure can appear concrete, although the substance of that structure is in constant flux.

Let us start with something tangible and identify its underlying processes before moving to more abstract biological structures. Solid, concrete things are surrounded by surfaces. The surface is all that we see of most things because it is the part through which the whole communicates with the rest of the universe. Although one might conceive of surfaces as passive and having nothing to do with dynamic processes, surfaces are places where the dynamic forces dominating the internal functions of an entity reach their functional limits. The skin of an organism corresponds with the furthest extent of the internal circulation system. The skin also coincides with the limits of many other fluxes.

Science looks for surfaces that define things with generality. We seek things relevant to systems according to many criteria, things that are detectable even when one looks at the system a different way. An entity that shows this persistence we call "robust to transformation." Processes held inside a stable surface usually reinforce each other. In social systems, language and commerce reinforce the limits in each other at an international boundary. Trade uses language and a common language facilitates trade. In the tree, growth puts leaves in the light, then photosynthesis provides material for growth; growth and light each refer to the limits on the other. These mutually reinforcing processes give a set of surfaces that can also be seen as mutually reinforcing, one surface for each process. For example, in lakes in summer, the surface between warm water above and cold water below is the surface at which oxygen and nutrient status coincide; the process of oxygen use

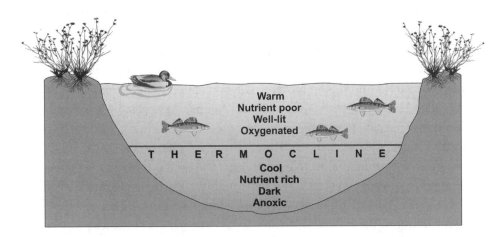

Warm
Nutrient poor
Well-lit
Oxygenated

T H E R M O C L I N E

Cool
Nutrient rich
Dark
Anoxic

FIGURE 1.7.   In northern temperate lakes in the summertime, warm water mixes above while cold water mixes below, separated at the thermocline where there is a very steep temperature gradient.

depletes nutrients, so there is high oxygen concentration above the thermocline and high nutrient status below (figure 1.7). The multiplicity of devices that can be used to detect the surface of an entity that is robust to transformation reflects the multiplicity of the processes associated with that surface. The thermocline is an emergent that then constrains processes in the vicinity.

Some plants grow and establish apparently new individuals by vegetative growth. A collection of individuals formed in that way has connections between genetically identical individuals. The aspen clones discussed earlier are a good example. The surface of a clone, as with many surfaces, is identifiable by sets of processes that press the surface outward. For example, the process of water transport within the clone is rapid. Internal processes reach the surface, but then attenuate rapidly. Fluxes associated with processes inside slow down at the surface; this can be used as a demonstration of the presence of a surface. For example, a radioactive tracer moves rapidly through a clone by virtue of the interconnections between individuals. The radioactivity moves rapidly up to the surface of the clone, but only very slowly across it into the outside world. Surfaces are places where signal is attenuated, stopped, or changed in some way.

We have come to understand that processes cascade upscale through fractal geometry (fractals get an extended treatment in chapter 2). The dominant processes can be recognized by the texture of the structures the processes leave behind. For instance, a naturalist recognizes the species of a tree by the texture and outline resulting from the growth processes for that species. That is how biologists know which species is before them without going into the technical details of fruits or leaf shape. The tall, narrow trees in French impressionist paintings are Lombardy poplars. Lombardy poplars are columnar, not bushy like most open-grown trees. Lombardy poplars are so common in a British or Continental European scene that the British Army has a simple classification of trees: fir trees, poplars, and bushy top tree.[16] We suppose that is enough information for aiming artillery, although a silviculturist would find it a bit austere. The poplars referenced there are just the variety called Lombardy, with other varieties of poplar being bushy top. Lombardy poplars have a mutated growing point that causes the branches to diverge more slowly than most poplars, giving a distinctive fractal dimension to that variety. They are reproduced by humans by taking cuttings, and so the mutation persists. The growing point cascades upscale in a fractal manner, leaving a characteristic form behind. Small events that involve stems branching and growing leaves eventually give large outlines. Behind the structure lies a particular set of processes.

As we have seen, surfaces disconnect the internal functioning of entities from the outside world. The disconnection is significant, but not complete. Therefore, the observer has to judge whether or not the disconnection is sufficient to warrant designating a surface. That judgment is what makes all surfaces arbitrary, even natural surfaces that are robust to transformation.

The inside of a natural entity is strongly interconnected, as in the case of the trees in the clone. The inside is relatively disconnected from the outside across the surface. The strong connections inside and the weak connections to the outside are a matter of relative rates of fluxes. This applies not only to tangibles but also to intangibles like functional ecosystems. Large terrestrial ecosystems are composed of the interaction and integration of plants, animals, soil, and climate. The surface of an ecosystem is not tangible, but is rather defined by the cycles of energy, water, nutrients, and carbon. If the ecosystem has the integrity to make it a worthy object of study, then it must have a stable surface in some terms that reflects that integrity. You generally cannot see an ecosystem edge with the naked eye, but you could detect it with the use of radioactive tracers and a Geiger counter. The surface of the ecosystem makes it an entity. Connections inside the cycles are strong relative to the connections to the outside world. Tracers put into an ecosystem will move around inside the ecosystem much faster than they move out of the ecosystem. The relative rates of movement define the ecosystem surface, even if the ecologist cannot see it literally.

Other intangible entities, like the regional cultures within the United States, are also a matter of relative movement. There are more social contacts within a cultural region than between the region and the outside world. Because of the separation by water, the islands off the Carolinas each had their own dialect, one switching the *W* for a *V*, as in some Dickensian accents (Sam Weller pronounced with a *V*). All of that was homogenized with the establishment of just one school to serve all the islands.[17] Even with the inroads of mass culture through television, the islands as a whole still maintain an accent that sounds more southwestern English than American. It is a maritime accent, like those in pirate movies. The reason is that the social contacts are strong locally but weak to the mainland. The ocean defines the cultural surface.

Similarly, ecological populations are held together and are recognizable less because one can see the whole population and more because of the mutually reinforcing processes that bind the individuals together. Where population members are on the ground is often not distinctive, but the breeding patterns are a force for unity and identity. The process of breeding involves finding mates in a shared habitat. The process of survival in a habitat is often heritable. The shared responses of individuals to habitat define and are defined by what is inherited.

The boundaries of organism are often distinct, the process that holds them together being a contained physiology. Biomes are often large and may have distinct boundaries. Allen was on a field trip to the inland forests of Washington State. His hosts took him through many miles of unbroken Douglas fir–dominated forests. Then they crossed a ridge, and all was different. A juniper scrubland presented itself. The boundary between these biomes was distinct, measured in meters, not kilometers. Allen's host, Tom Spies from the Andrews Forest site, told him that the juniper desert scrubland emergent on that ridge goes all across the great basin for almost a thousand miles. Physiology integrates organisms as blood circulates carrying energy and hormones. As to homogeneous vegetation across thousands of miles, fire and animals are the only things that move fast enough to integrate large biomes.

HIGHER AND LOWER LEVELS

A relationship between levels can be considered as the relationship between the internal functioning of an entity and the behavior of the whole. Lower levels are characterized by internal functioning, while the upper level relates to the whole entity. Many problems can be translated into the relationships between levels. Evolution by natural selection is an example of a concept that links levels. The lower level is occupied by the individuals that reproduce with varying amounts of success in between birth and death. A Darwinian view sees the aggregate of those successes and failures as determining the character of the upper level, the population.

We are now in a position to discuss the principles that govern the relationship between higher and lower levels. We recognize five interrelated criteria that order higher levels over lower levels.

1. *Bond Strength.* In the discussion of surfaces, we have emphasized that the connections inside an entity are stronger than connections across its surface. The surfaces that separate the entities of a lower level can be part of the bonding that unites those entities as parts of the upper level (figure 1.8A). Relative to the strong connections

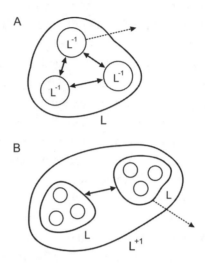

FIGURE 1.8.   *A.* L is the level in question; L^{-1} is the next level down, and L^{+1} is the next level above. *B.* The weak connections of L^{-1} to the outside world beyond L become the strong connections with level L^{+1}.

within the parts, only weak signals pass out through their surfaces to make a link between the parts. The parts can only communicate with the relatively weak signals that pass out through their own surfaces. The weak connections between the lower-level entities, between the parts, become the strong connections that give integrity to the entity at the upper level, the whole. This explains the principle of weaker bond strength. The higher the level, the weaker is the strength of the bonds that hold entities at that level together (figure 1.8*B*). Sometimes it is possible to see the bond strength at a given level by breaking the bonds and measuring the energy released. Note that breaking chemical bonds in a fire or an explosion releases much less energy than breaking the bonds inside atoms in an atom bomb or a uranium fuel rod.

2. *Relative Frequency.* Behaviors that might appear to be directional can be seen as cyclical if the set of observations has a wider extent. For example, eating can be described with much more generality as an activity that is repeated at a certain frequency, rather than as a series of items consumed. In fact, describing all sorts of human activities as recursions allows a completely new set of insights. Herbert Simon once noted that if a resource is to be found everywhere, we replenish it often. Simon, an economist, sees this as a matter of inventory; only keep a short inventory of readily available resources, but keep a long inventory of scarce material whose supply is erratic. There is no point in planning too far ahead with something as universal as air, so we breathe at a relatively high frequency and keep a three-minute inventory of oxygen. Water is common but not ubiquitous, so we drink many times a day and keep an inventory of a few days. Our biological requirement for food has been set by the

hunt, and so we eat somewhere between once and five times a day. We have an inventory of food in our bodies that could last a few weeks, at least. Time to exhaust supply should be a good measure of past selective pressures. If there is total system shutdown, the greatest urgency of renewal applies to the shortest inventory.[18] Humans are quite hard to replace because of food to reach maturity and time to learn how to be effectively human. There is therefore an inventory on each one that is a generation long. Long-term human knowledge has an inventory on it of threescore and ten years. Levels of observation can be ordered by the frequency of the return time for the critical behavior of the entity in question at a given level of organization. Higher levels have a longer return time, that is, they behave at a lower frequency. Thus, in a coherent account of ecosystems, many nutrients cycle once a year, but the whole upper-level ecosystem accumulates nutrients over centuries and millennia.

3. *Context.* Low-frequency behavior of the upper level allows it to be the context of the lower level. The critical aspect of a context is that it either be spatially larger or more constant over time than the lower level for which it is the context. It is not always transparent what the critical constancies are that the context offers. Sometimes the context involves change, but that change always happens; the constancy is the always of "always happens."[19] Seeds germinate based on a favorable growing season that has not yet happened but which is predictable from warming, moist conditions in the spring. Summer always follows spring, and that change is itself a constant on which temperate plants rely. If the environment is always changing, then what becomes dependable is the constancy of the change. Weedy plants thrive in disturbed habitats, and it is the constant upheaval that they find reliable and persistent.

4. *Containment.* Despite being unequivocally upper-level entities in the social hierarchy, dominant animals do not contain lower-level individuals. However, there is a special class of important systems where such containment is a requirement for existing at a high level. These systems are nested, where the upper level is composed of the lower levels. Organisms are profitably seen as nested in that they consist of cells, tissues, and organs. Of necessity, they also contain the parts of which they are made (figure 1.9A). In nested systems, the whole turns over more slowly than its parts, and the whole is clearly the context of its parts, so the nesting criterion aligns with the frequency and contextual criteria. However, the reverse is not true because the criteria of frequency and context do not depend on the system being spatially nested. For example, a reliable food supply behaves more slowly than the animals that depend on it, but it does not contain or consist of the animals it feeds. Unlike frequency and context, the bond strength criterion probably does apply reliably only to nested systems.

Nested systems are very robustly hierarchical in that the containment criterion corresponds to many other considerations. The nesting keeps the order of the levels constant even when the observer changes the rules for relationship between levels. Consider a hierarchy that goes from cells to communities. Plants may be seen as the synthesis of cellular interaction by physiological processes: cells to

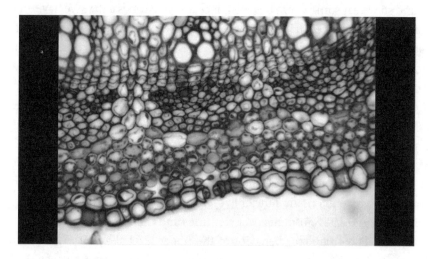

FIGURE 1.9A.   Cells aggregated to form a tissue are an example of a lower-level being contained by a higher level. (Photo courtesy of University of Wisconsin Botany.)

plant on physiological rules. However, the whole plant can be seen as a part of the ecological community on rules of assembly that bear no simple relationship to the physiological processes that built the plant as an individual. The switch from physiology to species associations causes no confusion because the nesting keeps things straight. Western medicine usually casts the human body in nested terms. As a result, the cells interact through membranes and pores between cells. But the organism of a whole human is joined largely through processes of hydrodynamics. The pumping heart makes the kidneys work, which is why heart failure engenders kidney failure. At a higher level, humans interact socially through symbolic schemes of speech that have little to do with either cellular or whole body physiology. But it is all kept straight by the nesting of the human hierarchy. That the nesting is a choice rather than a necessity is evidenced by Eastern medicine often using lines of force and critical points rather than nested body parts.

By contrast, in non-nested systems, the containment criterion does not apply, and so it cannot be used to keep the system ordered; accordingly, non-nested systems have to rely on some other criterion, and they must use that one criterion from top to bottom of the hierarchy. A change in criterion in a non-nested system creates a new hierarchy that only incidentally shares certain entities at particular levels with the old hierarchy. Pecking orders or food chains are non-nested in that the top of the hierarchy does not contain the lower levels (figure 1.9B). The relationship between the plants and the grazer is the same as the relationship between the herbivore and carnivore; once again, higher levels do not contain lower levels. If one takes an individual in a food chain and then considers that same individual's place in the pecking order, this amounts to a switch to a new non-nested hierarchy

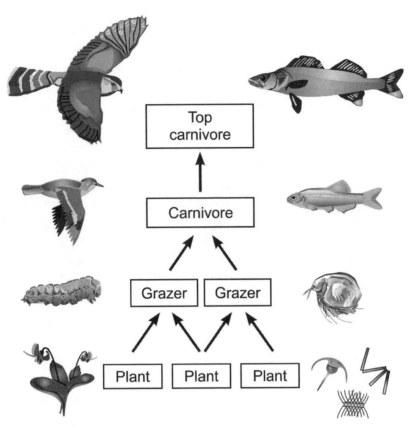

FIGURE 1.9 *B*.   Food chain hierarchies represent non-nested systems where the higher levels do not contain the lower levels.

that only incidentally articulates with the first non-nested hierarchy. From top to bottom of the food chain, the ordering criterion is always eat and be eaten; the social hierarchy of dogs at a kill is a different matter that needs separate consideration. The ordering in non-nested systems can easily be reversed, when "eats" is substituted by "depends on." A mapping of ungulate herds moving is slower and higher order than the movement of wolf packs that behave as satellite units. The wolf pack seen as moving around the herd as a lesser entity is working in the principle of "depends on" rather than "eating." Hierarchies must be used with precise criteria and for a particular purpose. Hierarchies are largely decisions about how to organize experience, not things independent of human concerns.

5. Constraint. For our purposes, frequency and constraint are the most important criteria for ordering levels. Upper levels constrain lower levels by behaving at a lower frequency. Constraint should not be seen as an active condition. Upper levels constrain lower levels by doing nothing or even refusing to act. For example, one should never underestimate the power of impregnable stupidity. High-frequency manipulations of elegant ideas can be held in the vice of persistent misapprehension. Contexts are generally unresponsive to the insistent efforts and communications of things held in the context. Constraint is always scaled to longer time frames than those used by that which is to be constrained. It is in this manner that deans constrain their university departments by doing nothing. A budget is announced perhaps once a year, and the departments must simply get used to it. Deans should not deal with every new circumstance, even those promoting worthy projects. Poor administrators change their minds when lobbied by their more powerful underlings. In such situations, there is nothing the unit as whole can bank on.

In this regard, consider an advancing dune system with plenty of sand still available for further advance and a suitable prevailing wind. Dune systems do not start easily from nothing; they mostly expand from established dunes that interfere with airflow. Dunes lose sand from their windward sides, and this sand is then deposited by the slack air on the leeward side. A flat terrain will allow the sand to blow away without forming self-perpetuating irregularities in airflow. If an intermittent river cuts the front of the dune system back, then that is a temporary limitation that is not a serious constraint. In time, the slow process of dune building will recoup the loss. However, if the river always returns before the dune system has had time to build past the riverbed, then the dune system is contained and constrained, and can never cross and establish a system on the riverbank on the far side. The individual floods are not the constraint; it is the "always" in the return of the river that makes the river a constraint. If the dune system could affect the course of the river, then there would not be constraint.[20] Note in this example that constraint can only be described at the right level of analysis.

Constraint is so important in the ordering of levels in ecology because it allows systems to be predictable. Predictability comes from the level in question being constrained by an envelope of permissible behavior. Predictions are made in the vicinity of those constraining limits. When a system is unpredictable, it has been posited in a form that does not involve reliable constraints. Any situation can be made to appear unpredictable, so predictability or otherwise is not a property of nature, it is a property of description. The name of the game in science is finding those helpful constraints that allow important predictions. Those are the models that George Box called "useful." Science would appear to be less about nature and more about finding adept descriptions.

In ecology, constraints are sometimes called limiting factors. An impossibly large number of factors could influence the growth of algae in a lake. Predicting population size is only possible when some critical known factor becomes constraining. Diatoms are microscopic plants that require silicon for their glass cell covering (figure 1.10).

Silicon is not very soluble in water, which is why we drink liquids from glass. But glass is soluble enough that soft museum specimens sitting in unopened jars

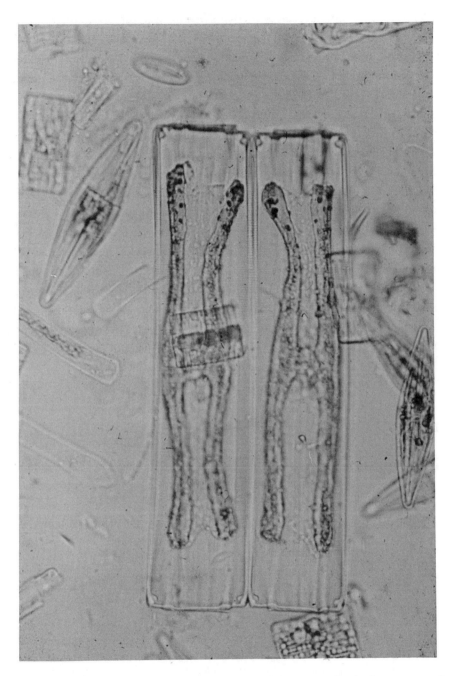

FIGURE 1.10. Diatoms develop a cell covering of glass. Accordingly, they are sometimes constrained by silicon supply in the water. (Photo courtesy of T. Allen.)

for a century come out just a bit crunchy from silicon of the container getting into the solution and precipitating on the specimen. If a lake has low silicon concentrations, then diatoms are reliably only a minor component of the plankton. Recent research on the great new dams in China indicates that the impoundments cause silicon to be removed from the water that flows through the dams because it is instead deposited at the bottom of the lake in the remains of dead diatoms.[21] The ecological result downstream and in the oceans and in the estuaries is a great depletion of diatoms, as well as the release of deleterious red tides in the estuary. Much ecological work is incidental recording of local considerations, but the new work on the effect of dams is an example of ecology that matters. When the constraint is lifted and silicon is abundant, diatoms may become abundant only if other factors are not limiting.

When control of a system changes, the situation becomes unpredictable. With a change in constraint, new factors take over the upper level. A good example is the switch in control that occurs in an epidemic of spruce budworm that can be seen as a switch in constraints. For decades, the budworms are held under the constant control of birds that eat them. Any increase in budworms will feed larger populations of birds that crop the budworm populations back down again (negative feedback) (figure 1.11). Budworms eat the young parts of shoots of conifer trees.

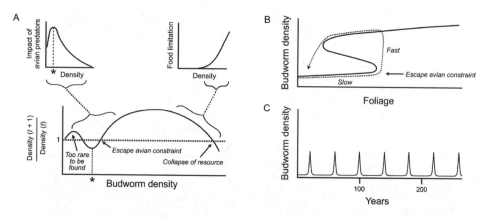

FIGURE 1.11.   A. When the density in the next time unit is greater than the density in the present time unit, the tendency will be for the population to increase in numbers. That is the tendency of the system when it is above the unitary line. It will proceed to higher densities until there is a constraint to hold the density in the next time unit the same as the present. The unitary balance creeps slowly toward higher densities as host foliage increases in quantity. At a critical stage, the population can reach a high enough density so as to escape predation as it slips above the unitary line to grow to outbreak numbers. The population increases uncontrolled until the pest eats itself out of resources. B. The increased levels of pest population with increased foliage are gradual until there is a short, rapid outbreak of the pest that finally destroys the resource base. C. Plotted on a time axis, the population accordingly shows intermittent outbreaks approximately every thirty years (Holling 1986).

The avian constraint on the budworms allows the trees to grow. As the trees grow, the budworms have more food and can handle the losses to bird predation more easily. The budworms are still constrained by the birds, but their population creeps higher. Eventually the trees are sufficiently large that they support a population of budworms that is still relatively small, but is almost large enough to saturate bird constraint. Any bird can only eat so much, and bird density has upper limits controlled by factors like nesting sites or territorial behavior that have nothing to do with bird food. Sometime during a critical period in the constraint of budworms, when bird populations are at their maximum, a chance event like a pulse in worm immigration can increase the budworm population. Being limited by some other factor, the birds cannot respond as before by increasing their numbers (figure 1.11*B*). The effect is to increase the budworm population a small but critical amount where the budworms can grow faster than the birds can eat them, a critically different situation. Budworms increase in numbers leaving bird control behind. The constraint on the budworm population is broken as the birds lose control of the increasing population of insect larvae (figure 1.11*C*).

Since there has been a breakdown of the constraint, there are problems of predictability. When exactly the constraint will be broken is unpredictable, although it can be generally expected during certain time windows. The whole system is no longer under the reliable constraint of the birds, and it takes on the explosive dynamics of the epidemic. The well-behaved system under the bird constraint is of no help in predicting what happens when that constraint no longer applies. During a budworm explosion, a new constraint takes over, the growth rate of the uncontrolled budworm population. Soon, yet another constraint ousts the budworm growth curve as the limiting factor, when all the food disappears as all the mature trees are killed. Pest starvation constrains the system and it returns to low budworm population densities.

The constraint of the birds gives the trees time to grow; the system behaves slowly. When the constraint changes, the budworms become the upper constraining level for the system dynamics. The system suddenly starts behaving according to the fast dynamics of the budworms. From the vantage point of the old avian constraint, the new constraint comes as a complete surprise. A reordering of levels to give control to a new upper level gives unpredictable behavior.

THE CONVENTIONAL FRAMEWORK: A POINT OF DEPARTURE

We now unpack our definitions that underlie the six criteria we use to order and unify ecology.

The Criterion for Organisms

We have already identified organisms as arbitrary (see introduction, figure 9). Botanists are more comfortable with this notion than vertebrate biologists because many plants regularly reproduce vegetatively and only gradually become autonomous and separate; however, vertebrates resemble ourselves in having an unambiguous physical boundary and obligate sexual reproduction in most species. Nevertheless, in humans, conjoined twins draw attention to the arbitrariness of even human organisms. Physical identity is an important point in the biology of placental mammals like humans because confusion between self and not self invites either spontaneous abortion or uncontrolled fetal parasitism. Both of these conditions reduce fitness.

Organisms are generally of a single genetic stock, although fruit tree grafting can lead to chimeras, branches where the outer layer of cells belongs to one partner of the graft while the core of the shoot belongs to the other. In botany, the formal definition of a new generation focuses on the reduction of the organism to a single cell, a fertilized egg, or a spore. The reduction to a single cell gives the genetic integrity of the individual. Even so, some population biologists view the parts of a plant, particularly grass shoots, as competing individuals. The shoots from a single root stock are called ramets,[22] as opposed to genets, which are plants derived from a single seed. Genets have genetic identity, while ramets have physiological integrity, as in a single stem. While the genetic and physical discreteness of organisms clearly fails in such conceptions, the individual grass shoot is adequately defined by its physiological autonomy.

Thus, to be an organism, a being should have at least one of the following: (1) genetic integrity, because it comes from a single germ line (egg or spore); (2) a discrete bodily form; and (3) physiological integrity within and physiological autonomy from other organisms. Normal humans are the archetypal organism; other beings pass for organisms, even though they may be compromised on one or more of these three general features.

The Population Criterion

Populations follow from individual organisms, for they are collections of individuals. Even so, there are individuals that could be defined as populations in their own right. Colonies of primitive animals like sponges and anemone relatives have all the critical characteristics that we use to define a single organism. For the most part, populations contain one species, but this is by no means a requirement.

Populations vary in size from a few individuals chosen ad hoc for a given purpose, to millions of individuals spread over a large geographic area. Spatial discontinuity between populations is often a helpful criterion, but discontinuity can be very much a matter of degree. Even the convenient tangible bounds of an island can be a matter of degree, in that influx from surrounding populations combined with significant emigration could make the population as transient as a collection of human beings in a crosswalk. The physical bounds of the island might be perfectly clear, while at the same time they may correspond to no significant biological limit.

Often there are biological underpinnings to populations like a certain genetic homogeneity. This too is a matter of degree. Pollen comes on the wind to insert genetic variation in plant populations. Newcomers may readily invade animal groups. Species are special populations of large size with a degree of infertility in crosses between groups. For example, woodland primroses and their close relatives in the adjacent pastures regularly produce hybrids of intermediate form.[23] In animals, the circumpolar species of birds blend into one another, and interbreed with their neighbors.[24] The ends of the ring often cannot interbreed because the two end species are too divergent. Despite exceptions in populations of all sizes, populations can often be defined on grounds of relative genetic similarity within. Populations are collections of individuals that can be delimited by many criteria, depending on the question the ecologist chooses to ask. Organisms in populations are united with something of a shared recent history. Thus, a flock of different species may be a valid population. Recent interbreeding is only one of the histories that might apply.

The Community Criterion

Here we take a stance possibly at odds with convention. We do not insist on our point of view, and others are free to use their own definitions of communities. Research on the way ecologists conceive of communities shows us that plant ecologists do not, for the most part, work on communities using populations as the parts.[25] An analysis of the proportion of research papers on communities involving particular organisms forces us to the conclusion that the majority of ecologists conceive of communities as consisting of individuals rather than populations. Population studies disproportionately focus on animals and small plants. Trees can be studied as populations, but the fact that they are bigger than humans presses their individuality on us. That makes it hard to see populations in a forest. Animals can be readily seen to herd into populations, and the spatial limits of the populations of small plants can be observed easily by standing and looking at them from above. That is why animals and small plants are the favorites for population but not community work.

The explanatory principles of communities may involve competition, but not the clean, clear competition measured in population experiments. Competition in communities is set in a variable environmental context that does not allow population competition to come to exclusion.

Communities are the integration of the complex behavior of the biota in a given area so as to produce a cohesive and multifaceted whole that usually modifies the physical environment. The forest is an archetypal community with its individual trees bound together as members of the community by a tangle of processes. Trees standing side by side represent an instant in a continuing complex process involving not only competition but also interference, accommodation, and mutualism, among other factors. Our community conception only allows us to study competition, mutualism, and the many other population processes in aggregate. Communities are held together somewhat by microevolution. Species do not so much evolve into community membership as they represent multispecies units that arise as relatively stable entities coming from the ready-made species present. The species are like LEGO bits. The bits in isolation do not determine the community, but there are combinations of them that are stable. We see the stable configurations. The stability therein often may not be a product of evolution.

Because of the radically different space/time scales of animals and plants, coherence is generally only recognizable in either animal or plant communities, but not in combinations. Plants are good at plant taxonomy, and animals are good at animal taxonomy, but each is not so good at the other. This makes coherent response as multispecies assemblages difficult between kingdoms. There are plant and animal mixtures, such as biomes and coevolved species pairs or sets. In biomes, the insistence on species is relaxed, and is substituted by plant form. The form reflects physical manifestations to a shared climate. Animals groom biomes. Coevolved pairs of species are very robust. Communities come and go in each interglacial period, but species pairs, such as pollination partners, evolve together in, for instance, moth genera and orchid genera. Neither biomes nor coevolved plant–animal dependencies have a straightforward relationship to communities.

The Ecosystem Criterion

The functional ecosystem is the conception where biota are explicitly linked to the abiotic world of their surroundings. System boundaries include the physical environment. Ecosystems can be large or small. Size is not the critical characteristic; rather, the cycles and pathways of energy and matter in aggregate form the entire ecosystem. Robert Bosserman's thesis research was performed in the Okefenokee Swamp in Georgia.[26] His work was part of a big push on ecosystem research at the University of Georgia in Athens, led by Bernard Patten and organized by Edward Rykiel. Bosserman's research was part of ecosystem studies like others at the site. The difference was his bold reframing of ecosystems independent of size. Most everyone else was measuring big cycles of nutrients and spanning the large swamp, but Bosserman's work dealt with a natural microcosm. They were clumps of floating carnivorous plants called bladderworts, the size of a few handfuls. That plant performs photosynthesis and so provides the primary production. But being in a swamp, it lives in a low-mineral environment and turns to insects for its mineral supply. When a water flea touches a trigger hair, small bladders that are under negative pressure open and water with the flea in it rushes in. The flea is digested. Bosserman measured all the major functions in the clumps and discovered that all the major compartments of a typical ecosystem were present. Primary production, trophic consumption, and detritus compartments that fed production were all

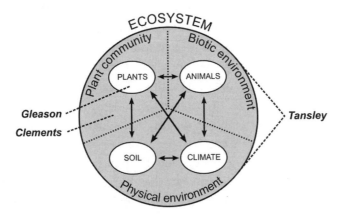

FIGURE 1.12.   Gleason, Clements, and Tansley looked at vegetation in three different ways. Gleason saw the plant community as a collection of individuals filtered by environment. Clements saw the plant community as an integrated whole, set in a physical environment. Tansley saw the plants, their biotic environment, and their physical environment as all components inside the ecosystem. The same material system is seen from different perspectives.

there. Ecosystems are defined by flows and cycle, with the bladderwort operating as a natural microcosm, showing that ecosystems can be of almost any size.

The origin of the term "ecosystem" goes back eighty years to when Arthur Tansley recognized the need for an entity that blended biota with the physical environment.[27] That original explication persists so that conventional accounts of the ecosystem include a large box, the ecosystem, with four boxes inside labeled plants, animals, soil, and climate (figure 1.12). Unfortunately, such a characterization refers to the intellectual history of ecology in the 1930s more than to the powerful ideas that flowed from Tansley's brainchild. The ecosystem contains plants, animals, soil, and atmosphere, but those names are not helpful categories for seeing how ecosystem parts are put together. The functioning subunits in the ecosystem consist not of plants, animals, soil, and atmosphere, but of process fluxes between them. Plants do not naturally separate from soil when the compartment in question is "below ground carbon."

Tansley's term met a need of his day, for ecosystem approaches did precede the name itself. A full decade before the term ecosystem was coined, Transeau measured the energy budget of a cornfield.[28] He subtracted outputs from inputs to identify net gain. It would have been more productive had ecology taken that study as the archetype of an ecosystem. Transeau's implied ecosystem emphasized fluxes and pathways that are hard to address in an organism-centered conception that insists on preserving plants and animals as discrete entities.

The critical difference between an ecosystem and a community analysis is not the size of the study, but rather the difference in emphasis on the living material.

In community work, the vegetation may influence the soil and other parts of the physical environment by the very processes that are important in ecosystems. In communities, the soil is explicitly the environment of the plants. The biotic community is recognized as having integrity separate from the soil. The same vegetation on the same soil may be studied as a functional ecosystem or as a community, depending on how the scientist slices the ecological pie. Ecosystems are intractable if the biota is identified as one of the distinct slices, particularly if separate organisms are allowed to be discrete parts.

Take the case of the cycling of nutrients for repeated use. The leaf falls from the tree; worms eat the leaf; rainwater washes the nutrients into the soil directly from the leaf and from the feces of the worm; fungi absorb those nutrients and convey them to the root to which they are connected; the root dies, leaving a frozen core of nutrients; in spring, new roots grow down the old root hole, collecting the nutrients; and the rest of the plant passes them up to the leaves (figure 1.13). Harris, Kennerson, and Edwards inferred two periods of vigorous, below-ground growth in yellow poplar, once in the fall and then again in the spring.[29] In terms

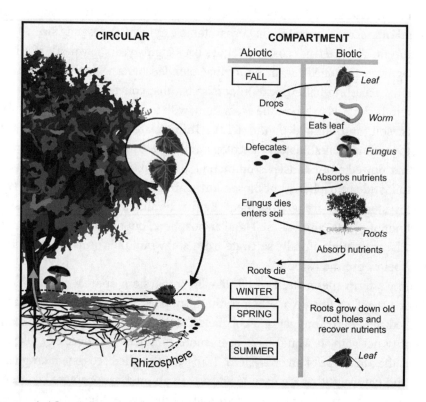

FIGURE 1.13.   For a mineral nutrient to pass once around a nutrient cycle, it must pass in and out of the biotic component several times in one year. Accordingly, biota are not simple subsystems of process/functional ecosystems.

of a well-specified ecosystem, the earlier description is a simple, efficient nutrient pathway. In terms of a model that insists on emphasizing living material as different from nonliving material, the pathway is horrendously complicated. In just once around the cycle, the nutrients pass in or out of the biotic compartment at least four times.

The Landscape Criterion

Landscapes occupy a range of scales comparable to ecosystems and communities, but according to their own distinct criteria. Although landscapes are generally accessible from commonplace experience—and represent one of the earliest criteria for studying ecological systems—they have lain neglected for most of the twentieth century, ignored by almost all ecologists except those studying wildlife. The landscape ecology of the nineteenth century was supplanted by community conceptions.[30] Community ecology has its origins in an abstraction of landscapes; one where the pattern of the patchwork on the ground is replaced by abstract community types defined by species lists and proportions of species abundances. Pursuit of the community abstraction left the landscape conception untended. When a new conception was deemed necessary in the middle of the twentieth century, the ecosystem and not the landscape prevailed. In a sense, Victorian country vicars did all that could be done with landscapes before the computer age. In the last three decades, landscapes have again come to command the attention they deserve.

Spatial contiguity is the ordering principle for landscapes. As the dynamics of species replacement in community patches became a focus of study some forty years ago,[31] the scene was being set for a reawakening of interest in landscapes. More than in any other subdiscipline of ecology, landscape ecologists are buried in data. A casual glance across a vista reveals an unmanageable mass of detail. To achieve anything more than the obvious, the landscape ecologist must have cheap, fast computation. Recent years have seen the development of several measures of pattern that formalize and quantify what heretofore has been subjective, informal assessment. Remote sensing from satellites has allowed the landscape ecologist to move upscale. In many studies, the student of landscapes has only to look; that which made landscapes so obvious as to be trivial early in the last century now makes them natural objects for study in a modern, computer-assisted world. With tools to identify spatial structure, landscape ecologists now turn to fluxes of material in spatial contexts. Landscape ecology now investigates the consequences of elaborate spatial structure.

The Biome Criterion

By definition under conventional wisdom, biomes cover large areas. However, it is worth considering what the essence of a biome is, aside from size. Biomes are defined by the dominant vegetation physiognomy, something not strictly scale-defined in itself. A biome also should have a critical climate component such that the other characters are responses to some meteorological consideration. Often, an assemblage of animals plays a central role in giving the biome its particular structure. Examples here would be the spruce-moose biome or the grassland biomes with their respective grazers.[32]

As the name suggests, biomes are characterized principally by their biotic components, although soils and climate are important parts of the picture. Biomes, at first glance, are a hybrid of community and ecosystem with a strong landscape reference. The distinctive character of biomes is revealed when the concept is applied to situations scaled smaller than usual. Small systems that are simultaneously physiognomic, geographic, and process-oriented might prove very helpful. Using the scale-independent biome concept avoids the confusion that arises when we try to use one of the other criteria to describe such situations. Landscapes, communities, and ecosystems used separately or in tandem cannot do the biome concept justice. However, they are often pressed uncomfortably into service because we lack a term for small biomes. A frost pocket is a patch of treeless vegetation set in a forest. The absence of trees allows cold air to collect and kill any woody invaders. It is not adequately described as a community because the species are incidental to their life form. A frost pocket has all the regularly recognized biome properties except size: physiognomically recognizable, climate determined, disturbance created, and animal groomed.

The Biosphere

The biosphere is the one ecological system where the scale is simply defined. Its scope being the entire globe, it occupies a level that is unambiguously above all the other types of ecological systems discussed so far. Considering it a large landscape is possible but awkward. The biosphere is more often studied as a macrolevel ecosystem. For example, students of the biosphere ask questions about global carbon balance, a problem involving the same sort of fluxes considered by an ecosystem scientist. There is, however, a subtle but unequivocal shift in the relative importance of Tansley's four parts to the ecosystem when we make a shift upscale to the biosphere. The atmosphere is definitely an overriding influence in the biosphere, whereas in subcontinental ecosystems, soil interactions with biota through water are at least equal ecosystem players with the meteorological components. The biosphere is not easily conceived as a large biotic community because there is not interaction of species on a global scale to hold the global community together as a working unit.

We have looked at other criteria independent of scale, and solved the problem by identifying the means of investigation. Systems dominated by three-dimensional diffusion as the means of connection give plant physiological studies something in common with biospherics. Plant physiologists, such as Ian Woodward in

Sheffield,[33] often move into biospherics. He considers the density of plant stomata as a surrogate signal for the past global atmospherics.

Another feature of the global system is that it is close enough to a closed system to make workable the assumptions of closure that physicists are wont to make. So the ecology of closed systems has biospherical qualities. This is a critical issue for projects that put life in space. As Peter Van Voris and his colleagues were developing the microcosms they used to investigate ecosystem function,[34] some of the systems they probed were closed (the final ecosystem microcosms used were in fact quite open). In the closed systems, they found that those consisting of mostly liquid would boom and then crash, ending with a few species existing as resting stages on the wall of the microcosm. If they included a significant soil component in their closed systems, diversity in an active state would go on more or less indefinitely. They came to understand that the slower dynamics in the soil acted as an anchor to the rich functioning of the system.[35] Systems need a context for their functioning; otherwise, fast dynamics come to crash the system. The escape of the budworm in Holling's model shows that phenomenon (figure 1.11). The fast dynamics of insect outbreak booms and then crashes the system.

Biosphere II was an experiment on a closed system. Although there were problems with some of the science, it was a bold move to find out patterns that follow from a system being closed. In Allen's visit there, he saw many examples of small closed systems in sealed glass vials with persistent life in them, but as cysts on the glass walls. In the big Biosphere II, there were major biomes of varied sorts, to see if the functioning of the Earth turns on the diversity of forms that it contains. Once the system was closed, there was a steady depletion of oxygen. The reason was that the new concrete cured over time, taking in oxygen. As a result, there was some introduction of various materials, such as oxygen. One introduction was a copy of the first edition of this book. The biospherians were excited to get it. Allen gave a talk to the staff and the biospherians inside, and they all resonated to what it said. While there were problems with the original science in Biosphere II, the vision of it has been lost in a takeover by mechanistic science in recent decades, and few care anymore. Closure is a way to access the functioning of whole systems.

A FRAMEWORK FOR ECOLOGY

Biology is not just complicated chemistry and physics, it is a different discourse. Biology cannot deny the second law of thermodynamics; in fact, it depends on it.[36] We can see biology as a set of rules that work as a special case of physical happenings. Some limits in biology are physical, but many describe things that are possible in physics but not allowed in biology (figure 1.14). Nerves depend on potassium crossing a membrane by simple diffusion. But before the nerve fires,

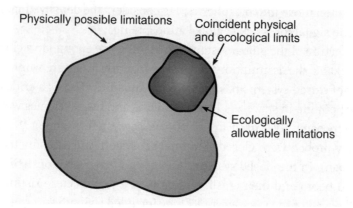

FIGURE 1.14.   Only occasionally is what is observed in ecology directly the result of a physical limitation. More generally, what is ecologically allowable is a subset inside what is physically possible.

potassium is not allowed by the biology to cross the membrane and go down the density gradient; instead, it is actively moved in the other direction to become more concentrated on one side of the membrane. In all this, physics is not violated; it is just that some of what unconstrained physics will do is not allowed.

Physics does not have precedence or privilege. In fact, Robert Rosen would say that the reverse applies. In his classic work in biology, *What Is Life*, Erwin Schrödinger pointed out that there was not a physics that applied to biology. And Rosen submits that such still pertains. The reason is that physics, as it is practiced, needs to make a set of assumptions. Now, those assumptions in physics—like all assumptions everywhere—are not true. Systems are not in fact closed, but in physical systems you can get away with calling them so. The point is not to find which assumptions are right, but what lies you can get away with. You simply cannot get away with assumptions of closure in biology. And there is on its face no life close to equilibrium. At equilibrium, life dies. Rosen says we should put open, far-from-equilibrium biology, such as organisms, in physics through analogy.[37] Then we will have physics for biology.

As scientists, we take a position that is distinctly data driven. When we recognize a type of thing, we are scanning our past experience for things that are like what we see. So type is subjectively chosen by us, but in the light of experience. Do not imagine then that type is only internal to ourselves, it involves things we have already experienced. Also, do not imagine that the scaling of things we see is imposed on us by nature, with no values on our part. Consider a pond skater and a big dung beetle. Put both on the surface of water, and one skates and the other sinks. It is a matter of scale; the beetle is too big to skate. Now put a metal bolt on the water; it too will sink. But somehow that is not a matter of scale; it is simply a different situation. We as biologists would dismiss the bolt as something else, not something heavier. But let us ask a physicist, who we would imagine saying, "Oh, it is a scale issue. The bolt is heavier than the skater so it breaks the surface tension, just like the beetle." Biologists might say that the beetle and the skater are both insects, and so are similar enough to have one scale applied to them. The bolt is not similar enough for our biological sensitivities. Thus, scale or not scale depends on the observer, and so scale has some of its origins inside us and our decisions.

The subjectivity of naming things ecological is crucial to ecological understanding. Calling something an ecosystem involves taking a point of view on a given tract of land that emphasizes something different than if one called that same material system a community. The framework we erect subjectively defines entities of an ecological sort and embeds them in a physical setting where scaling comes to the fore.

Ecological systems are complex and require careful analysis if the student of ecology is to avoid being lost in the tiered labyrinth of the material. The most important general point covered thus far is the recognition of the role of the observer in the system. Ignoring human subjectivity will not make it go away. Since one makes arbitrary decisions anyway, ignoring them abdicates responsibility needlessly. All decisions come at the price of not having made some other decision. By acknowledging subjectivity, one can make it reasoned instead of capricious. A real danger in suppressing the ecologist in ecology is to be bound by unnecessarily costly decisions. These could be exchanged for a more cost-efficient intellectual device if only the subjectivity of the enterprise were acknowledged.

One of the most important benefits of consciousness of the observer is knowing the effects of grain and extent on observation. The formal use of grain and extent gives opportunities to avoid old pitfalls and opens up possibilities for valid comparison that have heretofore escaped our attention. Also of great importance is being cognizant of patterns of constraint. Constraint gives a general model for couching ecological problems in terms that generate predictions. At this point, we have put most of the crucial tools in the box and are ready to go to work. It is time to erect a general frame inside which we hope ecology will become a more predictive science. It will certainly be more unified than the fractious thing that is generally practiced at the time of this writing.

Given the extended treatment of definitions of ecological criteria and the tools for linking them put forward in this chapter, the framework we suggest can now be adequately stated in a short space. The temptation we resist is to stack types of ecological systems according to an approximation of their size. Yielding to the desire for tidiness in the conventional ecological hierarchy is costly.

Up to this point, we have detailed our paradigm describing the difference between scale-based levels of observation from levels of organization. Our representation for this paradigm is a cone-shaped diagram containing columns to represent ecological criteria at different scales (figure 1.15*A*). Instead of stacking criteria one upon another in the conventional order, the diagram separates scaled ordering from an ordering of levels of organization. As a result, we can hold the scale constant and look at how the criterion under consideration is but one conception of a material system at a given temporal and spatial extent, for example, considering a forest as a patch of landscape, as a community, or as an ecosystem. We can put each type of ecological system at every scale. Organism, population, community, ecosystem, landscape, and biome systems will sit side by side at every scalar level. Not only can we compare the ecosystem and community conceptions

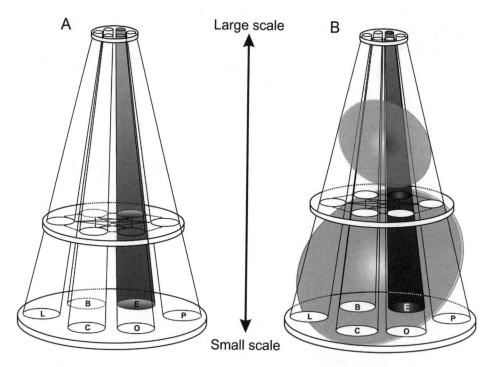

FIGURE 1.15.   *A.* The cone diagram metaphor for ecological criteria and ecological scale. The wide base indicates a large number of small entities; the narrow top indicates a small number of large entities. The cross section across the entire cone represents one middle-level scale. Although there is only one here, any number of cross sections could be inserted, each at its own scaled level. Each letter indicates a different criterion: O = organism; P = population; C = community; E = ecosystem; L = landscape; B = biome. The six lettered disks correspond to a comparison of scale-independent criteria that intersect at a given grain and extent. Individually, the columns represent a criterion for looking at the material system, for example, the abstract notion of community. The disk labeled C is an actual community with a particular spatiotemporal size assigned to it. In the C column, larger-scale contextual communities occur above that community, while smaller-scaled community subsystems would occur below. *B.* The large diagonal disks show an ecosystem, as an example, of a given size working as the context for a smaller community, while itself being held in the context of a larger community. Many other diagonal cross sections are possible, such as one that sees a cow's rumen as an ecosystem, obviously scaled below the cow itself. The organism is the context of its rumen ecosystem.

across a landscape of a given area, but we can do the same at larger- and smaller-scaled landscapes.

Our scheme is not limited to making comparisons horizontally across a given spatiotemporally defined level. We can move up and down the scale on the cone diagram and make comparisons between differently scaled entities of a single type of system. We could see how a given ecosystem contains smaller ecosystems, while being itself part of a larger ecosystem. As we move to levels below, we may find the mechanistic explanations of the behavior of the level in question. The levels above define the role of entities at the level in question in the functioning of larger systems. The upper levels define role, purpose, and boundary conditions. Predictions come most readily when the system to be predicted is up against a constraint imposed by the layer above.

Probably the most interesting—and certainly the most neglected—questions will involve slicing the cone diagram diagonally (figure 1.15*B*). Here, we change the type of system while also changing levels. In a sense, that is what the conventional hierarchy does, but we are not limited to the conventional order. Any type can be the context for any other type. Diagonal slices will allow us to see how community patterns at a given level are influenced by nutrient status maintained by a subsystem defined in ecosystem terms. With our ecological cone diagram, we can ask how low-level ecosystem function is constrained by population considerations of a dominant tree population. We no longer need to see a difficulty when animal communities carry nutrients around an ecosystem that sits across the boundary between two biomes. The biome boundary is clear by its own criteria, but it is leaky with respect to nutrients involving animals and ecosystems. We do not recommend a compromise that forms a general-purpose system designed by a committee composed of one population biologist, a community ecologist, an ecosystem scientist, and a biogeographer. Rather, we suggest a formal change in the type of system description every time a new explicit question or explanation demands it. We use the above scheme to organize the material in the chapters that follow. We see it as both flexible and encouraging consistency.

The cone diagram of figure 1.15 is our version of the Beaufort scale. We use it to fix order in rich ecological situations that are at least as complicated the seas that Beaufort assessed. Beaufort only separated situations on wind speed and its consequences. Our use of distinction between type of situation and its spatiotemporal scaling gives a richer device, one more suited to the many nuances of ecology.

2

THE LANDSCAPE CRITERION

CHAPTER 1 proposes that the major ecological criteria be seen as independent ways of viewing ecological systems, independent in that they need not be held in a rank order prescribed by the conventional biological hierarchy. Thus, there are two separate considerations for addressing any ecological system: scale and system type. Such a scheme opens up a formal approach to a whole complex of new relationships. There is something to be said for using this scheme to address first the most obvious and tractable ecological type. There are two candidates, landscapes and organisms; both are tangible. We employ the landscape criterion as the point of departure.

For all the separate criteria, we have tried to follow the same plan through this book. There is the option to hold the criterion constant and move up and down scale, looking at smaller and larger versions of the same type. In terms of the previous chapter, this involves moving up and down one of the columns in the cone-shaped diagram (see chapter 1, figure 1.15A). Alternatively, we can hold the scale constant and look at how the criterion under consideration is but one conception of a material ecological system at a given temporal and spatial extent. This cuts across the diagram horizontally to expose one scale. Finally, it is possible to see how a given criterion offers either a context or an explanation for some other criterion. This last option involves simultaneous changes in both the scale and the criterion used to address the material system. Here, we cut diagonal slices through the diagram to divide the different columns at different layers (see chapter 1, figure 1.15B). An overextension of moving diagonally is what happens in the conventional hierarchy. It is unlikely that moving more than a level or two diagonally will give coherent insight, but the conventional hierarchy fixates on changing criterion and scale, creating a straitjacket. In this chapter, we see how to move diagonally across scale and type with effect. More important, we look at landscapes in a flexible manner and in a variety of ways to provide a well-rounded point of view.

Landscape ecology goes back to early expeditionary naturalists such as von Humboldt,[1] but had lain quiet in America from the early twentieth century until recently. Only in wildlife ecology did an interest in landscapes persist. The study and application of wildlife management is generally an applied science, solving problems offered by some management issue in some space. This makes it unique in dealing with mobile organisms that work across ecological criteria at many scales. This is in contrast to plants that are largely stationary, and less-applied animal issues that often have a narrow range of spatial scales for any given study. Explanations of the science of wildlife ecology require a unique understanding of multiple criteria at multiple scales, usually in terms of landscape occupancy.

However, there has been a resurgence of activity on landscapes through the 1980s that promises to persist. The switch away from landscapes seems to have occurred when the community concept came of age, husbanded by Frederic Clements.[2] There was much debate about the proper meaning of community because there was conflict in the minds of early community ecologists between the community ideal and actual stands. They gave the name "community" to both the collection of plants living on a particular site on the landscape as well as the abstract archetype community defined by associations of organisms over millennia. At first, the roots of communities in landscapes refused to die. The tension in the minds of early ecologists was between community concepts as they apply today and the patches on a landscape that were the origins of the community concept from the late nineteenth century.

Landscape ecology has been neglected until recently, probably because landscapes were considered obvious. As we have argued with Beaufort's scale, obvious is a challenging thing to formalize. Certainly, landscapes are the most tangible of the large-scale ecological conceptions, for one only has to look to see them literally. The conventional wisdom is that landscapes are large, but some researchers in the spirit of this book are relaxing that consideration. Bruce Milne, for example, studies the landscape from a beetle's-eye view, so large physical size need not be a requirement.[3] In the plant kingdom, Allen performed Greig-Smith pattern analysis of algae on orchid leaves; so landscapes may be large or small.[4] Of course, the scaling of movement of algae and beetles are different, but both those studies are on areas much smaller than the classic landscape vista. There are sometimes unfortunate references to the landscape scale. That is just so much hand waving at about the scale of the Turner paintings discussed in the introduction, that is to say, landscape is about what humans can see from a vantage point. This is so even if ecologists are not explicitly aware of what they are doing. "Landscape scale" represents the generally slovenly use of language and concept in ecology that we wish to stamp out with this volume.[5] No, landscapes are types of systems, held together by strictly spatial criteria. "Regional scale" is just fine because it refers to the extent of a landscape.

Unlike landscapes, ecosystems and communities have never been considered obvious, probably because they both require special measurements before they become apparent. Modern technologies have suddenly given us access to landscapes in ways that press them upon us as nontrivial systems. We can now see landscapes over enormous areas while preserving remarkable detail. Remotely sensed images that distinguish between thirty-meter grid units using seven wavelengths are publicly available for ecological research.[6] The satellite images make these fine distinctions while scanning county-wide swaths around the entire globe. Computer processing of such masses of data allows one to identify quantitative signatures of landscapes. Signatures are general patterns of details exposed by mathematical treatment that collapses the details that would otherwise overwhelm human memory. Human brains are good at capturing their own version of landscape signature, but the summary treatments through computation open up new summaries with different implications. Landscape ecology is coming of age. In fact, landscapes were never trivial, but contemporary image technology has brought that point home.

THE LANDSCAPE CRITERION ACROSS SCALES

Landscapes can be related to other ecological criteria for organization, such that the landscape becomes the spatial matrix in which organisms, populations, ecosystems, and the like are set. However, landscapes are meaningful in their own right and so it is possible to consider differently scaled systems while using only the landscape criterion. Within large areas there are smaller subsets that are mosaic pieces. Patterns may be extended so that they become segments of larger patterns. Smaller patterns may be just details from larger patterns or they could be autonomous with their own distinctive causes. Let us first consider the landscape criterion by itself, and look at landscapes nested in larger landscapes, or landscapes composed of smaller landscapes. Let us focus on the landscape column of the cone-shaped diagram with its differently scaled vertical layers.

Through the last third of the twentieth century, mathematicians developed fractal geometry, a theory and method for dealing with the shape of complex entities.[7] These methods have been used to create intricate and strikingly realistic landscapes for science fiction films, and so have a place in popular culture. The fractal dimension of a pattern is a measure of its complicatedness, but in a remarkable way that gives a constant complicatedness across scales. Fractals are particularly pertinent at this point in the discussion because they are derived from qualities of the pattern over different grain sizes. In this way, a fractal dimension is an integration over levels while maintaining the landscape criterion. Moves across levels insert difficulties that cannot be resolved. In fractals, that manifests itself as the introduction of zero and infinity in the expression. Zero and infinity are undefined numbers and in fractals they show movement across scales to be undefinable. A critical property of complexity is that it is undefinable, so fractals are a way of moving forward anyway, even when the situation is undefinable. Complexity science in general turns on working when definition is not possible.

Fractal dimensions can be calculated in many ways, but there is one that makes intuitive sense for the purpose at hand. It involves the length of the outline of entities forming the pattern. Critically, this is a comment on measurement, as much as the thing observed in and of itself. As one traces the outline of a natural irregular shape like a coastline, there is always a degree of smoothing of detail, no matter how carefully the measurement is made. If one were to follow the outline in infinite detail, then the length would be infinitely long (figure 2.1A). On crude world maps, the outline of Great Britain might be made as a simple long triangle: one point for Land's End, another for John o'Groats at the tip of Scotland, and a third for the tip of Kent in the southeast. Note that the straight lines connecting the points are long. If the distance between the reference points were shorter, then some other coarse features would emerge. With slightly shorter segments, the top of Scotland could be seen as flat, not pointed, and Wales would appear as a rectangle stuck on the side of England. With still shorter segments, the outline would include a bulge for the Lake District, a rounded outline for East Anglia, and even a small notch for the Wash. As the outline is mapped with shorter segments, more details appear and the total length measured gets longer. The coastline of Britain measured to take into account the outline of every grain of sand on Brighton Beach would be of enormous length. The relationship between the lengths of the total outline to the lengths of segments used to make the measurement can give the fractal dimension of the shape.

Each shape has its own relationship between the length of estimators (chords for a circle) and the total estimated perimeter. Circles are not fractal because there is a limit to the circumference. Calculus works on systems that have limits (remember that "lim" in calculus). Mathematicians knew of systems without limits in the nineteenth century, but did not know how to deal with them, and called them mathematical monsters. The Julia set was one known then. Fractal systems have no limit; the coastline of Britain is fractal, and therefore has no limit. It is infinitely long. This does raise complications in that there is an island to the west (Ireland)

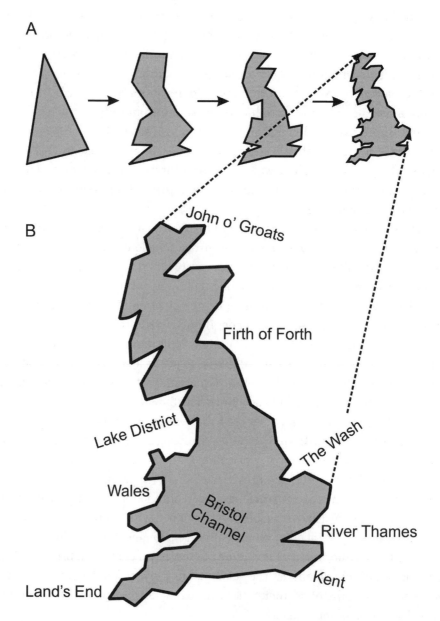

A

B

John o' Groats

Firth of Forth

Lake District

Wales

Bristol
Channel

The Wash

River Thames

Kent

Land's End

FIGURE 2.1. *A.* Maps of Britain capture more detail as the length of segments used for the outline are shortened (Mandelbrot 1967). With the longest straight-line segments, Britain appears only as a triangle. As the straight segments for the outline are made shorter, more and more detail emerges, while the length of the coastline increases. For infinitely small straight-line segments, the coast of Britain is infinitely long. The length of the outline of Britain is unbounded. *B.* As the measuring stick becomes shorter, more detail can be seen. The Wash appears as a notch between East Anglia and the North. The Firth of Forth, by Edinburgh, appears too. With shorter measuring segments, the Firth will go deeper in as a slash, as would the Humber in the dent just above the Wash. Anglesey, the knob on North Wales, would appear as an island once the detail of the Menai Straits can be detected.

that has a coastline only one-quarter that of England, Wales, and Scotland; and yet the coastline of Ireland is also infinitely long. Infinity one-quarter the size of another infinity is a quandary.

The more complicated the shape to be estimated, the longer the estimated perimeter as estimators shorten. The shorter estimator segments will pick up more elaborate details in a complicated shape. Figure 2.2 shows how the shorter segments pick up more detail in the complicated shape. The ratios of long to short estimated segment perimeters are larger in the more complicated shape. The shapes in Figure 2.2 are not fractal; they only show how complication is captured in a bigger ratio. But the same will apply to fractal figures that will have higher fractal dimension in more complicated patterns. Fractal dimension is a measure of complication. The complexity is found in the way the measurement transcends scale and invokes situations that cannot be defined. Fractals describe outlines that do not fit in integer dimensions, such as 1, 2, or 3. If a line is very wavy, it will come closer to filling a two-dimensional plane. A simple straight line is one-dimensional. So fractal wavy lines fit in a space that is too large for one dimension but too small to exist in two dimensions. Make a log-log plot of estimating line segment against the total measured circumference. If the line on the graph is straight, the outline is fractal, with the same degree of complication across scales. The steepness of the log-log graph tells how complicated is the outline across scales; it tells the fractal dimension of the space in which the pattern fits (figure 2.3A). A simple shape has much of its detail captured when relatively long line segments are used as estimators (figure 2.3B). Therefore, shortening the length of the estimators will not add much to the total length of the estimated perimeter. Long-line estimators poorly approximate complex shapes, so shortening the estimators picks up much extra detail. This greatly increases the length of the total track, indicating a complex shape and a high fractal dimension (figure 2.3A). The log-log plot is steeper for more complicated outlines that will only fit in a higher dimensional space. The steepness of the log-log plot indicates the complications of the outline. The fractal dimension of a system is the reciprocal of the slope.

In fractal systems, a plot of the log of the length of the measuring segment against the log of the total perimeter gives a straight line. In the world of mathematics of a given fractal equation, the line will be straight all the way to an infinite number of segments of infinite shortness giving an infinitely long perimeter. In the extreme case, these patterns have the same underlying process working across scales and would use exactly the same equation across all scales. Such patterns are not just fractal; they are self-similar fractals. It is clear in the Koch diagram that this is the case (Figure 2.4).[8] Not only does the same degree of complicatedness apply but the exact same outline is repeated again and again across scales in self-similar fractals.

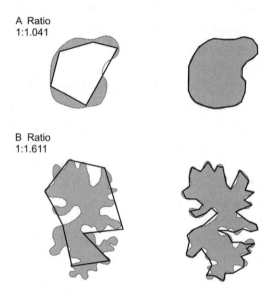

FIGURE 2.2.   The more complicated the outline of a shape, the greater the total perimeter will increase, as the line segments used to outline it are made shorter. *A.* The upper shape is simple and therefore has a small-ratio estimated perimeter with long and short estimator segments (1:1,041). *B.* The lower shape is complicated, therefore revealing a greater increase in perimeter with shorter estimator segments to yield a larger ratio (1:1,611).

The Koch diagram is self-similar and displays one set of activities that can be expressed as a single equation. The diagram starts with a triangle. The sides of the triangle are divided into three segments. Then a fourth segment is added. Now the line is one-third too long to fit between the corners of the triangle, and so the edge must buckle. The longer line is made to fit by inserting a kink that

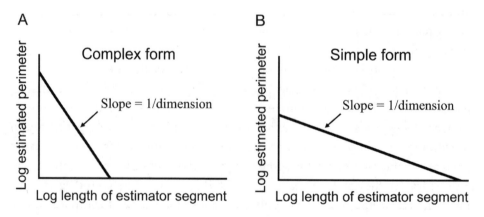

FIGURE 2.3.   If a log-log plot of the lengths of the estimators and the lengths of the total perimeter is a straight line, then the structure is fractal.

Self - similar pattern

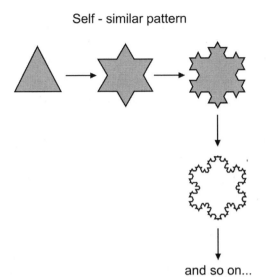

and so on...

FIGURE 2.4.   In the Koch diagram, straight-line segments are divided into four segments with a point merging in the middle of each straight line. Subsequent straight lines of shorter lengths can then be further divided into lines with a point in the middle. The system is self-similar in that successively finer-grained patterns follow smaller-scale versions of the original pattern. The fractal dimension of a Koch diagram is log 4 over log 3.

forms a new smaller triangular outline in the middle of the line. That, of course, now leaves four shorter lines, which are kinked the same way in the next round of elaboration. And after that second round of kinking, sixteen shorter segments showing triangular out-dentations now replace the original straight line. And the process can go on all the way down to an infinite number of infinitely short segments. The Koch diagram is self-similar. The fractal dimension of the Koch diagram is the log of 4 divided by the log of 3. The 4 and 3 refer to three segments being replaced with four.

Clouds were well known to have patterns similar enough at each scale so that one could not tell if the cloud was big and at a distance or close and small. Close and far views all look similar enough that you cannot tell which is closer. All look the same but with incidental differences; they are all cloud-like. Of course, clouds are material things, not mathematical creations. Clouds are fractal, and we can show that from scaled measurements. But they are not self-similar. The cloudiness persists across scales; each cloud and each part of each cloud are not exactly the same in outline, but they are all about the same. In the Koch diagram, patterns at different scales are not just generally similar, they are exactly the same. So self-similarity comes from mathematical processes that are technical descriptions.

To move our argument forward, we need to return to Rosen's modeling relations shown in the introduction (figure 1.5).[9] We have argued that you do not want a model to have size as one of its parameters. The error of insisting on a certain size occurs in biomes, which are mistakenly taken to be only large. We can now offer the technical basis of our position. We have pointed to formal models as being applied across scales, as the equations for aerodynamics apply to flying objects of different size. If two objects fit the same equation, then we can do experiments on one and apply the results to the other. The two observed (not calculated) systems together constitute a different sort of model; an analogy and compression down to what the two have in common. There is no symbolic or mathematically derived part of the compressed relationship. It is simply a material equivalence, an analogy.

So if we stay in a closed mathematical system, we might be creating a self-similar fractal. And they will apply to the infinitely large or small situations. However, when we move out to address the world as it is measured to be fractal, we are using a fundamentally different sort of model—an analog compression, not a mathematical cross-scale representation. Those measured patterns will have very small differences depending on the initial conditions of the situation when a particular scale was measured. Fractals are an expression of a chaotic system. In chaos we find one reason why biological systems repeat, but never perfectly. Take one electron out of one mouse in a mathematical experiment using a chaotic equation for growth and you can have a different number of mice in as few as fifteen generations. We can never get the initial conditions inside that level of precision, so even deeply controlled experiments come out differently when repeated.

Fractals that are not self-similar come from the measured side of fractals, not a mathematical equation. The very same process is in play when clouds form and grow, so they look the same across scales. However, each cloud starts with not exactly the same initial conditions. And the same applies to the creation of parts of clouds. The same general pattern of complicatedness arises because of the same underlying physical process, but the results are not exactly the same, as they would be in a self-similar fractal. And on landscapes, we encounter processes that in fractal ways cascade upscale until they run out of steam, or encounter some other process cascading toward the first. We therefore need to look at landscapes across scales to face the upscale rescaling. When we do so, we find that limits of the fractal dimension of one process are commonly observed to be overwhelmed by another. So while fractals apply across scales, they break down in practice. We need to deal with that fact.

When fractals come from iterating equations, it is possible in principle to go down to infinitely short segments giving an infinitely long perimeter. But in practice, measured fractals reach lower and upper limits, whereupon some other fractal

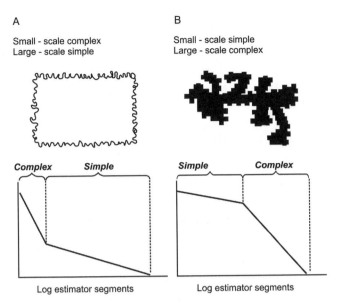

FIGURE 2.5.   *A.* The outlines are merely to show simple and complex patterns at different scales. The differently scaled patterns are so specifically scaled that the shapes are not strictly fractal, but are to illustrate the point that large- and small-scale patterns can readily exhibit different degrees of complexity. *B.* If the degree of complexity of the outline of a form is different at different scales, then the fractal log-log plot of the line segment against perimeter will bend at the scale at which the complexity of the pattern changes. At that point, the fractal dimension itself changes.

process with a different dimension takes over. Such is the case in the pedagogical devices in Figure 2.5. Sometimes the measurements fit inside a strictly fractal range, but usually measured fractals show their scaled limits. It is not uncommon for a measured pattern to have different degrees of complexity at different scales. Technically, the fractal pattern is denied if the benchmark is fractals coming from iterations of equations because the patterns change before infinite detail. But mathematicians with equations do not exclusively own fractals. Measured changes in pattern with scale would be reflected in deviations of the log-log plot from a straight line. These deviations would be significant if, rather than noise about a straight line, the log-log plot bent at a particular scale.

The indication would be that whatever family of processes was responsible for the slope of the line before the bend, a new set of pattern generators comes into play at scales beyond the bend. So in practice, in the world of measurements, fractal patterns do not reach infinity, the fractal breaks down. But fractals are still a very useful notion, even thus limited. To identify different fractal dimensions at different scales, one simply does a regression to find the slope over several shorter sections of the abscissa of the log-log plot.[10] The horizontal axis of the log-log plot goes from the shortest to the longest line segments used to estimate the perimeter, so the entire slope gives a fractal dimension that applies across a wide range of scales. Measuring the slope along only successive short sections of the abscissa gives a number of fractal dimensions. Each one of these fractal dimensions applies only to a limited range of scales.

Analyzing the fractal dimension of landscape patterns shows certain consistent behaviors of fractals. These have implications both for scaling the processes that generate patterns and for the difficulties innate in the observation of ecological systems. The observations have been of forest islands in satellite images of the United States. Across a range of relatively small scales, the fractal dimension is low. Some less complicated process is cascading upscale for a while. As the scale is increased, the fractal dimension suddenly jumps to a high value and stays high for further increases in scale (figure 2.6A). This indicates that some more-complicated process has taken over at large and very large scales. This result was consistent for many landscapes widely dispersed across the entire country. Two primary sets of forces seem to mold the landscape, one at small scales and the other at large scales. The explanation is that at small scales, human influences are the principal cause of pattern, while at larger scales, topography is the constraining force.

Anthropogenic patterns include fields that are usually rectangular. Tracks and roadways simplify landscape patterns by straightening corrugated edges. Human actions impose simple patterns on the landscape with accordingly low fractal dimension. The reason that the pattern is simple is that simple processes are moving upscale. When two waves of different periods are mixed together,

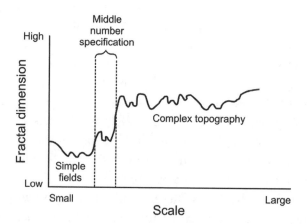

FIGURE 2.6A.   Krummel et al. (1987) discovered that the outlines of forests in the United States have small fractal dimension at small scale and a larger fractal dimension at large scale. Small scale appears to indicate human activity, while large scale seems under topographic control. At that point where the fractal dimension changes, the system has a medium number specification and is likely to be unpredictable. In medium number systems, constraints shift unpredictably—human versus biogeophysical, in this case.

FIGURE 2.6B.   Aerial photograph of the escarpment of the South Downs near Lewes, United Kingdom. This illustrates where fractal dimension goes up at large scale where simple human pattern gives way to elaborate geophysically caused pattern. Human patterns have similarly scaled reinforcing patterns. By contrast, the multiple geological causes are differently scaled. The South Downs–North Downs system is an anticline caused by folding in the Miocene, associated with the formation of the Alps. Weathering has exposed the underlying Jurassic deposits. There are two Cretaceous period chalk layers; the lower chalk is harder than the upper chalk, and one may be able to discern a terrace between the two strata. Beneath the pervious chalk lies impervious gault clay that gives rise to a line of springs with associated settlement at the foot of the chalk (e.g., Fulking, Edburton, Beddingham, and Firle, all villages on the north slope of the South Downs in East Sussex, five miles inland from Brighton). There always has been a shortage of water at the top of the downs that creates a problem even for feeding animals, and dew ponds were constructed and lined with clay to collect surface rainwater. Beneath the gault clay is a layer of greensand that gives rise to a low, sandy ridge with another line of settlements two miles further north (e.g., Hassocks and Ditchling, both villages in East Sussex on the North edge of the South Downs, seven miles inland from Brighton). Beneath the greensand is Wealden clay and beneath that are harder Jurassic deposits of limestones and sandstones with iron ore that give rise to the formation known as the High Weald, which is the former center of the Sussex iron industry. Beneath that is the Carboniferous with coal that was too deep to mine. All these strata terminate in sea cliffs between Beachy Head and Hastings, the oldest to the east. (Photograph, reproduced with permission, was taken on October 11, 2005, by Henry Law, whose many images may be seen at Elmer's Eye [http://www.pbase.com/pinoycup/profile]. Law also provided the technical details of the geology).

sometimes both waves peak together. Together they make a big wave. Other times, one wave is up and the other is down; then each cancels out the other. The aggregate wave is often complex and is the result of a pattern of interference between the component waves. Waves of the same length do not form complex patterns of interference when they are mixed. Most human enterprises that mark the landscape operate on a time frame of between a year and a few decades. Most things we do take about the same amount of time. It is therefore possible to think of human behavior as periodic waves of activity, most of which are

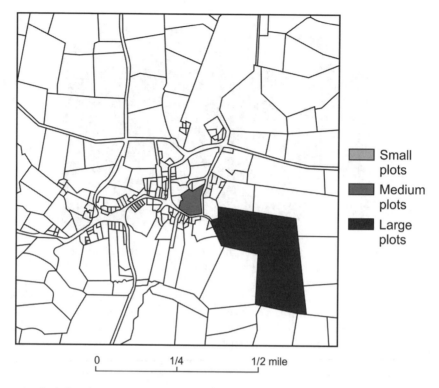

Small plots

Medium plots

Large plots

0   1/4   1/2 mile

FIGURE 2.6C.   The fractal pattern of human enclosure over time in Newnham, Northamptonshire, England. The small irregular enclosures are gardens. Next larger are the less regular but still uncomplicated enclosures of late medieval times. The biggest enclosures came with the parliamentary Enclosure Acts, with long straight edges. These enclosures drove yeomen into cities and new work like mining (Dury 1961:63).

approximately the same length. We discuss this further in chapter 9, regarding land management units of forestry and how policy changes modify the nature of land management unit size and texture. Processes with commensurate reaction rates are scaled the same and therefore in phase do not produce an interference pattern. Although human endeavors are various, they often share some common factor that causes the patterns left by humans to be mostly of one type. Property lines must be kept simple for record keeping purposes. Transportation is most sensibly conducted in straight lines. Because of this similarity, property lines and thoroughfares coincide rather than interfere. Forest land management units largely adhere to property lines but are less likely to do the same for thoroughfares. Humans have made impacts on the surface of the planet at geological scales, but most human activity is relatively small-scale. The small-scale pattern of forest mosaics in the United States has low fractal dimension because it is human-dominated. Mirroring the United States forest example, large-scale complicated outlines for the South Downs in Sussex, England, contrast with simple small-scale outlines for fields and roads (figure 2.6*B*). Dury (1961) shows how the English landscape moved from an open system through subsequent acts of enclosure as successively larger-scale fields were enclosed, but still with more or less the same straight line margins. At the scale of human enclosures in Britain, the system appears fractal (figure 2.6*C*).

By contrast, the pattern of topography is large-scale and complex. The reason for the complexity of topographic patterns is the obverse of the reason for the simplicity of anthropogenic patterns. The forces that give topographic relief like the South Downs are many, and each is remarkably independent from the rest. The independence of the causes for topography comes from their being disparately scaled. Glaciation, river scouring, rock compression, sedimentation under the ocean, and outcropping all contribute to topography, but do so over very different time scales. They therefore exert their influences independent from each other, except as each interferes with the work of the others. These forces are not contained at the small time and space scales of puny humans, and so topographic patterns are large. Because topography is the interference pattern between differently scaled processes, the form of topography on the landscape is irregular and has a high fractal dimension. At larger scales, the forest patterns are contained by topography. This explains the sudden jump to high fractal dimension as topography takes over from humans as the constraint on naturally occurring forests, and on the landscape of the South Downs.

There is other evidence of interference between human and natural forces on the landscape, but over time instead of across space. A study conducted by Monica Turner and her colleagues involved comparisons of various aerial photographs from 1935 to 1982 across the major vegetational regions of Georgia.[11] The areas studied were stratified random samples of the state, so the patterns can be taken as representative of more than just the places whose aerial photographs were analyzed. With changes in the economy, much farmland was abandoned and turned over to forest in the early 1930s. As might be expected, the fractal dimension of the farmland was low. Interestingly, abandoned farmland has a relatively high fractal dimension, indicating interference between disparate causal factors. Poor land is likely abandoned first, and that will often be the parts at the margin where the farmer had pushed up to the practical limits for plowing and other agricultural activities. These will have been the parts of the farm with irregular outline up against the hillside, river, or some other natural form. The edges next to the persisting fields would be straight enough, but the outer edge of abandoned farmland would be irregularly corrugated leading to a high fractal dimension.

In the 1930s, the fractal dimension of the forest was low. With successive decades, the proportion of forest increased. The most recent aerial photographs have larger tracts of forest with higher fractal dimension. The explanation is that in the 1930s, the forest outline was deeply constrained by the placement of fields. The first photographs were from the period just after the agricultural expansion of the 1920s. The low fractal dimension of forests then was because forest was merely the inverse of the cropland. By the 1980s, the constraint of agriculture on forests was considerably relaxed. The forests have expanded in the last fifty years to fill the landscape and are now under a new set of constraints. Forest constraints have become topographical and the fractal dimension of forests has accordingly increased. Forests in Georgia are now larger and show the greater complexity of perimeter characteristic of landscape patterns that are large.

PREDICTABILITY IN ECOLOGICAL SYSTEMS

Some predictable systems are specified so that they have a fairly small number of critical parts. We are not saying that the material system has in and of itself a small number of parts; it is rather how the system is specified (what constitutes a part is decided). A reliable equation can then be written for each part and its relationship to the others. Planetary systems would be an example here, where the respective masses and velocities of each planet are substituted one at a time into a single gravitational equation. These are called small-number systems.[12]

Sometimes the small number of parts involves using a small number of averages or representative values that subsume a very large number of parts. The number of entities contributing to the average must be large enough to make the average reliable. Equations are then written for the averages. These are called large-number systems, but they work the same way as small-number systems in that the number of reliable averages is small even though the reliability of the averages derives from a very large number of units contributing to the average. The gas pressure laws replace distinctive atoms or molecules with average "perfect" gas particles. The issue is not so much the number of parts, the name notwithstanding, it is the heterogeneity of the parts that matters. That in turn is an expression of the degree of organization complicating the constraints.

The problematic systems are called medium-number systems. These have too many parts to model each one separately, but not enough to allow averages that fully subsume the individuality of all the parts. Questions that cannot be answered imply a medium-number system specification. The reason why medium-number systems are unpredictable is that the constraint structure is unreliable. We simply cannot tell which of the too-numerous-to-monitor parts will take over a constraining role in the system. Note, we do not assert that nature is fundamentally unpredictable. It is the mode of system description that leads to the inability to predict.

In landscapes, the change in the fractal dimension at middle scales causes the system to be unpredictable over those middling scales (figure 2.6A). At small scales, the landscape is reliably influenced by human endeavors. At larger scales, the causes of pattern are reliably topographical. At scales in between, each part of the landscape has its own individual explanation. In some places topography might dominate, while in others the controlling factors might be distinctly human. In the middle scale, the landscape takes on medium-number qualities, as the quirks of local considerations jostle to produce patterns inexplicable in more general terms. Thus, a simple question, "What governs the form of abandoned agricultural land?" is in grave danger of invoking a medium-number specification of the landscape. The answer is not very satisfying: "Lots of unrelated things." On one boundary it is human endeavors, embodied in the land still under cultivation, while on another it is the local pattern of glaciation or erosion that defines topography. A favorite forest of ecologists in southern Wisconsin is Abraham's Woods, a plot of land owned by the Abraham brothers, who sugared the maples. There are two separate reasons for the edge of the woods. One is that the neighbor to the east preferred to crop his land, which is now agricultural across a straight-line fence. The other boundary is ecological, in that the woods lie on an upland horseshoe that backs to the southwest and opens to the northeast. Fires came from the prairies to the southwest, but could not burn downslope into the horseshoe. Over the top of the hill to the southwest, the maple forest peters out into eastern hop hornbeam ironwoods (*Ostrya virginiana*), and then to dry lime prairies. The boundary of Abraham's Woods has mixed causes. Weather and economics also interact to change the interaction between these two disparate sets of constraints. The scale implied in the question is the problem here.

There are many specific examples of medium-number failure to predict in landscapes. Dean Urban found that when woodlots are small, he can predict the number of birds with accuracy.[13] If the patches have room only for one territory, then he only needs to count the woodlots and estimate percentage of occupancy to know the bird population. If the land is forested in large tracts, then he only needs to know the area forested and the size of the average territory. Dividing the former by the latter gives reliable estimates of bird populations. If, however, the woodlots have room for a small number of territories, say two and a half, then the estimates are unreliable. The reason is that the orientation, shape, and degree of isolation, among other things, influence the number of birds. The differences between the woodlots become a matter not of simple area, but of many other considerations. The individuality of each woodlot makes a difference and so prediction is unreliable at best. Thus an innocent question, "How many birds are there in my system?" is fraught with danger. Temporal issues may also arise since, in the end, all constraints yield; when a prediction is made over too long a period, constraints fail along with prediction. It is well known that after seventeen days, local weather forecasts fail to the point of random because the trajectories and longevity on the predictive high- and low-pressure systems become deeply uncertain.

REPEATED PATTERNS

Although there is a difference in complexity between small- and large-scaled patterns in human-dominated systems, it is remarkable that the fractal dimension remains constant for such wide ranges of scale. Indeed, if it did not, we could not say that the patterns were fractal at all. Clearly there are different factors governing the processes that produce the patterns across the range of scales over which fractals are estimated. Therefore, the constancy of the fractal dimension is unexpected and requires an explanation. The answer is that there are remarkably few patterns used by nature (figure 2.7). As Peter Stevens notes, "In matters of visual form we sense that nature plays favorites."[14] He goes on to give us the reason "why nature uses only a few kindred forms in so many contexts."

The repeating patterns are spirals, meanders, branching systems, and explosive patterns. Additionally, many systems show 120-degree angles because they are both the consequence of space packing and what is left after stress release in a homogeneous medium. Cracks in soil and polygonal frost-heaving patterns clearly fall into these classes of landscape. The requirement for homogeneity is often not met at larger landscape scales; therefore, many 120-degree angles

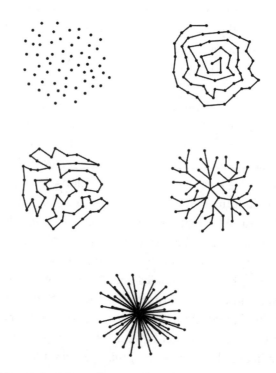

FIGURE 2.7.   More or less any collection of points can be connected to show a spiral, a meander, an explosion, or a branching pattern (Stevens 1974).

are at small landscape scales, with the gigantic exception of stress release when Pangea, Laurasia, and Gondwana land broke apart between 135 and 300 million years ago. Spirals, meanders, explosions, and branching patterns appear to apply across all landscape scales. Their ubiquity stems from four geometric attributes that have implications for processes: (1) uniformity, (2) space filling, (3) overall length of a linear system, and (4) directness of connection between parts of the system.

Let us connect the patterns to the geometric attributes. This will facilitate the linking of the processes common to landscapes (figure 2.7). Spiral patterns are beautifully uniform and fill space very effectively. They are also relatively short in total track, given the space they fill. Spirals appear when deep regularity is required. Seashells are spiral because of the directly competing exigencies of covering soft parts of the mollusk, as opposed to leaving an opening for food and expelling waste. An error in either direction means death. Random meanders are very similar to spirals, although they are not as uniform. Like spirals, they are not very direct in connecting arbitrary points. They apply when there is minimal energy available to fill the space but no particular need for regularity. A small rise in the landscape cannot be overcome by force, so the pattern must move around the obstruction, as small and local as it is. A river on a plain is a force that from time to time fills the space with water in very literal terms. Beyond the space filling of intermittent floods, the continuous processes of erosion and deposition tend to move the river across the space so as to fill it over time. Note that these processes and events apply to watercourses over a wide range of scales. Rapid transport of water from one place to another is not a critical factor on flat terrain, and so the lack of directness of meanders and spirals is not a problem.

Explosion patterns are very direct, particularly with respect to the center of the pattern. They overcrowd the center of the space and therefore do not fill space in the uniform manner of spirals. The angles between their tracks can, however, be very uniform. The total track of an explosive pattern is exceptionally long, and so in explosions, conservation of construction material cannot be a consideration. When the rapid dispersion of material is critical, the explosive star is the pattern to be expected. Some landscape patterns are of explosive origin in the literal sense, volcanoes being a case in point.

While explosion has a literal meaning, that definition only applies to the narrow and arbitrary time frame of primary human experience. Something that takes a whole minute is stretching the literal meaning of the word "explosion." If, however, we recognize the scale-relative nature of explosions, then the notion can be applied to other scales. An explosion involves movement of material so fast relative to normal system behavior that it occurs as an event. The context has no time to exhibit behavior during an explosion and thus appears static. If we now take a

FIGURE 2.8.   Erosion on a Colorado hillside shows how competition keeps channels apart. There are explosive patterns for at least three scales here: (1) the peak center left shows gullies coming away from its peak; (2) the gullied middle ground shows a smaller, more local explosion; (3) the circular pattern is lost by its proximity to the camera, but the regular striations on the slopes in the foreground show regularly spaced small gullies exploding down a slope of only several meters. Explosion is relative. (Photo courtesy of Gursel Kusek.)

system whose normal behavior is very slow, then something taking days or years could amount to an instantaneous event. In this way, patterns resulting from moderately paced dispersion of material can take on the explosion form, although there is not a literal explosion at the human scale; there is merely a functional explosion given the size of the system at hand.

The form of a mountain changes over millennia except in cases of literal explosions. A literal explosion, in the vernacular, happens very fast relative to the time frame of the observer. But using the time scale at which the mountain erodes away, the movement of water off a mountain is at explosive speed. The only way to disperse water that far that fast is with an explosive, starburst pattern. The regularity of the arms of an explosion pattern comes from competition between the parts of the moving front of the explosion. A part of the mountainside that happens to receive more runoff has a shallow channel cut in it. In subsequent runoff events, that channel takes water from either side, therefore cutting itself deeper with the water from adjacent regions (figure 2.8). Deprived of their water, the regions on either side of the channel erode more slowly than the channel itself. Channels without competing neighbors cut faster than those that have to compete for water with other gullies. Thus, regularly spaced channels grow faster.

Competition is a positive feedback in which nothing succeeds like success. Positive feedbacks require a supply of resources that can drive the positive feedback and feed the growth of the winners. Note that ink dropped a short distance leaves only a generally circular blob; while ink dropped from a great height forms an explosive splat. The blob is not explosive because the force behind it was insufficient to feed the winners of the positive feedback over the time period required to manifest the pattern (figure 2.9). For an explosion, the sides of the mountain must be steep to provide the abundant energy required. The movement of water needs to be forceful. Gentle slopes or even equitable and modest rainfall on a steep slope blunt the force of the positive feedback, so competing linear channels fail to form. Sheet erosion leaves no explosive marks on a satellite image.

Explosive patterns also occur on urban ecological landscapes. Given the relatively long time it takes for an urban center to grow, the requirements of movement of commuters twice a day are a matter of explosive speed. That is the reason for the explosion pattern of roads from major cities. Roads are expensive, but that cost is less important than the pressing need to move vast crowds in and out of the city quickly. Duplicating tracks of major arterial roads, each on a slightly different radius, is one of the costs of urbanization. For metropolitan road systems, the explosion is the pattern of choice. The Roman adage, "All roads lead to Rome," was true. A satellite image of the Italic Peninsula from the time of the Roman Empire would have shown the explosion pattern of Roman roads we might expect. The flow that caused that star was slower than the flux of modern commuters, but it was faster and larger than any other flow at the time. Note that the interstate system of high-speed freeways that connect the major cities of the United States does not form explosion patterns even when it serves New York City or Chicago. The interstate system is a network because it functions as a national transport system. It belongs to a higher level of organization where it functions as a network for the country.

Branching systems are common on landscapes. The geometric properties of branching networks are a compromise between the single circuitous route of the spiral and the heavily duplicated tracks of the explosion. In a branching system, the total track of the system connecting all points is exceptionally short. They achieve this economy of track material at the expense of some small indirectness of connection of individual points to the center. Branching systems are less uniform than either the even, constant curving pattern of the spiral or the uniform angles of the explosion. There is irregularity of detail in branching systems, although they do fill the space. When economy of track and relatively direct connection of all points to some central point are the important factors, then branching systems are the pattern of choice.

Note that the explosion of arterial roads from a city gives way to a branching system at a distance from the city center. This is because the urgency of moving people nearly home at rush hour is less, but the cost of building roads remains high. While duplication of major highways on adjacent radii is a necessary expense at the city center, costs are minimized in outlying districts by serving adjacent small

FIGURE 2.9. When ink is dropped onto paper, it does not leave the characteristic splat pattern unless it is dropped from a significant height. With a short drop, the ink only forms a blob because there is not enough energy to drive the positive feedback of the winning sectors of the perimeter.

communities with branches to serve both. Similarly, the explosion of watercourses from peaks soon forms a branching network of rivers once the slope eases. Animals like sheep free ranging on the Welsh mountains need to move to all parts of a space eventually so as to exploit resources. There is not an explosive flow because neither the sheep nor their resource base is centralized, and so traffic does not justify duplication of pathways. Therefore, grazing animals leave branching networks of trails on the landscape. Through Stevens's analysis of the ubiquitous processes that underlie spatial pattern, we see that the distinctive patterns on a landscape are composed of a small set of component patterns. The same general class of causes underlies branching systems of mighty rivers and the trails left by ants on a tree trunk. The explosion of Mount St. Helens is matched by the tracks of livestock to water tanks in the desert grasslands of New Mexico (figure 2.10). The gradient from distant parts of the range and the water tank with regard to water is steep. The animals implode to the tanks when they are thirsty.

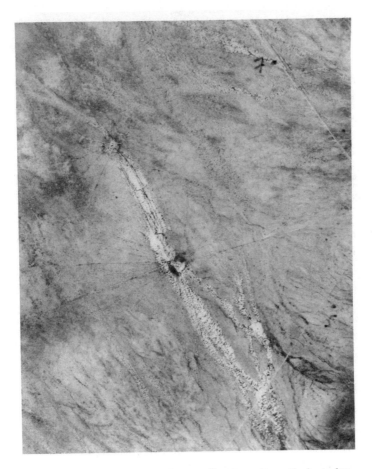

FIGURE 2.10.   The pattern here from the Sevielleta Long-Term Ecological Research (LTER) site in New Mexico shows an explosion pattern around water tanks, which causes grazing animals to move rapidly in toward the water resource. (Photo courtesy of NASA.)

LINKING TO OTHER CRITERIA

In chapter 1, we refer to the cone diagram of ecological levels. So far in this chapter, we have moved up and down the landscape cylinder (figure 1.15A). We now look at landscape in relation to the other ecological criteria (figure 1.15B), slicing across to other parts of the cone.

To test hypotheses about landscapes, we need to know the characteristics of a neutral landscape, one that does not reflect the constraining effect of topography, aggregations, disturbance history, and similar ecological influences. Such landscapes would need to be formed by a random process that propagated pattern in the absence of constraints. Such neutral landscapes constructed by Robert Gardner and his colleagues have distinctive characteristics.[15] If we consider a random landscape in terms of the percentage of the surface occupied by the habitat of interest, then the largest number of patches occurs when the habitat occupies 30 percent of the total space. At lower values, there are a few well-spaced patches. Above 30 percent, the number of separate patches decreases because the addition of more occupied habitat tends to connect existing patches more often than it creates new ones. The ecological processes on real landscapes seem to generate fewer and larger clusters than expected from the neutral random model with no constraints.

It emerges that changes in total habitat area do not have the same effect at all percentages of land occupancy. This is illustrated clearly in the case of the neutral landscape. There is a 10 percent decrease in land covered below the critical 30 percent level that will have negligible effects on the number, size, and shape of patches. Above 30 percent occupancy, the same 10 percent reduction will have dramatic effects. Thus, the effect of disturbance that removes habitats of a given type is not simple. The effects will be greatly influenced by the original quantity of the habitat in question, independent of the size of the disturbance.

In random landscapes, at values greater than 59.28 percent of the landscape occupied, the network of habitat interconnects so that there is a pathway from one side of the map to the other without leaving the habitat in question. There is percolation.[16] The rules for passage are edges not corners, like a castle or rook in chess, but not like the diagonal move of a bishop. This percentage for percolation is well known from a body of theory concerned with the percolation of liquids or passage of electricity through aggregates of material where assumptions of random placement are realistic. Additions of new pixels when the map is just below the threshold have very little room to avoid spanning. Well below the threshold, even aimed insertions cannot cause spanning. The map in Figure 2.11 is small and subject to the statistics of small numbers, with results influenced by map-edge effects. Student exercises in Allen's class used even smaller, 10 x 10 grids to make the exercise tractable. Even so, the students' random insertions created spanning clusters

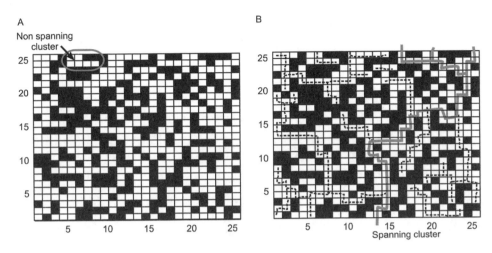

FIGURE 2.11.   *A*. Map of a 25 × 25 grid, with 625 cells. Sixty percent of the cells are white, and so 40 percent are black, all inserted at random. If white allows movement, and the rules are moves only across cell edges, not diagonals, there is a cluster of cells that spans the map. *B*. Sixty percent is just above the percolation threshold (59.28 percent), so it is not a surprise that there is a spanning cluster marked by a thick gray line. With a queen move allowed (including diagonal), the black squares almost percolate to achieve a spanning cluster, staying just below the spanning percentage of 40.82 for that rule for moving (Gardner et al. 1987).

still quite close to the 59.28 percent threshold. The track of the spanning cluster (figure 2.11*B*) is the gray line. There are many nonspanning clusters, like the one noted at the top of the map (figure 2.11*A*). The black squares on the map almost span under the permitted diagonal-move rule (a move like a queen in chess). Put another black square in place to take the black square sum over the threshold to 41 percent, and the black squares on diagonal-move rules would almost certainly span. It is no accident that the thresholds for the two rules for moving (58.28 + 40.82) sum to 100 percent.

Real landscapes, full of pattern-generating ecological constraints, usually allow percolation at a lower threshold of area occupied by the habitat in question than do random landscapes. In the next section, we consider the special parts of real landscapes that encourage percolation.

THE IMPORTANCE OF FLUXES ON LANDSCAPES

At the beginning of the rekindled interest in landscape ecology in America, there was a phase of naming and classifying landscape patterns. Forman and Godron reported on twenty-seven ways that landscapes shed water (table 2.1). This excessive terminology resembles the first stages in community ecology when, at the

TABLE 2.1   Natural Drainage Densities and Patterns

Drainage Densities	Subdendritic
Fine	Barbed
Medium	**Nonintegrated Drainage Patterns**
Coarse	Deranged
Integrated Drainage Patterns	Internal
Dendritic	Karst
Angular	Lacunae
Rectangular	Incipient
Annular	**Other Drainage Patterns**
Radial	Artificial
Trellis	Thermokarst
Centripetal 1	Dichotomic
Centripetal 2	Anastomotic
Pinnate	Yazoo
Parallel	Elongated Bay
Braided	

beginning of the twentieth century, Frederic Clements collected old terms and coined new terms for different aspects of community form and process. For example, Clements listed many words for nuances in plant establishment and names for different means of doing that in different places. He used no less than ten terms for ways a plant could invade. Such a large ecological vocabulary with so many fine distinctions may seem superfluous now, but underlying each word was a new conception. Most of Clements's community terminology has fallen from use now that the underlying concepts are fully integrated into the modern conception of communities. The same may come to pass with the burgeoning vocabulary in landscape ecology, but for the moment, the words draw attention to the richness of pattern and the meaning that the various forms might have.

Different geological and climate areas create pattern types. We can imagine what the patterns might be. The point for our argument is less the patterns themselves and more the large number of them that are named (Way 1973). For detailed images of each name in table 2.1, see Forman and Godron (1986).

In landscape ecology, the terms imply underlying processes. In some cases, these words are for structures and processes that link to other criteria for organizing ecological systems. A case in point is the term "corridor." A corridor is often identified as a one-dimensional passage that connects across a two-dimensional landscape area. We have a more general definition for corridor, suggesting it is something that is one dimension lower than the surrounding space that allows passage across the higher-dimension space.[17]. Corridors have dramatic effects on the capacity of the system to allow percolation of ecological material across landscapes. They are places of flux; they are communication channels. This has importance not only for the workings of communities and ecosystems and for the movement of organisms and populations, but also has particular significance for the scale-oriented approach that we employ throughout this volume. Let us weave together the more abstract scaling implications and the use of the simple tangibility of corridors to relate landscapes to other organizing criteria.

Corridors are the connectors across boundaries. They allow interaction of the parts of large-scale communities or ecosystems. Communities and ecosystems have integrity coming from a binding together of their parts. Those interactions are mediated by communication channels like corridors. Some of the exchanges between system parts can be by means of short-distance communication, not along corridors but by diffuse interaction. However, over longer distances, corridors are often involved because they facilitate flow of organisms and abiotic material.

An example of the effect of corridors is the westward migration of the eastern blue jay. It moved west across the hostile Great Plains environment along the watercourses such as the Platt River when humans changed the hydrologic regime of the rivers so cottonwood stands are more contiguous today than before Euro-Americans settled the West. That movement has affected the survival of other bird species in eastern Rocky Mountain forests due to eastern blue jay competition. The eastern blue jay crossed a major biome using that corridor. According to the Cornell Lab of Ornithology, "Recently, the range of the Blue Jay has extended northwestwards so that it is now a rare but regularly seen winter visitor along the northern US and southern Canadian Coast."[18]

The integrating role of corridors has an intuitive appeal. However, the functioning of corridors as the communication channels of communities and ecosystems is not a simple matter. Types of organisms encounter linear structures on the landscape differently. Only some types of organisms will use a given type of linear structure as a corridor. Large mammals might use the open strips under power lines or along logging roads as corridors for rapid movement, facilitating dispersal with burrs.

The fact that different organisms respond to potential corridors in different ways has important consequences for the scaling complexity of communities and ecosystems. Abiotic components are transported to various degrees down corridors and therefore could belong to alternative ecosystems depending on which types of corridor move them and which do not. For example, wind and water move different-sized particles, wind often uphill but water always downhill. Water-soluble materials will move down watercourses, but anything insoluble will move less. Different degrees of solubility change the rate at which water can transport deposits of various materials. The reason why lime is so important in natural and agro-ecosystems is that it changes solubility of minerals and therefore changes the way those ecosystem components relate to corridors of water. For both communities and ecosystems, inconsistent uses of corridors by different organisms and abiotic materials may not be the rule, but neither are they the exception; this inconsistency confounds simple conceptions of communities and ecosystems as places on the landscape. Cassandra Garcia looked at sewer sheds instead of watersheds on the campus of the University of Wisconsin, Madison.[19] Campus managers immediately recognized it as a new and profitable conception, because sewers work more forcefully as corridors.

This raises the question of whether a corridor for one purpose is not a boundary for another. While the Alaskan pipeline is a conduit for human purposes, there was deep concern that it might be a barrier for animal migrations. Many deep forest birds have an aversion to open spaces such that a logging road is a considerable boundary. In the primeval forest, these birds belonged to a system that was larger, occupying great tracts that were more or less continuous. The rescaling that has occurred through human intervention could have large consequences for genetic diversity and population persistence. By decoupling routes of reinvasion, narrow logging trail corridors do not allow local extinction to be reversed in small sections of forest.[20] Over time, this leads to global extinction. Meanwhile, birds that have more neutral responses to small open spaces are not rescaled at all by the corridors. Yet other species, those inhabiting forest edges, suddenly find themselves thrust into a much expanded and linearly connected network of forest edges. They become part of a new large-scale system. If such opposite effects can pertain within one order of animals, it is clear that difficulties in prediction across orders or kingdoms are to be expected. Conservation is suddenly a very complicated business.

There is some distinction in the literature of landscape ecology between corridors that are wide enough to have interiors and those that are so narrow they only have edge. Corridors with interiors can function as discrete entities in their own right. For all their fully apparent flux, stream corridors are a complex of corridors and can have permanent residents of their own. When corridors intersect, they often show unusually high diversity of organisms. In landscape networks, the flux is high and the alternative tracks at each intersection change the scale of the system.

A further complication is that landscapes are directional. They work like fish traps, where the fish can easily move in as they are channeled toward a hole, but once inside they are faced with surfaces that do not guide them back out of the hole. In classroom experiments using maps of the Madison, Wisconsin, area, students randomly move disk counters with throws of the dice. Lake Mendota is broadly triangular, with its bottom edge positioned as an east/west line. The other two sides are roughly diagonal. The randomly moving counters easily move south along the diagonal edges of the lake. But coming up from the south, the counters run straight into the south edge of the lake and get stuck somewhere on the University of Wisconsin campus.

THE SIZE OF ANIMALS ON LANDSCAPES

Some of the most important work done in landscape ecology, since the first edition of this book, is from Holling's lab. It pertains to levels in hierarchies and suggests from where some of the discontinuities between levels come. Holling and his last batch of students (Craig Allen, Garry Peterson, and Jan Sendzimir) looked at log temporal versus log spatial range of various ecological phenomena.[21] Peterson united three other works to plot salmon life cycle onto atmospherics and human activity on the landscape. Adult salmon in the Columbia River basin move up to headwaters to spawn in sectors of the rivers so local that they are measured in meters.

The young fish then work their way down to the ocean and cross the Pacific Ocean during a period of years. Thus, different stages of an animal's life are differently scaled in time and space. Holling shows how the food choice, home range, and migration differ for three animals—moose, beaver, and deer mouse (figure 2.12). While the three biological facets of the organisms differ in several orders of magnitude for all three, the differences between these animals of very different size vary only one order of magnitude for each of the respective food choice, home range, and migration activities.

This scaling of animal sizes led Holling to rank all of the animals in a region (everglades and other regions). He noted the size of all the species of animals from data in the literature. Then he ranked them on size. In his inventory (not sample), he noticed that certain sizes were disallowed. Instead of a continuous increase in size across the ranking, he found jumps in size in restricted places on the rank (figure 2.13). If he plotted not the size of animal but the difference

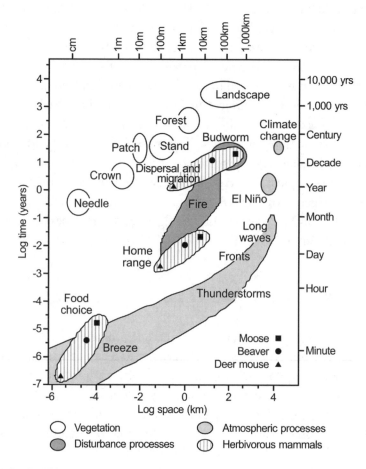

FIGURE 2.12.   In a log-log plot of time against space, we see that atmospherics, fire, and biotic placement on the ground all can vary in their own ways by many orders of magnitude. Animals' (a) food choice, (b) home range, and (c) dispersal and migration differ by several orders of magnitude. Three animals, orders of magnitude different in body mass, are plotted on the big log-log plot of time and space. Remarkably, the different species cluster within just over one order of magnitude for each of the respective a, b, or c activities. This is suggestive of universals of landscape occupancy (Allen and Holling 2008, with permission of Columbia University Press).

between an animal and the next one up in size, the disallowed sizes stood out as peaks of difference in size. The groups of allowed sizes are called lumps in animal size distributions.

There was great criticism of the work, complaining that Holling had statistically insignificant data on the lumps of species of a given size. But that criticism can be largely discounted because the critics were arguing with sampling statistics, when Holling was not sampling but dealing with an inventory of all the species. Inventory statistics are very different. The work is difficult to prove or disprove with particular tests, but Holling and his students took a series of triangulations on the issue. There was significant reinforcement of the ideas from several directions. This is one of the few ways to validate ideas in complex situations.

The lumps were not universal. They were different for different regions. Holling was in Florida at the time, so Florida was one of the places. But Holling is Canadian, and the Boreal Forest was another place he calibrated. The lumps were different in number and the break points were at different places, depending on the region. The lumps were different for different major taxa. Birds had one set of lumps, while mammals had another. The lumps of reptiles and mammals were the same, and bore the same relationship to the lumps of bird species. Cleverly, they tested bats against bird lumps, and unlike other mammals, bats had a one-to-one mapping onto bird lumps. This suggests that the functional dimensionality of the environment is critical. Reptiles and mammals live two-dimensionally, while birds and bats live in a three-dimensional environment. Nobody tested tree-living animals, but one might expect a fractal dimension between that of birds and ground-living mammals.

Other patterns were telling. The lumps in urban parts of the region were the same as the rural parts, but in urban settings some lumps disappeared into the gaps between their neighboring lumps. Urban is simply a special environment that is situated in the region. Holling and his students also noticed that invasive species were at the edges of lumps, presumably where they could get a footing. They found that the edges of the lumps were disproportionally occupied by rare and endangered species.

There were several hypotheses about the origin of the pattern, but one that relates to landscapes and corridors has the best support. Animals of a certain mass require a certain amount of food. Mass relates to linear length of the animal through a cubed function. Larger animals have a longer stride. Stride relates to transportation costs. Not only do horses walk, cantor, and gallop, but mice do also, and the speeds at which they switch gaits are scaled to be equivalent across species.[22] One of the remarkable things about humans being bipedal is that their special gait makes them very efficient at traveling great distances. Over the literal long run, humans can run down a horse, which seems to be our hunting strategy. It appears as if Holling's lumps are the interaction between three factors: the amount of food needed by the animal once it makes it to a patch of resources, the speed with which the animal can traverse a landscape, and the distance between favorable patches (figure 2.13).

The functional distance between patches could be modified by climate, as when permafrost melts into a bog. That is one of the theories for the megafauna extinctions at the end of the last ice age. We also see changes in the size of animals, especially on islands; when the Mediterranean filled with seawater and Crete became an island, the predators did not have space enough to survive. Rats go down holes, beavers live in water, and elephants are too big to be attacked. In the absence of

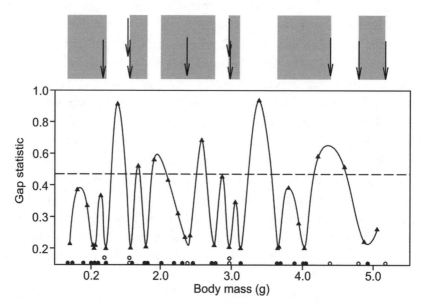

FIGURE 2.13.   When mammals of the Everglades are rank ordered in body mass, there appear to be disallowed sizes. The gap statistic on the graph show as peaks certain sizes with a big gap between adjacently ranked organisms (triangles are differences, or gaps). The upper part of the image shows blocks of allowed size, with gaps of sizes that are disallowed on that landscape. The arrows show the sizes of invasive and endangered species; note that they occur next to lump-size edges (about three or four times random expectation). Something about the ecology of marginal size is going on at the edges. There are suggestive narratives. Open circles along the abscissa show listed (endangered) species (Holling, Peterson, and Allen, in Allen and Holling 2008, with permission of Columbia University Press).

predators, beavers got much bigger. Mammoths got smaller as that size was suddenly allowed by the absence of predation. Big fossil rats and massive sloths are well known from special ancient island environments.[23]

Viable sizes of animals and those sizes that are prohibited come from an interference pattern between the periodicity of the landscape mosaic and the transportation costs across a given mosaic that offers patches of a certain quality. We see interference patterns as a common cause in ecology under several criteria. Communities are made of species with periods of occupancy. The common community configurations as to species associations come from interference patterns between species periodicities. For instance, the intertidal animal communities appear held in certain patterns of species association by the competitive superiority of mussels over other shellfish and the periods of mussel removal by starfish or storms.[24] We would also argue that the stable configuration in biomes is a wave interference pattern between the periods in plant life cycles (germination times, seasonality for reproduction, droughts, and frosts). Patterns that are found commonly are often a matter of interlocking frequencies.

ENERGETICS OF MOVEMENT ON LANDSCAPES

Holling's work on lumps opens up a new field of the energetics of living on a landscape. Some recent work on return on investment showed that different consumers respond differently to sheer quantity of resources as opposed to increases in efficiency of refining resources.[25] Profligate consumers, called high-gain users, can be predicted by flows in thermodynamic studies. Efficient users are called low gain, and they can be predicted by degrees of organization, which has no rate, even if it influences rates.[26] Low-gain, efficient users take advantage of the great mass of low-quality resources. Hot spots are very beneficial, but run out quickly. High and low gain capture a lot more than standard population biology vocabulary of r and K strategy, so gain deserves its separate treatment. But r strategy maps to high gain, while K strategy maps to low gain to an extent. In the old parlance of population biology, users of high-quality resources were called "r strategists" after the r in the logistic equation, the growth term. When only low-quality resources are available, profligate users cannot refine what is there and are therefore functionally out of resources altogether. Species using the r strategy move in fast and crash the resource. Low-quality resource users were called K strategists in the old parlance, after the K in the logistic growth equation for growth. K strategists are prudent and focus on consuming close to the capacity of the system to provide renewable resources over the long term. In general, humans are K strategists, moving close to the carrying capacity. As animals go, humans invest much in children, so their survival rate is high. Crashing a resource would kill children in whom investment has been made. But at a larger scale, human technology has released huge resource bases that have allowed a temporary switch to an r strategy in reckless increases of consumption. Our mutualism with dogs is a technology that moves us up-scale on landscapes in terms of Holling's lumps. Our mutualism with cats moves us down-scale and lets us go down cracks to remove vermin, another resource expanding technology.

There are three stages to efficient use: capture, refine, and burn. Engineers working on engines tend to focus on the efficiency of the burn, the transfer of fuel into motive power or profit. They focus on efficient use. Energy engineers focus more on how to make fuel out of low-quality resources such as sunshine or wind. In this way, we see the flexibility of human ecology across a range of strategies.

There are ants that eat food, and go straight for the energy in food. They are high gainers. Other species raise fungi on feces and flowers, converting waste into food. They are relatively low gain. The *Atta* leaf-cutting ants are the most efficient, as they go for very low quality but very abundant resources, leaves. Not much in a leaf, but the fungi used by *Atta* ants are some seven times more efficient at extracting benefit from that particularly poor resource than the fungi of primitive ant species that collect better material. The hyperefficient *Atta* ant colonies are much larger, and they explicitly build corridors hundreds of meters long, which they keep clean with "street cleaners." But we should note that, efficient as they are, *Atta* colonies can be agricultural pests (akin to the dual strategy of humans). *Atta* ants are so efficient that they suffer pests of their own that consume their fungus resources. Just like us, they make pesticides. They raise a second species that debilitates the pest. The general rule appears to be; if you are too efficient, you are going to need things like pesticides. As K strategists ourselves, our population has become characteristically large. The complications of high and low gain being relative appear at a global level. At the global scale, we as a species are approaching an r strategy, with which we are in some danger of crashing our global resources.

All of this makes it easy to transfer ecological knowledge from nature to human ecology. For instance, the Roman Republic is well cast as a high-gain consumer of gold by looting. By contrast, the Roman Empire had to garrison places where the gold had been taken. They moved into low gain, and taxed the peasants. Not much in a peasant, but there are lots of them, and they deal in renewable resources.[27]

In this same vein, Peter Allen's (2009) thesis at the University of Wisconsin looked at a small town in Wisconsin, Evansville, where he considered switches in the occupancy of the landscape (figures 2.14 and 2.15). An idea, elaborated in following chapters about management, looks at ecological issues in terms of profit. We contrast high-gain use of resources with low gain. High gain is pressed by steep resource gradients (think r strategy) and can be predicted by rate and strength of processes (rate-dependent). Low gain uses shallow gradients (think K strategy), and uses what it can get efficiently, using structural constraints (rate-independent). Peter Allen was able to characterize Evansville in the nineteenth

FIGURE 2.14. Evansville, Wisconsin, was named in 1842 for Dr. J. M. Evans. Land records indicate that Amos Kirkpatrick purchased land on May 29, 1839, and many historical accounts report his double-log cabin as one of the first built in what is today Evansville. The grid pattern of the streets was established in 1855 (Montgomery n.d.).

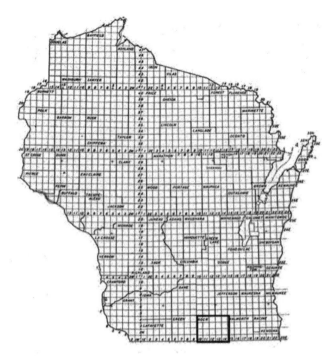

FIGURE 2.15.   Wisconsin was platted into townships in the 1830s. Rock County (outlined in bold here) is on the Illinois border. "Rock County consists of twenty townships. In 1833, Lucius Lyon surveyed the southern boundary, which is the Wisconsin, Illinois state line. In 1836 the final township was surveyed" (Rock County Wisconsin Surveyor 2012). Evansville is in Union Township, in the northwest corner of the county.

century as a low-gain consumer of a diffuse resource, sunshine. It was a market town and was the first collection point of agricultural produce. As quality fossil energy came online, local industrial towns such as Janesville grew under high-gain consumption, while low-gain Evansville stagnated. Evansville relied principally on railroad, but Janesville grew more with road transportation. Rail is planned and low gain. Roads are not planned; they are simply straightened in response to more traffic in high-gain mode. The fractal patterns of both road and rail are the same, but rail is planned from on top (low gain) and serves diffuse resource use. Big lines come first and side branches come later as structure is created downscale.

Wisconsin roads started as local networks, and they cascaded upscale (high gain), not according to plans but in response to traffic flow, a steep gradient (figures 2.16 and 2.17). Part of the slump in Evansville is that the railroad is now closed. But as fossil fuel starts to present problems of supply and consequence of use, we are now considering growing fuel. Biodiesel is a diffuse, low-gain strategy. In 1987, Evansville was declared the soybean capital of Wisconsin, and the city instated Soy Bean Days in 2008. And as we might expect, there are halting plans to reopen the

FIGURE 2.16.   The original surveyor's record (1833–1836) of Union Township. Evansville is now situated in the 26th and 27th sections, but would not be settled for another three years and was named Evansville some six years later. The lines are section lines bounding each square-mile section. Much of the land will have been settled in forty-acre parcels (T. Allen's parcel in the next township north was unusually homesteaded as twenty acres). Most roads will have been on the quarter-square-mile sectors around 160 acres in area.

1871

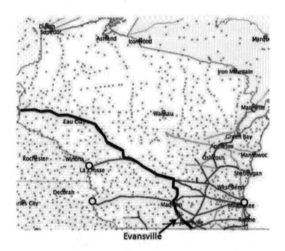

FIGURE 2.17.   The railroads in Wisconsin in 1871. The map shows roughly the same fractal pattern as the roads at present. Most of the rail lines are now closed, but they grew from the top down, first big rail lines with branches later. Roads were bottom-up, starting with the local grid of Figure 2.16, and straightening out corners as traffic demanded direct connection between hubs. The early lines (see open circles) starting in 1850 were only from Milwaukee, but crossed the state by 1858. In 1860, a line intended to connect Chicago to Minneapolis reached Evansville, which was then the railhead until 1864, whereupon Evansville boomed. The line was a top-down plan that reached Minnesota in 1871. The small dots are extant small towns. Note their linear pattern in the north, as they reflect railroad towns on fractal branch lines into the pine forests as the Robber Barons cleared the forest around the turn of the twentieth century. (Map created by Peter Allen.)

railroad in an extended low-gain pattern of concentrating resources. This is given closer attention in the applied resource chapters that deal in ecological economics. But at this point, the landscape implications deserve mention in terms of creation of corridors.

TOPOGRAPHY AND DISTANCE AS SURROGATES FOR INTERCONNECTION

The linking of the landscape criterion to the other ecological criteria often involves flux, but not always along corridors. Sometimes ecological material passes across the landscape diffusely. Landscape topography may slow down that diffusion. In this case, topography operates as a filter on ecological signal. The wider the expanse of unfavorable terrain, the more the high-frequency signal from low levels of organization is filtered. Landscape topography changes the functional scale of the communities on the ground.

Bruce McCune's study in the Bitterroot Mountains reveals how topography can constrain communities.[28] The Bitterroot Mountains run north and south on the Montana–Idaho border for some fifty miles. At regular intervals, there are valleys that cut into the range from the east. Since they are so close together and are geologically homogeneous, one might expect them to be replicate systems. If communities are the result of filtering of biota by competition mediated by the environment, then the communities in the valleys should be the same in composition. The unexpected observation is that they are not the same. Furthermore, the differences do not align themselves with any environmental factor, not even a spatial ranking from north to south. After years of tedious measurement of the environment, McCune was forced to presume some other cause (figures 2.18A–C).

The valleys are completely separated from each other by the mountain ridges. The ridges filter out propagules so that only what is in the valley already is available to colonize. In a more open system, we might expect the constant pressure of invasion to produce a full suite of competitors from which the environment could select. Selection over millennia should produce an environmentally determined outcome, probably one with the same vegetation patterns in all the valleys. However, the mountains constrain the valleys so much that they are too small to reach any sort of unique equilibrium between vegetation and environment.

Ilya Prigogine, Nobel laureate for chemistry, studied the emergence of higher-level order in chemical systems far from equilibrium.[29] We have already introduced the notion of emergence associated with his work. The gradients underlying emergence come from the system being held far from equilibrium so the second law of thermodynamics can drive the system. The particulars of that high-level

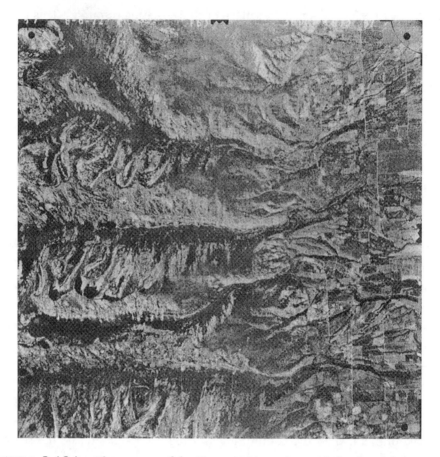

FIGURE 2.18A.   The canyons of the Bitterroot Mountains are isolated and, being small, each has its own vegetation arising from a unique disturbance history. This photograph shows Fred Burr and Bear canyons. *Larix* does not occur in Fred Burr. Fred Burr and Bear lack *Thuja*, but Mill Creek canyon (on the bottom margin of the photograph) has it. Bear has *Pinus monti-cule*, but other canyons in the image do not. Note how the Bitterroot valley running north and south up the right side of the image isolates the canyons from each other vegetationally. (Photo courtesy of the U.S. Forest Service.)

order are the consequence of the particular configuration of the system as lower levels of order went unstable. For example, the detail of the nucleus that starts a snowflake determines the particular form of the high-order symmetry in the fin-ished snowflake; the instability is of liquid water at temperatures below freezing. Freezing does not start at zero degrees Celsius. It starts with a gradient between zero degrees and the temperature of supercooled water. The heat of crystallization warms the water as freezing consumes the gradient. Complex systems are formed by successive reorganizations where a series of instabilities cause the emergence of a series of structures at higher levels. That is why complex systems require sev-eral levels of organization for their adequate description.

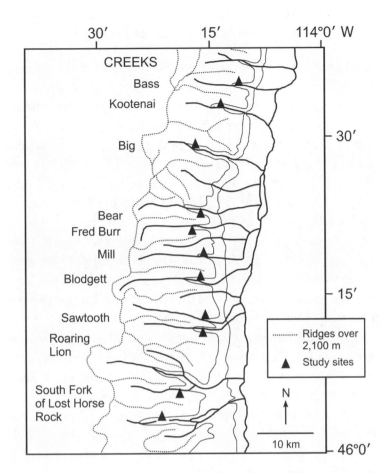

FIGURE 2.18B. A map of the entire Bitterroot Mountain range, showing McCune's study sites (McCune and Allen 1985b).

FIGURE 2.18C. A view from the trail to St. Mary's Peak looking down on the Bitterroot valley with the ridges between the canyons marching into the distance. (Photo courtesy of B. McCune.)

Prigogine has stated that complex systems contain past disturbances in their lower levels. A low level is disturbed and collapses up against a more global constraint of the new higher level. What makes the system unstable becomes an integral part of the working of the new higher level, for example, the nucleus of a crystal determining crystal form. An ecological case in point is fire-adapted vegetation (high level) where individual fires destroy and disturb the susceptible community (low level). Fires change the vegetation over time until it is fire-adapted. At that point, long-term survival of the community requires fire to remove invaders that are not fire-adapted. Fire-adapted communities have incorporated fire as part of the system. Fire is a friend of the emergent higher level, but its destruction of biomass indicates its historical role as disturber of the primitive low-level system.

The emergence of higher levels is dependent on the system being large enough in temporal and spatial terms so that the disturbance can be put comfortably inside. In the case of the Bitterroot valleys, they are too small to incorporate local disturbances. With the mountain ridges filtering out the propagules of many species that are in the Bitterroot range at large, each valley is dominated by the happenstance of what was available at crucial times of establishment. Each valley is controlled not by a process of orderly maturation of community structure through environmentally monitored competition, but by historical events.

Note here that what might be a historical accident at a large scale could be seen as environmental context at a smaller scale. Being somewhat isolated from seed sources from other valleys, the processes at work in each valley are isolated and see the local single valley as global. Within that narrower purview, what occurs across one valley is environmental context, even though seen in the context of the mountain range at large, what occurs across one valley is historical accident.

LANDSCAPE SIGNATURES ACROSS BIOMES

While there is much to be said for putting effort into linking particular patches on remotely sensed images to ground truth, the power of remote sensing for broad comparisons needs to involve a single, all-encompassing signature for each entire landscape in the comparison. The generalized unified patterns can be compared as wholes rather than as a set of local patterns with local causes. Nontrivial comparison at large scales have to be conducted at a level of great generality; otherwise, the sites are merely completely different. It is possible to generate a single number that describes one site such that comparison between several sites across several biomes is possible. One compares the single numbers for each site. We used Haralick texture analysis to get the single number that characterized each landscape.[30] Pixel grayness was compared across a landscape and entered as hits in a grayscale matrix (figure 2.19). White next to white, gray next to gray, or black next to black appear as hits somewhere on the diagonal of the grayscale matrix (figure 2.19C).

By contrast, different degrees of adjacent grayness would appear as hits nearer the black/white or white/black corners of the right-hand matrix in figure 2.19C. The first step is to get a summary number for each grayscale matrix. The summary number is a weighted sum of the grayscale matrix with weights applied to a part of the matrix. For instance, a higher weight close to the gray/gray diagonal gives bigger sums for locally homogeneous landscapes. The trick in this study was that different directions were used to create directional grayscale matrices for a single landscape, so there were four grayscale matrices created for the landscape in question, one for each direction of comparison (figure 2.19B).

The size of the sum does not matter, but relative differences in sum on the same landscape with different directional comparisons does matter. Landscapes composed of generally circular patterns will give the same sum, independent of direction used in pixel comparison (figure 2.19D). Sinuous landscapes will have big differences in grayscale sum because sinuous patterns will show homogeneity in the direction of the sinuosity but heterogeneity at right angles to the sinews (figure 2.20).

The difference between matrix sums for a given landscape in different directions appears to be a quantitative estimate of the general significance of biological processes in generating pattern. The argument is as follows. Pattern on remotely sensed data can be helpfully divided into just two types: (1) fossil pattern that is a scar left by some ancient event, and (2) pattern that is reinforced by a continuing process. The difference between reinforced pattern and fossil pattern is a matter of temporal scale. Anthropogenic and biologically significant patterns both fall into the self-reinforcing category for the most part. For example, a cornfield generates a crop that is sold, and the money is used to put the field back in short order the next year. Geological pattern is of the ancient scar variety, generally a temporally longer pattern. A different sum for different directional matrices says the landscape is sinuous. Small differences for directional matrix sums across direction indicate that the landscape is dominated by circular, or at least isodiametric, patterns.

The relatively large amount of water in eastern U.S. ecosystems makes them biologically driven. Therefore, the eastern United States is given to self-reinforcing patterns over scar patterns. The opposite pertains to dry western sites. Since biology involves growth about a point in space, biologically driven patterns should be isodiametric and should smear the edges of fossil patterns. Fossil patterns will be relatively sinuous and have sharp edges.

When this method was applied to different Long-Term Ecological Research (LTER) sites across the United States, the different sites could be ranked according to the coefficient of variation of the different directional summaries.[31] The Okefeno-kee Swamp in Georgia is a system driven by biological processes. There is essentially no topography, so the pattern should be blotchy, not streaky. Computer processing

A

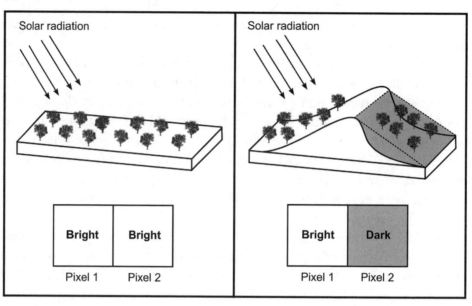

B

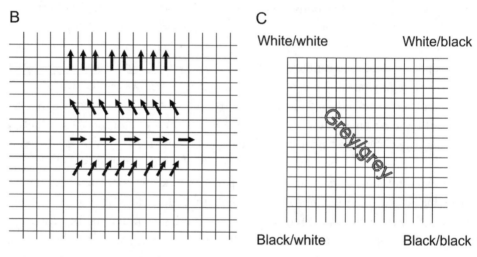

C

White/white   White/black

Grey/grey

Black/white   Black/black

D

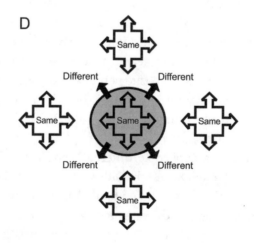

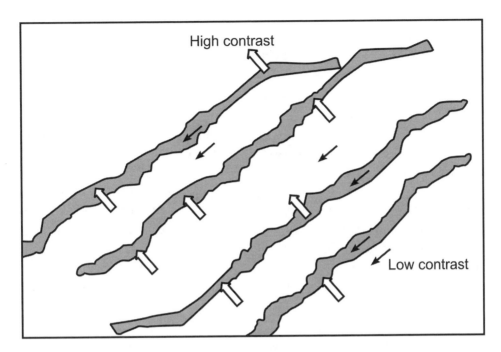

FIGURE 2.20. Sinuous landscapes will have low contrast locally in the direction of the sinews (thin black arrows), but will be highly contrasted at right angles to the sinuosity (thick white arrows). The four directional matrices will have different sums depending on the direction used to generate the grayscale matrix.

FIGURE 2.19. A. Haralick texture analysis depends on the similarity or difference in grayscale. The method compares one pixel with its neighbor and enters that difference as a hit on a matrix of grayscales. The hit is scored in the matrix on one side for the grayness of the first pixel, say a horizontal ordering, and the other side for the grayness of the second pixel, say a vertical ordering on the matrix. B and C. Different ways of sampling a matrix. The right-hand matrix is filled by comparing adjacent pixels. The left-hand side of diagram A would score a hit close to the white/white corner of diagram C. The right side of diagram A would score near the white/black corner. Similar grayness will score on the trace diagonal of the matrix as white/white, gray/gray, or black/black. The hits in the scores in the matrix are summed according to some weighting. For instance, higher weights for the hits on the trace would give a big sum for a relatively homogeneous landscape, but low values for matrices with contrasting adjacent pixels. The left-hand matrix here (B) shows pixels on the landscape, and the arrows show the different directions that can be used for comparisons in creating the grayscale matrix on the right. Thus, four separate grayscale matrices were created. Their weighted sums were calculated and compared. D. Isodiametric patterns will have the same grayscale matrix in all directions. Only the edge of patches will have high contrast. Inside the patch and outside the patch, which is most of the landscape, will be locally similar. That indicates a dominance of biotic, self-reinforcing patterns on the landscape.

of the image indicates generally radial patterns because of a low coefficient of variation across direction with respect to pattern contrast. The Jornada research site in the deserts of New Mexico is very different. It has a sinuous pattern and gave high coefficients of variation between directions, indicative of a fossil pattern.

The more radially symmetrical a pattern, the more important the biology is as the principal constraint. Since biological processes are constantly reworking material, isodiametric pattern indicates a landscape dominated by high-frequency constraints. This is certainly true for the Okefenokee Swamp that is as dynamic as one could expect a large landscape to be. Decomposition creates bubbles in the muck at the bottom of the swamp. This raised the swamp bottom up to make islands. Trees grow on these islands. Meanwhile, the alligators worry the sides of the island until it breaks away, floating around with its trees in place. Thus, relatively large landscape units in Okefenokee Swamp are uncommonly mobile. Not all biologically driven systems show this extreme high frequency of movement, but in general, the more biology constrains the system, the shorter the significance of historical events (figure 2.21).

In desert terrain, scarce water constrains biological effect, which results in persistence of abiotic scars on the landscape. In the Utah desert, the marks of General Patton's World War II tank-training maneuvers are still fully apparent and will probably remain for centuries (figure 2.22).[32] To an extent, fossil patterns fall prey to the forces that drive reinforced pattern. Biological processes soften fossil patterns and therefore lower the contrast in sinuous patterns from steep topography. The persistence of Patton's tracks is helped by the destruction of the biocrusts by the tanks. Biocrusts stabilize soil and retain moisture and nutrients, but have not recovered from compression by the tank treads. In dry lands, biology is not such a potent force, so scars persist (Prose and Wilshire 2000). If biological processes became significant in the desert because of a change in climate, making it wetter, the reworking of material in biological processes would fidget Patton's tank tracks to oblivion in a few years.

LANDSCAPE POSITION

The general issue for landscapes is the need for clever but massive data handling. We have just seen this in the previous section on landscape texture analysis. In

FIGURE 2.21.   *A.* The Okefenokee Swamp in Georgia (photo courtesy of B. Patten). *B.* The Northern Lake site in Wisconsin (photo courtesy of Northern Lakes Long-Term Ecological Research, or LTER). *C.* The Jornada Long-Term Ecological Research site in New Mexico represents a successively drier series (photo courtesy of Jornada LTER). *D.* Konza Prairie LTER, also studied for texture by Musick and Grover (1990). With a drier landscape there is less biological pressure on the system, less continually reinforced pattern, and more linearity to the pattern on the ground (1941 photo courtesy of Konza LTER).

A

B

C

D

FIGURE 2.22.   Patton's tank tracks in the U.S. western desert, where they traverse cryptobiotic crust at Camp Granit after seventy years. (Photo by William Hereford, Prose and Wilshire 2000.)

business, the new issue is "big data." Landscape ecologists have been in that business for a quarter of a century. Sometimes whole new structures emerge out of such analyses. For instance, we had no idea that the fractal dimensions of forests jump up between small and large scales. It is particularly impressive that this occurs across all regions in the United States.

A similarly surprising pattern occurred in the landscape position studies in northern Wisconsin. The Northern Lakes Long-Term Ecological Research site had data for about a decade on many facets of a set of seven lakes. The measures were of physical, chemical, and biological data. The lakes were spread across about a five-mile range (figure 2.23). The lake highest on the landscape was some fifty feet (15 meters) higher than the lowest. But how could the lakes be compared on so many measurements in so many different units? For instance, how could you compare and sum data on phytoplankton diversity alongside pH?[33]

The trick was to rank the lakes on the variance of each of the lakes. The clever part is that ranks are ranks, no matter on what data, and so they are comparable. Accordingly, all the lakes were ranked on the variability of all sorts of data. It was then possible to work out the average rank for a given lake. When the results came out, the researchers were amazed. In only a fifty-foot height difference there was a perfect rank order corresponding to position on the landscape. The lake highest on the landscape had the greatest variation. This was explained by the lake at the top only

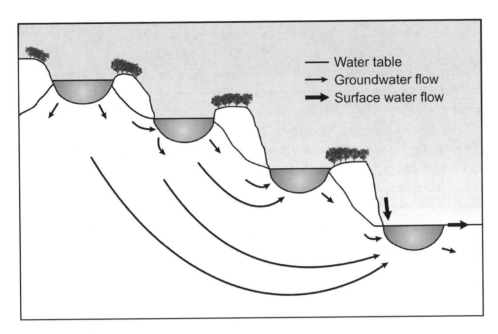

Water table
→ Groundwater flow
➡ Surface water flow

FIGURE 2.23.   A schematic of the Northern Lakes Long-Term Ecological Research (LTER) site showing that the highest lakes are connected to lower lakes via groundwater. It takes hundreds of years for the rain reaching Crystal Lake at the top of the string to reach Trout Lake at the bottom. This smoothes out the variation of the lower lakes in a sequence as the lower lakes have more of their input from groundwater. The lakes were perfectly correlated to the degree of variation and height of their position on the landscape (adapted from Webster et al. 1996).

having rain as input. Those inputs are pulsed with storms. All lower lakes received underground inputs from higher lakes that buffer the water quality. For instance, Crystal Lake high on the landscape was only yards away from, and only inches higher than Muskellunge. And yet the variation was less in the measurements of the lower of the two lakes. Remarkably, the lakes were various in size and type. Some were bog lakes; others were oligotrophic or mesotrophic. And none of these distinctions applied to landscape position. Calculations are that water in Crystal Lake takes about three hundred years to reach Trout Lake at the bottom of the landscape.

The Northern Lakes site was one of a couple dozen LTER sites spread across the nation and even around the world. The North Inlet LTER site near Georgetown, South Carolina, is estuarine, opening into the Atlantic Ocean. On inspiration of the Northern Lakes results, North Inlet performed similar analyses. And they got results as good as Northern Lakes. The difference was that their most variable measurement site in the estuary was the lowest on the landscape. The reversal comes from their source of pulsing in the system being the ocean. Tides, storms, and the like drove the variation. Higher on that landscape, away from the sea, the effects of tides were ameliorated. Apparently, small variances in landscape position make all the difference.

CONCLUSION

Landscapes are the most tangible of the ecological criteria. Therefore, we tend to study them at conveniently human scales. There are, however, small and large scales at which we can profitably study landscapes. Despite the wide range of scales of ecologically interesting landscapes, there is a remarkable unity to the landscape criterion. The reason is that many of the patterns at the scale of an unaided human experience, say in a landscape painting, are remarkably universal. There are indications that the fractal dimension of the landscape experienced by insects is lower than the environment in which we function. However, many processes like surface compression and tension, explosions, or meandering flow recur at scales from the landscape of a leaf surface as seen by a mite, all the way up to remotely sensed images of continents taken from twenty-two thousand miles above the Earth. The reason is that Euclidean space has the same geometric properties no matter what the scale. Much of the world rests in whatever state requires minimum energy for persistence, for example, tight packing to minimize stress. These minimum-energy states are generally limited to certain geometric configurations, for example, we have already pointed out that meandering occurs when there is no energy to override obstructions. Another general notion is that branching systems arise when material for track is expensive. We see these and a few other patterns repeating over again at all scales. On Titan, the largest moon of Saturn the temperature is minus 179 degrees Celsius. Water there is like a rock, and it will make rounded boulders as it suffers erosion. There are rivers on Titan, but they are made of liquid methane.[34] It is such a different place from Earth, and yet there appear to be rivers showing the standard branching patterns with some meandering.

There is a remarkable unity to landscape processes. Stevens, in his "Patterns in Nature," shows an image that applies to anode/cathode relationships on surfaces, magnetic fields, and stress patterns on a plane, as well as in fluids flowing around an obstruction. His explanation for this unity is that space is space for all of those processes and relationships. The spatial ordering criterion is powerful and remarkably general in its application. Complaints about the lack of mechanistic explanation for fractals in metabolic ecology can be dismissed in these terms because we are dealing with much more fundamental processes. For all sorts of reasons, fractal patterns emerge because of the nature of space and distance in which the processes are developed. Many very different material systems all show the same pattern. There is no mechanistic reason for this consonance because magnets are not fluid, and neither is stress in a solid, but the analogy to fluids flowing around obstructions holds because of the general nature of space as a context (figure 2.24).

The power of the landscape approach is in its intuitive appeal. We can use our unaided senses to great effect on landscapes. The familiarity of the landscape

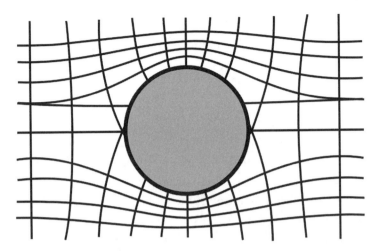

FIGURE 2.24.   The pattern presented in Stevens (1974). It emerges the same for anode/ cathode relationships on surfaces and in magnetic fields, as well as stress patterns in fluids flowing around an obstruction. All show the same pattern.

criterion makes it a mode where interesting hypotheses are most often generated, even hypotheses that pertain to other criteria like populations, communities, or ecosystems. For example, the pulsing epidemics of spruce budworm are popula- tion and ecosystem problems, but the manifestation of the event on the landscape is what first grabs our attention. Spatial patterning is so intuitive and powerful that graphic expression of trees in forests and patterns in rivers are created in U.S. Forest Service research so patterns can be understood. Remote sensing of water temperature has allowed the creation of movie sequences of temperature down a river. The movie looks as if the image of temperature is created as if a temperature- sensing helicopter were flying down reaches.[35]

Although many interesting phenomena, rightfully belonging to other ecological criteria do appear on landscapes, the relationship of landscapes to other ecological conceptions is far from simple. The essence of landscapes can be captured by only a few patterns, but communities, for example, contain many species. Worse than that, the primary units of communities are not scale-independent as are the units in landscapes. While big landscapes can be close to self-similar, employing the same few patterns as small landscapes, there are great differences between large and small populations, communities, or ecosystems.

As with organisms, human perception of landscapes is probably the result of selective pressures on our species. It is reasonable to suppose that we have been selected to perceive the world in a way that allows prediction. It is possible for humans to use landscapes to effectively geo-navigate from one point to another. Prediction comes easier in familiar circumstances. Since changes in scale alter perception radically, it would be of advantage to perceive in a way that recognizes patterns that occur at multiple scales; then the world remains familiar even under changes in scale. The scale-independent patterns of landscapes are meanders, spirals, explosions, and branching systems.

Humans like fractals, presumably because there is a repeating pattern at different scales. Often things and appearance both change with scale change. The beauty of fractal patterns is that they stay the same. Therefore, principles that apply to small things also apply to large things. It is a reflection of Western scientific thinking that fractal geometry has come about only in the last few decades. Recent work by Ron Eglash, a cultural mathematician, reveals that African villages have been constructed on fractal patterns from well before Mandlebrot's 1960s mathematics on the issue.[36] Eglash's TED talk is well worth a listen. This fractal pattern of villages is restricted to Africa. In South America, there are commonly repeated patterns within a village, but they are not the same pattern at different scales. In Africa there are roughly circular villages with similarly roughly circular houses inside. The chief's house is bigger and circular. The spaces inside the big house are circular too. There are even smaller circular devices inside, too small for living people. They are for the spirits of the dead. The model is not just the same pattern repeated everywhere, for different villages have their own individual fractal patterns, not necessarily circular. So the culture is not just copying, it is an explicit understanding of fractals in principle.

It is remarkable how late Western science and mathematics discovered fractals. It was probably writing alphabetically that slowed us down. Marshall McLuhan said that alphabets work below the level of meaning.[37] He pointed out that pictographic languages never go below the level of meaning. And oral cultures would find it even harder. Alphabetic cultures can suspend belief. They also tend to linearize thought. Western science is very spatial because it can afford to neglect time and narrative; they can be written alphabetically. Westerners have time to invest in irregular spatial relations, about which they also write. But if one is in an oral tradition, keeping track of time is harder and is achieved by telling stories. A reading of The Odyssey shows that it is an oral presentation that has been written down.[38] Odysseus has an adventure, but before moving the story forward, he ritually kills a beast, wraps its loins with fat, roasts it, and drinks wine. The narrator needs time to regroup before another cycle of action. The comedian Larry the Cable Guy has a catch phrase, "Git-R-Done," which works the same way in his oral tradition. It is a spacer. "Git-R-Done" is his version of wrapping loins with fat. African bead games are hard for linear-thinking Westerners because the endless cycles within cycles do not come easily (figure 2.25). In Africa, Allen would lose at those games all the time. So, in the absence of writing, Africa developed cyclical patterns within cyclical patterns to keep things straight. Stories repeat, and structures repeat in a culture where they do not write things down. African drumming can carry rich information in its fractal, cyclical patterns, which was a point of consternation for imperial whites. Fractals order elaborate spaces.

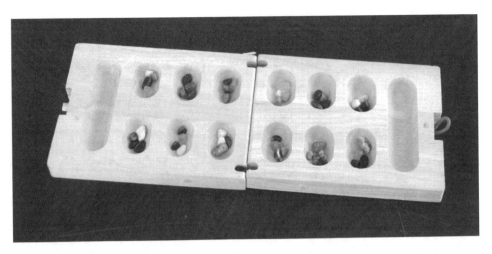

FIGURE 2.25.   An African bead game called Ayo in Nigeria, but also known as Mancala. Beads are dropped following a path that loops up one side and down the other. The complicated set of games involves jumping in and out of cycles.

Mapping communities or other ecological criteria onto landscapes is not easy because they are not simply spatial. However, the ease with which we see space often means that problems under other criteria are detected first under the landscape criterion. The problem with lead pollution in ecosystem function studied at Oak Ridge was first manifested in a spatial pattern of deeper leaf litter next to the smelter. In classical Greece, the Roman traveler Pausanias made note of groves of trees whenever he saw them.[39] By implication, most of the trees must have been gone by then. And this explains failed ecosystem function with the silting of harbors also reported in classical documents. The landscape is the first place to look for lead indicators. It is no accident that we have started our comparison and contrast of different ecological conceptions with the tangible landscape.

Throughout this chapter, a recurring theme has been the need for massive computation to deal with landscapes. This is the same point of tension that we captured in Beaufort and his scale. Landscapes are rich and tangible, and that is what makes them so hard to countenance. The same applies to the rest of ecology, even if not so obviously. The mathematics of landscapes is as elevated as any all across ecology. Clever summary is the hallmark of ecology, particularly of landscapes. While Beaufort would appear to have done something fairly obvious, it only appears that way in retrospect. The same applies to ecology. There were millennia of mariners who surely could have used his scale, if only someone had invented it. They had not; it took Beaufort's genius to do it. And so it is with ecology. Like mariners of old, those who study in natural settings need powerful summary devices. We have seen this in strategies for landscape ecology. The whole subject of ecology has arisen to make the richness of nature manageable. Tangibility makes ecology a challenging endeavor, all the more so for being so obvious.

3

THE ECOSYSTEM CRITERION

BECAUSE LANDSCAPES are tangible, they are particularly useful for putting into practice our scheme for comparing ecological observational criteria. The scheme for looking at ecological comparisons and explanations laid out in abstract in chapter one, and then implemented in chapter 2 is: (1) keep the criterion constant and change the scale (see chapter 1, figure 1.15A); (2) hold the scale constant and change the criterion using that same tract of land; and (3) change the scale and the criterion together so that, for instance, landscapes can become the context above or the mechanism below ecosystems, communities, organisms, or any other type of ecological structure (see chapter 1, figure 1.15B). Mode of observation 3 can use any ecological criterion for observation above or below an instance of any other criterion for looking at the system. Each hierarchical ordering of the system can use any mixture of criteria for order in the hierarchy. Now we are in a position to test the general usefulness of our scheme with something more challenging, the intangible ecosystem.

But first we must deal with potential confusion in terminology. Robert O'Neill and his colleagues, in "A Hierarchical Concept of Ecosystems" (1986), mean something broader by the term "ecosystem" than we have in mind here.[1] Allen was one of O'Neill's coauthors, and has not changed his mind between that book and the present volume. The point is that these are decisions relative to a purpose, which is different between the books. The difference between the deconstruction of O'Neill et al. and that of Allen and Hoekstra is not material or substantial; it is simply an expression of a different purpose. All naming and organizing is specific to a purpose. Their process/function type of ecosystem is what we mean by the term

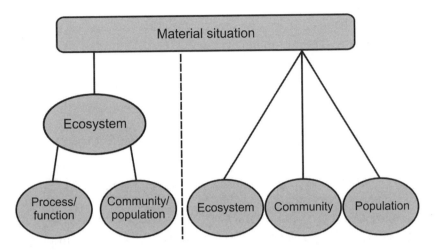

FIGURE 3.1.   Two conceptions for the term ecosystem. O'Neill et al. (1986) is shown on the left side of the figure; Allen and Hoekstra is on the right side.

"ecosystem" here. They also include population/community in their definition of ecosystem; we reserve the term "community" for that type of system and give it equal status with "ecosystem" rather than make it a type of ecosystem. O'Neill and his colleagues, like us, were at pains to discourage ecologists from requiring that communities be contained within process/function ecosystems. That is why they included community as a type of ecosystem, so it could not be taken as a default component of the process/function conception of ecosystem.

The material situation is undefined. From that, O'Neill et al. define "ecosystem" to include species and population conceptions as well as flux and process into which organisms and species melt. Here, Allen and Hoekstra take the same undefined situation and restrict the term "ecosystem" to just process/flux aspects of what is observed. Community, oriented toward species, organisms, and population, is for Allen and Hoekstra a separate consideration of equal standing to, but different from, ecosystem (figure 3.1). The points of view differ much in the manner of the inertial frame in physics wherein the sun may either be taken as a relatively unmoving body, as opposed to the sun seen as moving across the sky. There is no contradiction in the sun moving or not, and neither is there in the two usages of ecosystem. Each simply indicates a relative reference system proposed for different purposes of emphasis.

However, O'Neill and colleagues, in their broad definition of ecosystem, depart significantly from conventional usage. Ecologists who call themselves ecosystem scientists would cleave themselves away from ecologists who study "population/community" systems. Conversely, students of "population/community" systems would not often call themselves ecosystem scientists. Rather they would answer to the name "community ecologist" or "population biologist," depending on how they prefer to approach multiple-species assemblages. We have invested much effort elsewhere in this book to make sure that communities are not seen as necessary parts of ecosystems. The terminological device of O'Neill and his colleagues is superfluous for our treatment. We do not need to pay the price of using an unconventional vocabulary on this point, so we choose not to do so.

THE HISTORY OF THE ECOSYSTEM CONCEPT

The process/function approach to ecological material, the subject of this chapter, goes back to the early decades of the twentieth century. In a recent summary, Sterner shows that there were several early close passes at an ecosystem conception set in aquatic ecology where connections between biota and nutrients figured large.[2] Water invites a conception of flux. In a distinctly terrestrial setting, Transeau focused on energetics as early as 1926: "Let us now examine the energy budget of hypothetical acre of corn in heart of the corn belt in north central Illinois where corn achieves yields as great anywhere, and not far from Madison Wisconsin, one of the stations at which solar radiation has been studied."[3]

Transeau looked at energy flux in photosynthesis, respiration, and transpiration, casting the system in terms of energy flux. He looks back to early scientists citing who identified that there was something in air (in fact, oxygen) that is needed to sustain life. Transeau referenced Brown at the beginning of the twentieth century for the energy budget of a green leaf, and worked his way up from there.[4] Transeau reckoned that nutrient status pertained to his study, and based his thinking on work from previous centuries in that regard. Transeau's was a very self-conscious effort that realized there was something new here, classically based and finally brought into ecology and made cohesive.

The word "ecosystem" did not appear for another decade, when Sir Arthur Tansley was trying to clean up the terminological mess at that time in community ecology.[5] Tansley noted that the plant community has an environment with which it interacts: animals, soil, and the atmosphere. The interaction goes both ways and so the community is in a sense part of the environment of its own environment, in Tansley's conception. At that point, Tansley folded in the animals, climate, and soil to make something more inclusive, the ecosystem. Tansley's agenda was to make the plant community a more rational conception. It was not ecosystem science, to which his creation in fact led. By folding in the biotic and physical environment, Tansley had, perhaps unwittingly, moved the discussion into the arena of mass balance in a black box. That tension between biotic and thermodynamic principles echoed over some decades until Odum in 1969,[6] after whom ecosystem ecologists finally stripped away the biotic ordering principles of organisms, species, and Darwinian adaptation from the thermodynamic principles.

We notice that the modern ecosystem concept took some time to emerge. The critical player in bringing together energetic (Transeau) and mineral nutrient flux was Lindeman, whose 1942 paper put a lot together.[7] It was sufficiently new that Lindeman's mentor, G. Evelyn Hutchinson, had to put pressure on the editors to get the work out. He tracked a departure from "(1) the static species distributional view point," the view of species in places that goes back to biogeographical posture that was established early in the nineteenth century with von Humbolt. Lindeman then lists "(2) the dynamic species distributional viewpoint, with emphasis on successional phenomena." That is community established as a collection of species changing over time in the works of Clements in the early twentieth century. It is that which had become the mess that Tansley (1935) was trying to clean up. The third stage of intellectual development that Lindeman identified was his preferred view, "(3) the trophic-dynamic view point." He considered a lake as the material reference, and worked through how the three intellectual postures, Transeau's, Tansley's, and his would play out. Lindeman's mature view updated earlier aquatic interest in nutrient flux around species and genera with decay as critical in the works of Shelford (1918) and Strøm (1928) (figure 3.2).[8] Diagrams of these and other earlier syntheses are readily available in Sterner (2012), along with an affectionate account of Lindeman's life. Lindeman is always cited as the one who put "the trophic-dynamic viewpoint" together. Lindeman's "Ooze" is central to his diagram, the aquatic version of soil in Tansley's conception. Critically, Lindeman (1942) put the energetics and nutrient recycling together, even though he does not cite Transeau (1926). Ooze allows the synthesis between the aquatic and terrestrial postures of previous decades.

Lindeman still does not completely step away from the Clementsian view in his "(2) the dynamic species distributional viewpoint, with emphasis on successional phenomena." There is still implicit presence of organisms as organisms and something like succession is a big player. He introduces the unity of energy and nutrient flux, but does not abstract the cycles as the parts of the system. The parts are still plants and animals with some "ooze." Teal's (1962) classic paper addresses energy flux in and around a salt marsh, with significant recycling and a strong focus on explicit and quantitative energy flux seen in the field.[9] In one last great effort to accommodate Tansley's original idea and Lindeman's type 2 and 3 conceptions, Odum (1969) looked at energy and material flows in ecosystems, which for him, seen in a contemporary light, are a hybrid of community and ecosystem conceptions. This is something of a rehabilitation of the unity that Clements saw, with energetic and flux arguments as justifications for that wholeness. It is much more than "Clements warmed over," as Allen once heard it called in the 1970s (personal communication with Orie Loucks at an Ecological Society of America [ESA] annual meeting). E. P. Odum, with his principles of development and his identification of humans as critical components gives the last grand view; he was not completely successful, but made a brave effort.

A Shelford, 1918

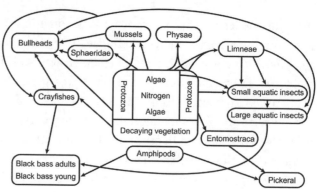

B Strøm, 1928

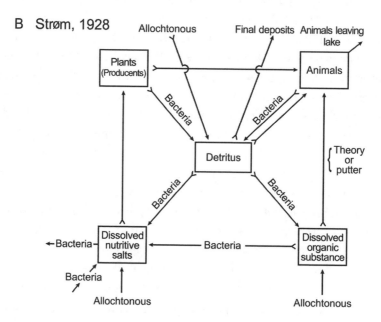

C Lindeman, 1942

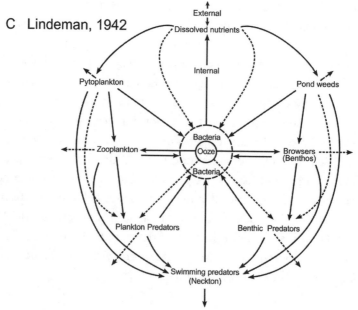

After Odum, computer technology gave much impetus to ecosystem and community studies as separate enterprises. On the one hand, computers allowed huge simulations to drive the ecosystem flux and process approach. On the other hand, there was abundant analytical power to address great masses of interrelationships between species summarized with multivariate statistics to press forward the community/population view considered in our next chapter on communities. After 1970, Oak Ridge National Lab had a cabal. They called their sessions "Labac," a reverse spelling of cabal. These were brilliant, computer-savvy young ecologists (for instance, the authors on Van Voris et al. 1980) backed by a Cray supercomputer. These scientists took energy and material flux to such levels of sophistication that a study of ecosystems strode off on its own, separate from communities and succession. The University of Georgia, Athens, and various other centers for the biomes of the International Biological Program (IBP) were part of that effort.[10] Ecosystem science formed a self-conscious invisible college at that point, where all the players across North America collaborated and knew each other well.

A parallel invisible college for community ecology formed in a network centered at University of Wisconsin and Cornell, in North America, with strong connections to the Australian Commonwealth Scientific and Industrial Research Organisation (CSIRO), and the community work of Greig-Smith and students in North Wales. Some, Allen for instance, played in both the community ecology and ecosystem camps, but changed posture as necessary across the divide (figure 3.1). It was time to separate community conceptions from ecosystem insights. In this ecosystem chapter, we follow the development of those new ideas on the flux and process side.

THE ORGANISM IN THE INTANGIBLE ECOSYSTEM

There are several meanings for the word "ecosystem" in the ecological vernacular. Even when the word is used informally, a process focus is implied. There is a difference when an ecologist speaks of a forest ecosystem as opposed to a forest. We do not insist that ours is in any way the right one, for all definitions are arbitrary.

Organisms are as tangible a set of entities as one could want. Watching organisms melt into the intangible pathways of the ecosystem can bring home the essential intangibility of ecosystems. The failure of organisms to offer ecosystem

FIGURE 3.2.   The conception of ecosystem dynamics in the first half of the twentieth century. Lindeman's conception introduces the "ooze," where detritus brings the link between biotic and abiotic compartments. Lindeman notes Shelford in 1918, with food chains but no link through detritus (diagram A), but corrected in Strom in 1928, with detritus holding a central position (diagram B). Lindeman's massive contribution (diagram C) was the quantification that he laid down on top of Strøm's correction of Shelford (Sterner 2012).

explanations and predictions comes from their lack of discreteness in ecosystem function; organisms are more crosswalks across which ecosystem function passes. The pathways in which organisms are subsumed are the functional parts, not the organisms themselves. The parts might be the nitrogen cycle or the carbon cycle. Another part might be a pathway that starts with overgrazing, which removes vegetation excessively, therefore changing the surface albedo of the soil, which changes climate, which affects plants further, perhaps in a process of desertification. The pathway is the ecosystem part. Thus, animals present in ecosystems are usually not parts of ecosystems, they are only connectors within the parts.

Organisms as discrete entities in ecosystem models are at odds with organisms as conduits (our preferred model). But in the narrative at the end of the study, organisms can be both, as the narrator slips between definitions. Ecology is not the study of the material system undefined. It is the study of ecological material discussed in terms that the ecologist decides in the opening narrative on a process that refines the story. Banning of organisms as organisms in ecosystem depends on the situation to be explained, because salmon, for example, do function as organisms in ecosystems.

ORGANISMS FOR ECOSYSTEM
AND COMMUNITY SCIENTISTS

Certain special approaches to animals necessitate balancing calories in, say, locomotion physiology. However, most organism-centered biologists of a taxonomic ilk spend little time dealing with fluxes of matter and energy, and for the most part are unconcerned with keeping track of the mass balance in the system. The notion of mass balance focuses on conservation of matter and energy in terms of the first law of thermodynamics; certain quantities enter the system and must remain if they do not come out. It is not that communities violate conservation principles; it is rather that such principles do not predict community structure or behavior. Evolution by natural selection works through principles like competition, mutualism, and predation. It is a structure-focused point of view where we notice that some structures are selected and others are not. Communities are ordered on evolved organisms, and the abovementioned principles are used to explain the workings of the accommodation between community members. Evolution does not violate conservation of energy or matter, nor does it violate the thermodynamic principles of increasing global entropy. However, insights into relative fitness are not often gained from knowing that organisms respire or otherwise expel material and energy at a rate commensurate to their consumption minus their growth. Conservation of matter and energy is, for the most part, irrelevant to the community ecologist. By contrast, the ecosystem scientist could not do the most elementary bookkeeping in ecosystems without invoking conservation and principles of mass balance based on the first law of thermodynamics.

Thus, important predictors in one type of system are of little use in the other. For example, the community structure of forests in the southeastern United States was radically altered by the blight at the end of the nineteenth century that removed the American chestnut as a critical component of the canopy of the eastern deciduous biome and its communities. Meanwhile, the contemporary record at that time gives little indication that the ecosystem function in those same places was altered one jot, even at the height of the epidemic.[11] The chestnut, as indicated by simulation studies, seems to have been merely one of many equally workable alternatives for primary production and energy capture. This notion that something can matter a lot in one framework but not in another is an important consideration when, at the end of this book, we turn to strategies for basic research as well as management issues.

The relationship of ecosystems to communities is called a many-to-many mapping. That is to say, neither conception invokes with any regularity the entities that pertain to the functioning of the other. To see the consequences of many-to-many mapping, consider ecological succession where a dominant tree species is replaced by another. Expressed thus in community terms, one sentence has laid out the situation, and the consequences of that successional event for the community can be readily described. For the ecosystem, there may or may not be a change in the rate of recycling, in the carbon budget, or in mycorrhizal efficiency, to name just three uncertainties. Worse than that, take any one of those many ecosystem factors related to the one community factor, and we find that it relates back to many community factors; for example, mycorrhizae could have much the same ecosystem consequences while representing a whole suite of different fungal community assemblages. Only occasionally does the same chunk of the world relate across the community and ecosystem conceptions in a one-to-one mapping.

One notable exception that does achieve such a mapping is the case of Pacific salmon in the rivers of the western United States and Canada, where we find that the fish belongs as a discrete entity to both types of system. Salmon are distinctive in that they spend the bulk of their adult life cruising the ocean but return for reproduction with astonishing accuracy to exactly the same place they were spawned. One of the reasons salmon are such a sporting challenge for the dry fly fisherman is their lack of interest in feeding during their spawning run. Salmon eat almost nothing as they return to spawn. Both the male and female fish have all the energy they need for breeding, so their guts are superfluous and in some subspecies are disintegrating. Once having laid or fertilized the eggs, the salmon in the western rivers of North America die.

Heavy fishing during recent decades has caused the salmon run to fail. The first impression was that not enough salmon had survived the catch to return and lay an adequate number of eggs. It seemed that the small number of fertilized eggs had not offered a large enough population of hatchlings. This deficient-year class supposedly manifested itself as the missing adults of years later. The remedy was thought to be restocking the rivers with the missing hatchlings from human fish hatcheries. Unfortunately, this was only moderately successful. The problem was apparently not missing hatchlings seen as community members, but an ecosystem variable that mapped onto the adult salmon.

The error had been easy to make, for it mistook salmon for components of the fish community. Of course they are members of the fish community, but that was not the role they played in the system's sickness. Missing were dead adults who had just spawned, not live hatchlings. Rivers constantly flush away nutrients, an ecosystem consideration. Dead adults rot, release their nutrients, and stimulate algal growth. The algae are food for microscopic invertebrate animals, the food of the hatchlings. The adult salmon were the source of eggs, but more significantly, they were the source of nutrients. The fish were a critical ecosystem property, a nutrient pump upstream, one fish at a time. The solution to the fishery management problem was to put mineral nutrients into the headwaters of the rivers at the critical time.

Telling here is that eastern rivers in North America have salmon that swim upstream to breed, but then swim downstream, back to the sea, to get ready for next year. Eastern ecosystems are wooded and provide nutrients to the rivers there. But the western ecosystem headwaters are largely in deserts, in places like Idaho, Montana, and Wyoming. There is little biological production there to churn mineral nutrients that might go into the rivers. Western rivers are accordingly nutrient poor, so production at the base of the food chain is small, and the fingerling salmon do not have enough to eat. Therefore, the dying adult salmon are not needed in eastern North America. In the East, there is natural selection at the individual level, where salmon are selected to survive to mate next year too. The death of salmon in the western headwaters is a clear example of group selection overcoming individual selection.[12]

The deal is clinched in nitrogen isotope ratios.[13] Most nitrogen atoms have seven protons and seven neutrons to give an atomic weight of fourteen; hence, nitrogen-14. The only other stable isotope of nitrogen has one more neutron to give nitrogen-15. More than 99.5 percent of nitrogen is nitrogen-14. Because of different processes of use and reuse in the ocean as opposed to the land, the terrestrial ratio of nitrogen is different from the oceanic ratio. In the terrestrial ecosystems beside the Columbia River, nitrogen is closer to the oceanic ratio. The data clinch the salmon's crucial role. Bears eat salmon and defecate in the woods. The salmon as organisms are the source of nutrients.

Another example of a species-specific role of animals in ecosystems works in the opposite direction of the salmon. In the canvasback duck, it is exporting nutrients such that shallow lakes in Minnesota are kept clear. It is species-specific to the canvasback. Recent studies of the poorly drained, shallow glacial lakes in the prairie regions of southwestern Minnesota shows the difference between species of ducks. Different species have opposite effects on the lake ecosystems. The specific lake studied by Steve Thomforde and Peter Allen was Huron Lake, a shallow lake of some 800km². The canvasback duck is the largest of the diving ducks in North America. It is a migratory species that feeds on the tubers of wild celery (*Stuckenia pectinata*). The plant grows widely distributed, but plays an important role in the shallow lakes. They and the duck used to keep the lakes clear.

In the old days, the wild celery was abundant and took up a quantity of nutrients. These plants are the favorite food of the canvasback, which would gorge themselves before their migration south. The ducks exported nutrients so the water quality was high. The name canvasback has competing etymologies: one is from the light brown back feathers on the bird; the other comes from the huge hunting pressure on the duck to the point of wholesale shipping of dead birds to Eastern cities. They were packaged in canvas for travel, which was to be returned, since the bags were labeled "Canvas Back." In 1920, Lake Huron in Minnesota became eutrophic, and remains so despite ninety years of restoration effort. The lake in Clear Lake, Iowa, is not clear anymore. With more farming in the region and no canvasbacks, the nutrient export system broke down. The ducks and geese that replaced the canvasback eat in the surrounding cornfields. They then return to the lakes, where they defecate. A system of nutrient import has thrown the switch to eutrophication. The turbid water quality works against the wild celery too, so reestablishing the old healthy ecosystem is not going to be easy.[14]

Examples of species-specific ecosystem processes are fairly uncommon. Quite often those species that fit the bill, usually animals, are called keystone species. They may anchor the community in place, but the keystone name appears to apply to species that usually lock into an ecosystem function. Beavers have been called a keystone species, or ecosystem engineers. In engineering a favorable environment for themselves, ecosystem engineers take on some specific ecosystem function. There are not many keystone species because community function does not commonly map onto ecosystem function.

THE SIZE OF ECOSYSTEMS

Having shown that the organism is only occasionally a discrete part of an ecosystem, now let us see how the ecosystem criterion relates to the other tangible criterion, the landscape. Area is a landscape criterion, but ecosystems are not readily defined by spatial criteria. Ecosystems are more easily conceived as a set of interlinked, differently scaled processes that may be diffuse in space but are easily defined in turnover times. Processes pertaining to very differently scaled areas encounter each other in the full functioning web of the ecosystem. Thus, a single ecosystem is itself a hierarchy of differently scaled processes. This should not be confused with the hierarchy of differently scaled ecosystems, where bigger air sheds belong to larger-scale ecosystems. There are differently scaled processes inside a single ecosystem, as well as sets of differently scaled, differently inclusive, whole ecosystems. Which one of these is which depends on the intellectual posture of the scientist and the concomitant measurement regimes that are imposed.

Attempts to specify a particular area for an ecosystem raises difficulties. The problems stand out clearly for the meteorological part of ecosystem pathways, but boundary issues are by no means limited to the atmospheric part of the system. A place on the ground does not adequately delimit the climatic aspects of ecosystems. It rains most afternoons in the Great Smoky Mountains in the summer; plants cause the precipitation through their transpiration. Thus, the rain is not ecosystem context; it is part of a pathway that is itself a critical component inside the ecosystem. There is an appropriate air shed that is partner to a given watershed, but it covers a much bigger area than the watershed. With the atmosphere as part of the ecosystem, the spatial boundaries of the ecosystem move every time a new weather system passes through the region. So ecosystems generally cannot be pinned to a place. It is not that parts of ecosystems lack a boundary in space; it is that such a dynamic boundary is impractical for most investigations. There have been attempts to use averages of the position of air masses, but that is different from defining the explicit boundary of the ecosystem at an instant (figure 3.3).

Some ecosystem parts, like surface water flows, can be mapped to a restricted, stable site, a watershed. Nutrient retention processes of a particular ecosystem could be associated with a particular watershed. Watersheds are bounded in space. Even so, the entire ecosystem regularly violates the watershed boundary on the ground. Such violations come, for example, from the ecosystem's internal airflow, which moves in and out of the watershed. In addition, animals move in and out of watersheds, causing their associated nutrient loops to spill out over the edge.

If an entire ecosystem cannot be said to be in any given spatially defined place, then we need some other way of telling ecosystems apart from each other, or a single ecosystem from its context. Put another way, how big is an ecosystem? The search is for natural surfaces, boundaries that coincide with a large number of limits. Later, we show how to identify and specify intangible surfaces of ecosystems using some of the devices that biochemists use to separate their intangible pathways. For the moment, we deal with limits associated with tangible surfaces in ecosystems, like watershed boundaries, but we do this without prejudice against diffuse surfaces that give the intangible bounds of an ecosystem. Preliminary discussion of ecosystem boundaries can move forward more easily using tangible surfaces, so that is where we start.

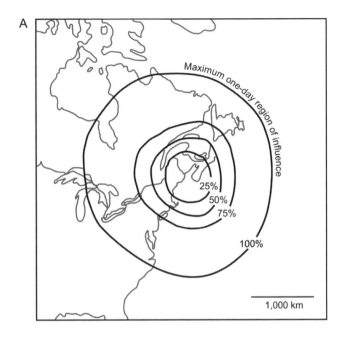

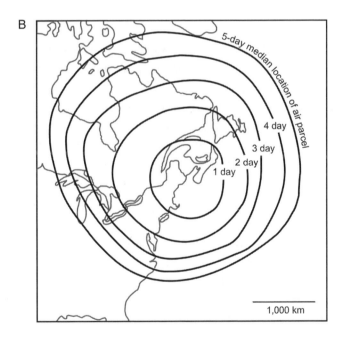

FIGURE 3.3. Not only is it possible to identify the watershed of given ecosystems but it is also possible by calculations of average wind speeds and direction to identify air sheds for a given site in North America. The two maps show: A. the percentage of the time a one-day air shed is contained within the circumscribed area; and B. the median air shed for one to five days for Kejimkujik, Nova Scotia (Summers 1989).

The watershed makes a robust natural boundary because a number of critical local processes reach their limit at the edge of the one catchment area. That does not mean that the watershed defines the whole ecosystem. Watersheds are simply measuring devices just like a meter rule. Some precision is needed in defining measurement devices away from the measurements themselves and the thing being measured. Nutrient deposition input to the ground and surface water flow patterns are both held within the drainage basin. The coincidence of a boundary with many processes indicates a linkage between those processes. The system connects nutrients arriving through deposition to all parts of the ecosystem by the flow of water. Both diffuse nutrient input and water flow coincide at the edge of the watershed.

It is important to work across the entirety of the system so that we encounter most of the pathways that give critical system behavior. We need to know the size of the system so we can study it all and then go on to find how it is different from its surroundings. The size of an ecosystem is given by the largest extent that only just contains the definitive pathways of the system. The ecosystem is relatively homogeneous inside its boundary. Watersheds are not as much the boundaries of real ecosystems as they are devices for convenient bookkeeping. Water falling in a watershed comes out in one place, where it is possible to situate a weir for measuring output. Inputs of water and nutrients can be reasonably assumed to arrive out of the air in a diffuse way across the watershed at random. Studying a whole watershed facilitates a connection between diffuse input to an explicit point of output. As a result, there is a big difference between increasing the size of the study site while remaining inside the system boundary, and increasing the extent so as to cross the system boundary. Rain falling on just the other side of the hill runs away from the local system center instead of toward it. The integrity of the one watershed is broken. But the problem is a measurement issue, not something that matters to ecosystem function. We can move up the column of ecosystems within ecosystems (shown in chapter 1, figure 1.15A); we can hold the criterion constant and increase ecosystem size.

In figure 3.4, we see a set of images of the Kickapoo watershed in southwestern Wisconsin. Image A shows all the details of water flow across the approximate four hundred square miles of the whole river basin. Any water falling in that basin, if it flows out of the system, leaves into the Wisconsin River at the bottom of the figure. But there are smaller watersheds within the basin, and images B and C show the watersheds of fifth- and sixth-order branches. Moving over the ridge between watersheds can be considered in different ways. One might be moving from one local watershed to another. But one could equally well note that the movement is upscale to a higher-order watershed that encompasses both watersheds on either side of the boundary. This illustrates the complications of using the watershed conception.

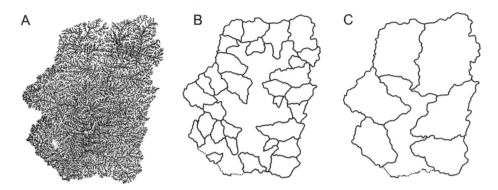

FIGURE 3.4.   In a detailed study of the environmental impact of a dam on the Kickapoo River, the Center for Biotic Systems, University of Wisconsin, Madison, collected maps of the fine detail of the stream patterns in the entire watershed of approximately fifteen by twenty-five miles, and then aggregated these small rivulets into successively larger watersheds within (A) the entire Kickapoo watershed system; (B) fifth-order watersheds; and (C) sixth-order watersheds.

While the watershed example has an intuitive appeal, the same arguments apply to less tangible, temporal aspects of ecosystems.

THE CRITICAL ROLE OF CYCLES

An ecosystem resembles a complex of biochemical pathways in that cell metabolism and ecosystem function both mix and interdigitate with neighboring systems. Different ecosystem processes share their respective spaces to a degree. The various pathways shunt nutrients and energy around the ecosystem very similar to the way biochemical pathways shunt molecules and chemical energy. The Krebs or citric acid cycle occurs in mitochondria, but does not occupy a particular place. It is better considered as an intangible but discrete part that is defined by its role in cell functioning. It keeps its identity by maintaining strong connections between its parts dispersed across the cell (figure 3 5A). Despite its intangibility, this biochemical pathway has very real consequences. Figure 3.5B shows how the Krebs cycle of figure 3.5A is set in a higher-order system that eventually connects respiration to photosynthesis. Thus in biochemistry, we are dealing with non-nested systems where connections are a matter of level of analysis, rather like the watersheds of different order seen for the Kickapoo River in Figure 3.4. So the same boundary issues apply in ecosystems; they are consequential but are for the most part intangible. We need special procedures for measuring them. Like biochemistry, ecosystems are process-oriented and more easily seen as temporally rather than spatially ordered.

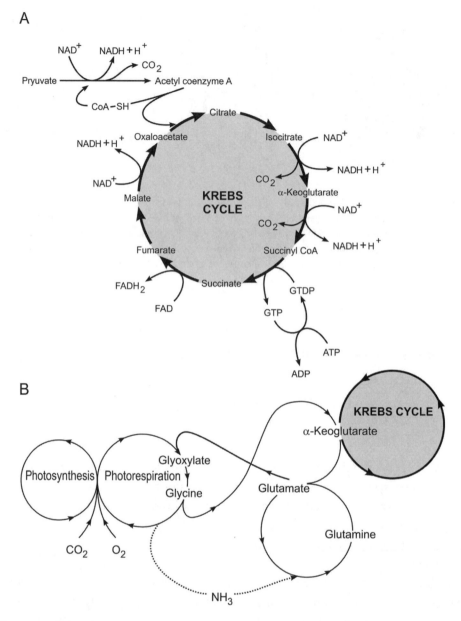

FIGURE 3.5.   *A.* The Krebs cycle. *B.* The Krebs cycle is but one cycle in a complex web of pathways. Radioactivity fed into the Krebs cycle eventually leaks into other parts of cell function.

Consider the way biochemists plot the course of biochemical cycles. They add radioactive material in a chemical form that belongs early in the pathway, and then let the system run for successively longer periods of time. Molecules that occur early in the cycle show radioactivity quickly. Later parts of the pathway take successively longer to pick up the radioactivity. In doing this, biochemists engage in a temporal version of our looking for the areal extent of the watershed. They look for the sequence of events that takes the longest time to occur while still reflecting an orderly pathway; they seek the extent of their system to find the limits of the cycles they investigate. They take note of pathways that appear to turn around and eat their own tails, for it is in recurrence that the pathway achieves closure and therefore becomes a discrete identity.

The boundaries of pathways may not map onto space, but are temporarily defined. There is often reference to life emerging as the first cell. Before the first cell there were enzymes, and the increase of speed of reaction with enzymes separated protolife from its mere organic chemical environment just as much as a cell membrane might. Cell membranes simply speed up interactions within and delay reactions across. What the object is and what its environment is can be defined well in temporal terms, which only sometimes map into a coherent space. Richard Levins gave a talk in the early 1970s on the campus of the University of Wisconsin. It was a telling moment when he said that "ashes to ashes and dust to dust" only means that you sink into a sea of commensurate reaction rates when you die. Your tangible corpse and molecules may persist, but you are gone.

Note that the smaller ecosystems may be as distinctive as the individual biochemical pathways in a cell. Therefore, the move to larger ecosystems is no more a simple summation than is the move to a cell from its biochemistry. Big ecosystems can be as different from small ecosystems as cells can be from biochemical cycles.

The methods of ecosystem ecologists resemble those of biochemists. Ecosystem experimentalists commonly pulse their systems and wait for the signal to pass through the entire network. Sometimes they even use radioactive tracers like biochemists; therein lies the history whereby Oak Ridge National Laboratory has become one of the world-class centers for ecosystem research.[15] Radioecology really started there. Radioactive tracers tell the ecologist what are the important connections in the system.

Many experimental pulses in ecosystems do not involve radioactive tracers and might take the form of a clear-cut of the forest with or without removal of the downed timber. This was the approach of Gene Likens and the team of researchers studying the Hubbard Brook watershed (figure 3.6).[16] It is fortunate that at least some aspects of the diffuse ecosystem do map onto the ground because that allows the experimentalist a means of pulsing the ecosystem in a meaningful and tractable way. Clear-cuts performed across entire watersheds facilitate the mass balance calculations. Clear-cuts on portions or mixtures of watersheds are difficult to calibrate.

One often measures materials that flow around inside the system in a complex fashion. However, relatively few measurements may be needed if one is satisfied to let the ecosystem integrate the signal. This is done by making the measurements

FIGURE 3.6.   The Hubbard Brook site, where some of the first ecosystem clear-cutting work and nutrient cycling calibration was performed. The weir in the foreground allowed measurement of water and nutrient fluxes. (Photo courtesy of Gene E. Likens.)

at the ecosystem outflow in the stream. The inputs often can be calculated by measuring precipitation or deposition at a few local sites in the ecosystem and making the reasonable assumption that input is diffuse and relatively even. The Hubbard Brook study found that the system leaks mineral nutrients when the trees are removed, but reseals itself very quickly as soon as the woody vegetation achieves significant cover again. By measuring system loss and input through deposition, ecosystem scientists calculated how long it would take the system to achieve the same nutrient capital as before the clear-cut. The clear-cut allowed the ecologists to assess the role of woody vegetation in recycling resources. Woody vegetation in combination with soil and its microbes comprised the nutrient retention part of the system. As with many critical ecosystem functions, nutrient retention is performed by a cycle.

Like biochemists, ecosystem scientists are very interested in cycles. If ecosystems were only conduits through which material and energy passed, they would not be profound conceptions. It emerges that ecosystems cycle material to a considerable degree. Cycling causes the ecosystem to achieve a significant identity. Without cycles, the system would quickly run out of resources and be directly constrained by the physical environment. If we think of rivers as ecosystems, the least interesting aspect of them is the flux of water through them. Their distinctly ecosystemic properties come from the way mineral and organic materials spiral downstream, passing through the biota many times before leaving the system.

What makes living systems such as ecosystems distinctive is the way they escape the constraints of the physical world by recycling resources.

James Kay identified that with living systems' coded programming, organisms are relatively closed. A frog's leg twitches whether jolted with electricity and acidity or if its nerve is simply poked. It is the same response every time. He also suggested that physical systems are information open. A metal bell conducts electricity, heats up if warmed, or rings if struck. The response is different for each stimulus. As for ecosystems, Kay saw them as being in the middle; they are information ajar.[17]

Recycling does not come free; it demands dissipation of energy. The order of life comes from energy dissipation in driving the critical cycles. Ecosystems, like organisms, expend much energy in repair and use of their cyclical pathways. In holding nutrients in the system, the ecosystem invests energy in microbial growth in the summer and in root growth in the autumn and again in the spring. The fungi die wholesale once a year, while fine roots die and must be replaced twice. In organisms and ecosystems, the cycles are integrated to become the system itself.

The mean residence time of carbon in a deciduous forest is on the order of fifty-four years, according to simulation models. Although carbon moves rapidly in and out of leaves through photosynthesis and respiration, some of it becomes locked into tree trunks for centuries. For most of the fifty-four years, the carbon does nothing and is not in the process of cycling. Nitrogen, by contrast, moves around the system from biota to soil and back again on an annual basis. Nevertheless, the mean residence time of a nitrogen atom in the forest is on the order of eighteen hundred years.[18] There is not much carbon by percent in the atmosphere, but it is reliably there in workable amounts and therefore the ecosystem keeps only a two-hundred-year inventory. Most ecosystems are nitrogen-starved and so the inventory for nitrogen is about ten times longer than for carbon. The difference in turnover time decouples these critical elements. This is Herbert Simon's idea that the notion of inventory applies to biology.[19]

BIOTA IN ECOSYSTEMS: PLANTS AND PRIMARY PRODUCTION

Plants and animals relate to a generally different set of ecosystem functions. We now contrast the different roles of plants and animals in ecosystems. First we consider plants. The roots of trees are crucial parts of the rhizosphere, but it is the rhizosphere, not the root, that is the functional unit in the nutrient retention subsystem. There are other biota—the fungi that also allow the rhizosphere to do its job. Note that the fungi play their part not as discrete entire organisms, but rather as a guild of fungi whose members individually work interchangeably (see chapter 1, figure 1.15A).

Along with Curtis Flather, we analyzed the ecological literature to see which organisms were used for work on which major concepts.[20] We performed a computer search of BIOSIS, a literature-searching database, to retrieve papers by their paired use of types of organism in the study and the ecological concept investigated. Hundreds of thousands of paired hits were found and tabulated. The numbers of each organism/topic pair were expressed as proportions of total effort devoted in terms of each type of study and to each type of organism. Knowing the amount of research on a given class of organism and the amount of research on a given concept, we could calculate how much research would be expected to be devoted to both at the same time. The actual amount of research in a given area was sometimes more and sometimes less than we would have expected. Those items in the matrix scoring above expectation indicate where there is a natural match between organism and concept. The results gave insights into the way ecologists think about all sorts of ecological entities, particularly ecosystems (table 3.1, figure 3.7).

We found that the archetypical organism in ecosystems has the life form of a tree. Trees are also the prime entities for community and succession studies. There are, of course, ecosystems and communities that are much smaller than human size, but the intuitive feel we have for a community or ecosystem is that they are both typically large systems. We are talking here of the gestalt, not the materiality of ecosystem size. When something is larger than ourselves, we cannot readily see it all at once. Because in a forest the dominant organisms are larger than us, the intangibility of the upper-level ecological entity is not only acceptable but also expected. In ecosystem and community concepts, there would seem to be an unspoken integrity of an unspecified whole; there are unifying processes, but ecologists do not feel confident to tease them apart in communities and ecosystems. Trees comprise the primary production compartment and are also the site of carbon storage. They live a long time and are therefore helpful, long-term capacitors that smooth variations in fluxes. By behaving slowly, they constrain other ecosystem parts and therefore lend the ecosystem stability. Their behavior is so slow that it passes unnoticed at the scale of human primary perception. This slowness fits in with the intangibility of function that is implicit in the ecosystem.

Our study in BIOSIS showed that herbaceous plants are not generally considered as ecosystem parts. The principal exception is grasses, which appear to readily lose their identity as discrete biotic entities. The reason for this is that grasses all look the same to the untrained eye. They are also often seen playing their ecosystem role of supporting secondary production in grazed systems. Organisms that can play a role unambiguously in an ecosystem function are favorites for ecosystem research.

TABLE 3.1 Binary matrix when 0 means less than and 1 means more than expected effort in research on that concept for that organism. Columns and rows are ordered to show blocks of concept/organism concurrence.

TAXON	COMPE-TITION	DISTUR-BANCE	POPULA-TION	SYMBIO-SIS	EVOLU-TION	CAPTURE	BIOGEO-GRAPHY	NICHE	HABITAT	ECOSYS-TEM	COMMU-NITY	SUCCES-SION
Lichen	0	0	0	1	0	0	1	0	1	1	1	1
Algae	0	0	0	1	0	1	0	0	0	1	1	1
Ericaceae	0	1	0	0	0	0	0	0	1	1	1	1
Trees	0	0	0	0	0	0	0	1	1	1	1	1
Bryophyte	0	0	0	0	0	0	1	0	1	1	1	1
Conifer	0	0	0	0	0	0	1	1	1	1	1	1
Gymnosperm	0	0	0	0	0	0	1	1	0	1	1	1
Pteridophyte	0	0	0	1	1	0	0	1	1	0	1	1
Basidiomycete	1	0	1	1	1	0	0	0	1	0	0	1
Ascomomcete	1	1	1	1	1	0	0	0	0	0	0	0
Roseaceae	1	0	1	0	0	0	0	0	1	0	1	1
Graminae	1	1	1	0	0	0	0	0	0	1	1	0
Composite	1	1	1	0	0	1	0	0	1	0	1	1
Angiosperm	1	1	1	1	1	0	0	0	0	0	0	0
Mammal	1	1	1	0	0	0	0	0	0	0	0	0
Fish	0	0	1	0	1	1	0	1	0	1	0	0
Insect	0	0	1	1	0	1	1	1	0	0	0	0
Bird	1	0	1	0	0	1	1	1	1	0	0	0
Amphibian	1	0	1	0	1	1	1	1	0	0	0	0
Reptile	0	0	0	0	1	1	1	1	0	0	0	0

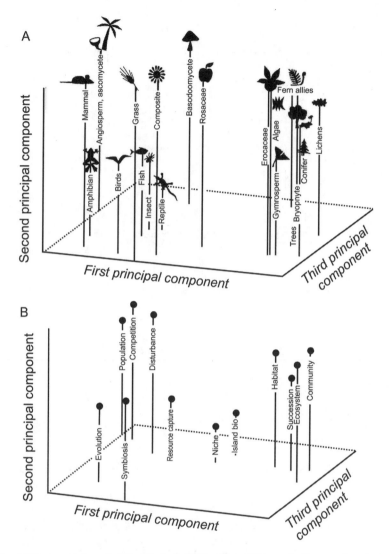

FIGURE 3.7. *A.* Principal component analysis of the relative use of organisms as research objects for certain ecological concepts. *B.* The inverse analysis, where ecological concepts are placed relative to each other depending on the organism that is used to study them.

BIOTA IN ECOSYSTEMS: ANIMALS AND PRIMARY CONSUMPTION

An animal eating means different things at different levels. Becky Brown has considered primary consumption in ecosystems in a way that leads to insights about the importance of scaling in measurement protocols.[21] This is a clear example for showing how level of analysis changes what ecologists see and can discover, so we give it an extended treatment. These scaling effects determine which processes in herbivory emerge as explanations. There is a school of ecologists, including Sam McNaughton, interested in the way that plants grow more vigorously after grazing than if they had not been grazed at all. The term used for the process whereby plants grow to make up the loss due to grazing is "compensation." When the plant more than fully recovers the loss, it is called "overcompensation." The literature is, however, confused because the scale at which the compensation to grazing occurs is often unstated. As a result, the theoretical point of exact compensation, against which observed conditions are compared, becomes equivocal.

Although herbivory is a process, it is often measured as the percentage of leaf damage at a given moment. Clearly, there is no time for compensation to occur in such data. Measuring the status of damage to leaves relates two processes: animals eating leaf material and leaf growth. The two processes occur at very different rates; damage to individual leaves is a fast process, much faster than leaf growth.

Some plants, like grasses and their monocotyledonous relatives, do not lend themselves to estimation of leaf damage at an instant in time because they keep growing their leaves from the base, below the point where the animal nibbled. In these plants, there is the possibility of compensation at the level of the leaf. Note that grasses give their name to the process of animal primary consumption, grazing. Grasses appear to have accommodated to grazing more than other plants by rescaling their parts so as to make the individual leaf the unit of response. This takes the natural unit of ecosystem primary production below the level of the whole plant and fixes it at the leaf. This biological rescaling appears to give grasses great selective advantage in that they are the dominant plants in ecosystems where animal pressure on plants is greatest.

The problem for other herbaceous plants, forbs, is that not only are the leaves damaged by primary consumption but so are the growing points of the plant stems. Leaf capital is destroyed and the means of replacing the capital is disorganized. They lose capital and the means of production. The unit for compensation to grazing in plants other than grasses is the entire shoot, plant, or clone, not the leaf. The growing point being nipped off, a new growing tip must become active to take over replacement of losses.

Let us move upscale and measure removal over time. With removal seen as a continuous dynamic instead of an event, there is time for the plant to respond to grazing, and we can now investigate compensation. Extending the period between defoliation and measuring plant recovery allows time for (1) further grazing to occur, and (2) growth patterns to depart from linearity. Recovery could easily involve a lag period in which the plant reorganizes its resources, followed by a period of rapid growth that eventually asymptotes to normal growth rates. Many protocols for measuring recovery from herbivory suffer from one of two limitations. If they attempt to measure long-term compensation, there are many assumptions introduced that may well be unwarranted. Otherwise, they pertain only to short-lived aspects of ecosystems and are likely to miss some of the higher-level aspects of compensation.

Only through sophisticated data-collecting protocols is it possible to move upscale and leave time for some of the more interesting patterns of compensation to be captured in the measurement. When we leave time for compensation, there is also an increase in the spatial aspects of the system, since we now have to take into account belowground parts of the plant. After defoliation, the plant can compensate the leaf compartment by transferring material from roots. At the level of the whole plant, the response of the roots is what an economist would recognize as deficit spending to stimulate the plant's economy. It is a form of compensation. However, deficit spending takes time to have the desired effect, so there is a considerable increase in temporal scale if we wish to measure compensation at the whole plant level. That increase in temporal scale is large enough to violate the assumptions of a single grazing event and linear growth of the experimental plant and the control.

The proponents of overcompensation have shown that saliva from grazers does stimulate growth of leaves. This not only encourages compensation by more leaf growth, but more leaves also alter the means of production. This might be the key to overcompensation, but it is set in a hierarchy of measurement schemes. The fracas that is the contentious literature is significantly a set of semantic arguments. Different measurement schemes change what compensation would be, and therefore changes what overcompensation might be. The error is to give privilege to an arbitrary level of analysis.

Let us move higher in the stack of nested ecosystems. Over still longer time periods, changes occur because of grazing operating at a level higher than the plant itself. The plant may grow faster than it otherwise would have done because it lives in a grazing regime. The critical change wrought by grazing over the long term is not on the plant directly but on the context of the plant. If grazing opens space by damaging other plants, grasses that can recover faster do so. They now live in a context altered by grazing to favor their survival and growth. It is over this long term that grazing impacts ecosystems at the high level and large scale at which we normally apply the ecosystem concept. Clearly, grasses gain advantage under at least moderate grazing. That is partly why we have grassland ecosystems. Perhaps stimulation with saliva could play a role there, but there is also a web of other processes at work. Livestock grazing over a couple of hundred years removed plant material in central New Mexico to the point that the ecosystem was carbon deficient. Cattle were nutrient exporters. Restoring carbon with Albuquerque sewage sludge produced dramatic positive effects.[22] At that point, humans become part of the compensation equation.

Thus, even the simple question, "Do plants compensate for grazing?" invokes a hierarchy consideration, each level introducing nuance as to what compensation means. Almost all the players in the literature assert that some levels of compensation are more important than others, to the point that the whole discourse is significantly semantic more than discrimination of material happenings.

BIOTA IN ECOSYSTEMS: PLANTS, ANIMALS, AND NUTRIENTS

There are many examples where herbivory has changed nutrient cycling rates. Such an increase in nutrient cycling can in turn increase primary production. Steven Carpenter and Jim Kitchell showed that phytoplankton is negatively correlated with zooplankton when observations are separated by two days. This is because the higher numbers of zooplankton crop down the floating plants. Nevertheless, Carpenter and Kitchell found that phytoplankton are positively correlated with zooplankton over a ten-day window.[23] Ten days is long enough to allow the nutrients consumed in the cropped population of phytoplankton to reemerge in the water and be expressed in increased production. Correlation changes completely with the time frame of the context. The extra days allow a larger-scale process to manifest itself, a nutrient-based phenomenon that is only tangentially related to the original grazing phenomenon. The irony is that collections of samples are often once a week, exactly in the middle between the two correlations. In chapter 2, we refer to medium-number systems where the constraining influences shift, not allowing prediction. Sampling once a week—the worst possible choice for sampling frequency—invokes a medium-number system.

Animals often play a role in complicating phenomena related to nutrients. As they cross the system boundary, animals play an important role in the nutrient budget. Their rapid movement sometimes allows them to be nutrient importers, as in the case of salmon returning to the stream of their birth. They are particularly important in systems of that sort, where nutrient loss is a critical factor. Insects are important parts of the nitrogen import into bogs where carnivorous plants trap them and digest the insect bodies on sticky leaves. Pitcher plants trap insects in chambers in their leaves. Bogs are particularly dependent on these nutrient imports because their acid water keeps nutrients in solution, so allowing the nutrients to be exported with water. In unequivocally terrestrial systems, A. D. Bradshaw has pictures showing how dogs bring nitrogenous waste into reclaimed urban landscapes.[24] The explanation for patches of bright green, vigorously growing grass in parks created from leveled slums are the spots marked by dogs in

FIGURE 3.8.   Patches of bright green healthy grass on this restored urban site are attributable to the ritual urination of dogs bringing nitrogen into the system. Although the restoration
used topsoil, the essential absence of nitrogen-fixing fungus plants led to low nutrient status,
allowing the signal from dogs to stand out clearly. According to A. D. Bradshaw, the smaller
clumps of green are "one dog once," while the most vigorous patches are many dogs many times.
(Photos courtesy of A. D. Bradshaw.)

their territorial ritual urination (figure 3.8). There are not enough dogs to support
the nutrient budget of the reclamation so as to compensate fully. Nitrogen-fixing
plants can offer sufficient inputs to support their ecosystems. Despite the input
of nutrients from insects, bogs remain nutrient poor. Thus, unequivocal input of
nutrients by animals with local effects may not change the nutrient status of the
system at large, but there are exceptions.

Animals can play an important role in cycling nutrients inside ecosystems, as in the case of the zooplankton cited earlier. Also, soil arthropods break down leaves in a timely fashion so that the nutrients can return to the plants. Most ecosystems receive only small amounts of nutrient input. The slowest part of the system constrains all the others, and thus painfully slow nutrient import might be expected to deeply constrain most ecosystem components. However, many ecosystems escape this nutrient constraint by cycling and building up significant nutrient capital. On the short-grass prairies of Colorado, Bill Parton and his colleagues sensibly divide the nutrient budget into a fast, middle, and slow nutrient cycles.[25] They calculate that the system is very resilient to even long-term abuse because only the capital changes as the slower pathways continue to deliver nutrients for the system's primary production.

In summary, biota in ecosystems may not be readily identified as discrete organisms, so the naturalist may not have an intuitive understanding of the role of biota in ecosystems. Nevertheless, biota does play a crucial and readily identifiable set of roles in ecosystem function, particularly with respect to nutrients. The ecosystem concept opens a whole new window on the activities of organisms.

THE SPECIAL CASE OF AQUATIC SYSTEMS

Returning to our BIOSIS investigation on the proportional use of various types of organisms for various types of study, fish were the only exception to the rule of underutilization of animals in ecosystem work (table 3.1, figure 3.7). They are the only animal taxon used in ecosystem research more often than expected. There are two reasons for this. First, fish live in an environment alien to us. Without scuba equipment, we cannot see them clearly in their native habitat. Out of sight leads to out of mind, as fish become a component inside a black box. Black boxes are studied from outside; input to and output from the box are compared and the state of the system is calculated from conservation principles. Being unseen as autonomous organisms in their natural habitat, the image of a fish with head, tail, and fins does not interfere, in the minds of ecologists, with the ecosystem functional role that fish play.

The second reason for fish being preferred ecosystem attributes is that fish play some of the same roles in lakes that trees play in forest ecosystems. They are nutrient storage units that bridge across times of paucity or destruction. Fish integrate nutrient input to the system as they grow larger and persist for years at a time. Lacustrine ecosystems appear to operate faster than terrestrial ecosystems, so years in a lake are probably equivalent to centuries in a forest. A fish holding nutrients in its body for a few years plays the same role as a tree locking up carbon and minerals in its trunk for centuries. In springtime, it is fish-nitrogenous waste that keeps the phytoplankton primed and ready to lock onto the nutrient inputs when the ice on the lake melts.

Algae are favorites for ecosystem work. This is because, like fish, algae perform discrete ecosystem functions. They are the primary producers and nutrient-capture compartments of the lake. In terrestrial systems, many types of plants are primary producers, and their role of nutrient capture is less conspicuous than that of algae in aquatic ecosystems. In addition, their size makes unicellular algae invisible to the naked eye. Accordingly, they sit easily, unseen in the ecosystem black box.

It is probably no accident that much of the pioneering work on ecosystems was performed on aquatic ecosystems.[26] Lakes represent one of the few examples where a landscape boundary clearly coincides with most aspects of an ecosystem boundary. Water operates as a vehicle for exchange in aquatic systems, and so aquatic ecosystems are particularly well integrated. This means that the physical size of a body of water determines the type of system. The size of a lake is more influential than the size of a tract of land in determining the type of ecosystem that can be supported. The difference in size between lacustrine and oceanic systems leads the systems of each type to be driven by different factors. This weaves together notions of size, patterns of constraint, and prediction. John Magnuson proposes that it is useful to think of lake fish assemblages as being related to extinction, while the species list in an equivalent volume of the ocean is driven by invasion.[27] If a lake is subject to winter kill every five years or so, a happenstance fish invasion might do well for only a few years. In such lakes, the invasion is unpredictable. However, we can predict with confidence that after only a few years the whole population will be killed. The size of oceans changes the scale of the causal factors and reverses the patterns of prediction. In oceans, invasion is a reliable driving force. Oceans allow invasion to reliably override the happenstance of local extinction. Lakes experience invasion and parts of oceans experience local extinction, but neither is predictive.

COMPLEXITY AND STABILITY WARS

One of the patterns that have caught the attention of ecologists has been the trend in species diversity across different biomes from the pole to the equator. There are fewer species found in the arctic tundra than in the boreal forests to the south. As one moves south through the temperate deciduous forests, diversity increases further. The exception is desert diversity at 30-degree latitude. In the high tropical moist forests, the diversity is very high indeed, reaching levels of hundreds of tree species per hectare. From this pattern have come assertions that increased diversity increases stability. Robert May has disparagingly called this notion part of the "folk wisdom" of ecology, the idle opinion of naturalists.[28] What matters for stability is the hierarchical nature of the connections (figure 3.9A and *B-1*). One of the arguments for orthodoxy is that the multiplicity of species provides alternative pathways should any community member suffer a local extinction.

Loss of stability involves changes in system structure. New system behavior involving the old structure can be confused with change in system structure itself. The difference between behavior and change in structure is fundamental. Behavior is rate-dependent, while structure is rate-independent. Once the system is structurally defined and observation protocol is fixed, behavior follows independent of the observer. However, structure is always observer-dependent, because structure is a matter of arbitrary definition. A change in structure, as in going unstable, follows from the observer finding that the old system specification is untenable or contradictory. Of course, how a system behaves depends on what the structure doing the behaving is defined to be in the first place. The observer asserts system structure, but that still does not make behavior observer-dependent in the way that structure is observer defined.

After specification through mathematical formalism, constants and the manner of change of what can be changed are fixed. That is the boon of quantifying ecology, but it does not come free. Because mathematics facilitates proof, the temptation is to "prove it," which slows things down a lot. The real power of mathematics in ecology comes from laying out what we are saying in unequivocal terms, and that opens the door to grounded intuition. The insight is not always in the pattern that is quantified.

In assaults on the folk wisdom of ecology, Gardner and Ashby (1970) and May (1972) both showed that an increased number of parts and connections, and an increase in strength of connections all tend to make systems unstable. But there is one critical proviso. The big assumption is that connections between parts are made at random. We are not concerned with that assumption being true since all assumptions are false; so the issue is not that random connection is false in nature. One does not make assumptions to find which are true; you make them to see if you can get away with them. From Gardner and Ashby and from May, it appears we simply cannot get away with an assumption of random connection. Whatever else these authors say, both publications show that random connection is on its face an untenable proposition in almost any discussion of diversity.[29] They are explicit about that and should not be blamed for any misapprehension that diversity necessarily lowers stability.

The bottom line is that both studies say conventional wisdom is not a general case; and that is indeed true. The nonrandom patterns of connection that must apply need complexity theory and hierarchy theory. Both those theories go far beyond the parameters of the graph theory used by May and others. May's work shows that there is no straightforward relationship between diversity and stability, and that there is no reason to expect it. Our conversations with ecologists indicate

A

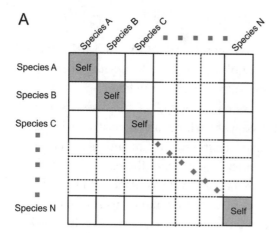

	Species A	Species B	Species C	■ ■ ■ ■ ■ ■ ■	Species N
Species A	Self				
Species B		Self			
Species C			Self		
■ ■ ■ ■					
Species N					Self

B-1

	Fox	Rabbit	Lettuce	Farmer
Fox		eats		competes for rabbits
Rabbit	is eaten by		eats	competes for lettuce
Lettuce		is eaten by		is eaten by
Farmer	shoots	shoots and eats	tends and eats	

B-2

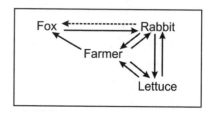

C-1

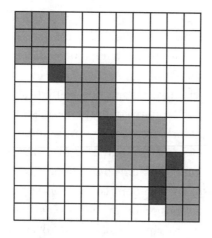

C-1'

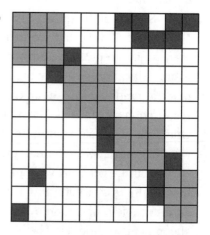

C-2

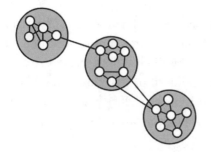

C-2'

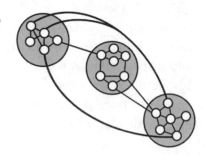

that this is not generally understood across ecology. A further complication is that diversity is a measurement for which there are many causes.

An ecological system becoming unstable is such a dramatic event that it is hard to remember that stability is a relative matter. Stability comes from the observer's specification of the system as much as it comes from the system under observation. Like complexity, instability is not a property of the system itself, but is instead an aspect of the mode of system description. Consider a system showing unstable behavior. If the system is specified to be more general, then what was unstable behavior becomes subsumed by a coarser system specification. A tree crashing to the forest floor can be seen as either the local tree exhibiting instability or as a healthy, normal process of replacement in a stable forested landscape. What is unstable over a narrow range of tolerances becomes normal system behavior if the observer uses a broader system specification.

The level of system description is not always transparent. If it were, there would be no need to consider scaling problems more deeply than engineers, who calculate complicated but straightforward scaled outcomes. Accordingly, aspects of system description that might not appear to have anything to do with stability can re-specify the system so that it stays within or travels beyond the bounds of stable behavior. The length of time a system is studied determines whether it passes through unusual states. Study it for a long time and it passes through unusual states often enough for them to be considered normal; this implies a relatively large system. Study it for a short time and the unusual state may arise only once and be seen as a system aberration; this implies a smaller system than the one that was found to be stable. It is in these considerations that hierarchy theory gives us a chance to understand diversity, which is really a matter of connection of parts. Patterns of connection determine stability; diversity does not.

FIGURE 3.9.   *A.* Interaction matrices have columns of species often arbitrarily arranged, but the rows are in the same order. Thus, the diagonal trace refers to self-interactions, while other values away from the trace are filled with values reflecting the direction and strength of the interaction between species. *B-1.* A simple interaction matrix in an agroecosystem. *B-2.* The corresponding species interaction diagram. *C-1* and *C-2.* A stable, hierarchical system expressed as an interaction matrix and as a set of systems and subsystems. The interaction matrix with all connections close to the trace (top left to bottom right diagonal) is stable. The order of the rows and columns is arbitrary, but if the system tends toward instability, it is not possible to get all connections close to the trace; no arrangement will make the trace dominant. Note that the subsystems are not strongly connected to each other and so no one subsystem can bring the whole down. *C-1'* and *C-2'.* These are equivalent to *C-1* and *C-2,* but there are strong connections between subsystems. This would tend to make the system unstable. Note that the long, strong connections appear off the trace of the interaction matrix. Rearranging the columns and rows to get those strong connections to the trace would disorganize the arrangement of the subsystems on the trace. Connections far off the trace of the interaction matrix are the hallmark of instability.

There is a body of work called food web theory. It remains suspect because it has never solved the problem of bounding the system or level of specification of the parts. Data of species of algae are sometimes reliable at the level of species, when the plant has a recognizable form. But many times species identification is difficult. *Staurastrum* is an algal genus where two generally tetrahedral forms are joined together at a constriction, a waist. But sometimes a division of *Staurastrum* produces a half cell that is not tetrahedral but is flattened, putting it in another genus. Two halves of a cell from different genera! Worse still are algae that are not much more than little green spheres. Only a specialist in that part of phycology stands any chance of getting those algae accurately down to species. Collectors of most ecological data give up and present only data down to the genus, or worse, only to the family. So here are data that underlie food web research with widely different levels of specification. It is fair to ask what food web theorists are studying.

The level of bounding such systems is also a problem. Lakes are much better bounded than most terrestrial systems, except islands. So here is another reason to question what food web research studies. There are apparently patterns that arise regularly as scale-independent in mathematical treatments of these data. In the absence of a clear definition as to what is being studied at what level of analysis, great caution should be taken in interpretation of those patterns. We can expect some very general explanations with little meaning, probably some expression of central tendency. Central tendency says that, with change over time, more likely configurations will arise. That is why the second law of thermodynamics applies so universally. Throw two ordinary dice repeatedly, and a total score of six or seven turns up most often, simply because there are more ways to achieve them. The values of two and twelve are most rare, because there is only one way, respectively, to make them.

There is a hidden assumption in relating diversity to stability, which is that the new species entering to increase diversity are connected to the rest of the system. This assumption comes from the nineteenth-century idea of the economy of nature, that everything is connected to everything else. If May has his jab at the folk wisdom of ecology with diversity increasing stability, ours is that "everything is connected." In fact, that view is mistaken, in that most things are not connected to any but a few others. True, there would appear to be a gravitational interaction between all things, but that is practically immeasurable as well as irrelevant at ecological scales. When we find ourselves wrong, or see some wrongheaded idea in someone else, the temptation is to say, "You are wrong." It is much better to ask, "What is that person trying to say?" When we say, "Everything is connected to everything else," we are trying to say, "There are unexpected connections that will change the outcome. Expect to be surprised."

Ecological theorists build matrices of interaction to describe many ecological situations, which include terms for the strength of interaction between the column and row entities (figure 3.9A). Most of the values in these matrices are zero, except in restricted cases. Most pairs of species chosen at random do not interact in any meaningful way, except for highly focused matrices (figure 3.9B). Several theorists in the 1970s, including Robert May, Robert MacArthur, and Richard Levins,[30] pointed out that if indeed there is significant interaction between all parts of an ecological system, or even if the connections are random, then diversity would be held low, as we discussed earlier. This is because destabilizing positive feedback is bound to arise with increasing diversity if the increases come with unorganized connections. The theorists' point was not that diversity decreases stability, but as we said earlier, that ecological systems are connected relatively weakly and very nonrandomly.

The critical feature of stabilizing elements added to a system is that they be only locally connected to just part of the system. In large stable systems, there are local knots of connection that amount to relatively discrete subsystems, and stable additions must be to members of only one subsystem or, at most, to members of adjacent subsystems. At the outset, the columns and rows of a community interaction matrix have no particular order, except that the rows are ordered the same as the columns so that the self-reference of species is down the diagonal of the matrix. It is, of course, possible to change the order of the columns as long as there is a corresponding change in the order of the rows. That way, self-reference continues to be down the diagonal, the trace. In an appropriately reordered interaction matrix, the previously mentioned subsystems occupy local blocks astride the principal diagonal of self-reference in the matrix. Stabilizing additions are connected inside these blocks and generally nowhere else (figure 3.9C-1). Destabilizing additions to the system form bridges between otherwise disconnected major sectors or blocks in the system. In the interaction matrix, these entities would have large interaction values far off the principal diagonal (figure 3.9C-2). The effect of such bridges between quasi-discrete subsystems is to create long loops of effects in the system. The delay in these long loops often causes the signal to return at a time when it amplifies system change, a destabilizing positive feedback. It creates a delayed negative feedback, which will at least oscillate and may well amplify.

Random connections in the system are only close to the principal diagonal by chance, and that is why randomly connected systems become readily unstable. Randomly connected systems are complex because there are many long loops in the system, most taking their own unique period of time to return a signal fed into the loop. Coherent organization is lost.[31]

THE EMPIRICAL STRIKES BACK

Complexity is a matter of the number of system parts and the strength and patterns of connection. Diversity of species by itself appears not to have a simple relationship to complexity. There is a large literature on diversity and stability, but the striking feature of it is the paucity of empirical data to support the myriad of hypotheses. Many of the experiments done recently are too large and therefore lack the focus that gives unequivocal results. A notable exception is the empirical test of complexity against stability by Peter Van Voris and his associates.[32] They used microcosm ecosystems, they made an empirical observation that implied complexity, and they related complexity to system stability.

They took eleven plugs out of an old field and enclosed each one in its own chamber (figure 3.10A). Ten plugs were used for the test and the eleventh was the control. Although they were from a single community and had levels of diversity that were in the same range, the microcosms were not replicate sample communities. This is because they were too small to contain the same taxa, rather like the valleys in the Bitterroot Mountains as studied by Bruce McCune, discussed in chapter 2. The differences in the particulars of the flora and fauna were large enough to be consequential. For example, some had ants in them while others did not.

Despite the clear differences in sample communities, it did emerge that the microcosms were replicate ecosystems. There were ten exact frequencies of behavior displayed by all the microcosms (figure 3.10B). To investigate the ecosystemic aspects of these microcosms, Van Voris and his associates turned away from species composition and toward measurements of fluxes.

Each microcosm was watered once a week, and carbon dioxide was sampled hourly for six months. The two measures of stability they chose both related to calcium leached from the system after the watering. One was relative system resistance, the capacity of the microcosm to retain calcium after an insult of cadmium; the second was a measure of relative system resilience, the speed with which the microcosm bounced back to patterns of calcium leaching that existed before the poisoning with cadmium. In the results, the rank order of microcosm resistance was the same as system resilience. Thus, an ordering of the system on stability was achieved.

The relative system complexity was assessed by looking at the patterns of carbon dioxide released from the microcosms. Complexity is an attribute of the total system and it therefore does little good to look at a collection of the parts, as does diversity. Carbon dioxide considered in the appropriate way is an integrator of the system, much in the way that water quality coming out of a watershed is a good indicator of many aspects of a total watershed system (figure 3.10A). Van Voris and colleagues took the time series carbon dioxide data and transformed them into a power spectrum.[33] To obtain a power spectrum, a sine wave is run through the data and the amount of variance in the data given account by that wave is recorded. If the data show a periodicity at the length of the wave, then the wave will account for much variation in the raw data. If there is no periodicity in the data at the length of the wave, then when the data show a high value, the sine wave may be high, low, or in the middle. The wave would show no relationship to the data and would

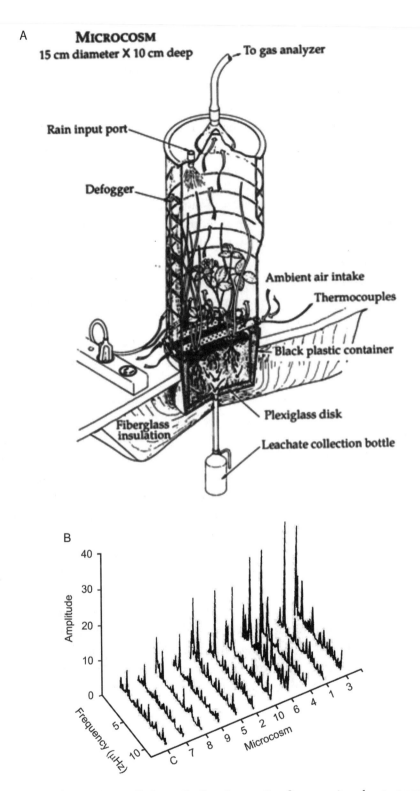

FIGURE 3.10   A. The small plugs of soil and vegetation from monitored output ports are shown. B. When the Van Voris et al. microcosms were ranked according to the number of peaks in the power spectra, those to the right with the larger number of peaks appeared the most resilient and resistant to disturbance ("C" was the control) (Van Voris et al. 1980).

therefore give little account of variance in the time series data of carbon dioxide. After one wave has been considered in this way, a slightly shorter sine wave is passed through the data. The variance that the wave extracts is noted. The process is repeated until a graph can be plotted of length of the successive sine waves against the variance extracted from the time series data by the sine waves.

An intuitive example might be a data set of temperature of most places in the temperate regions of the Earth for a couple of years. The first long sine wave would pick up the seasonal differences. Hot in summer, cold in winter, and then hot again in the following summer. But a shorter sine wave of three months length would be out of sync with the summer winter periodicity, and would therefore pick up little of the fluctuation in the temperature. But as the sine wave becomes as short as twelve hours, the cyclical change between day and night would be picked up. We can plot the period of the sine wave against the variance or signal picked up by the sine wave. For the Earth going around the sun and the Earth rotating on its axis (figure 3.11A), we have a graph with a big peak on the left end for the annual cycle, and a flat, low signal for all shorter periods until the smaller peak that indicates a twenty-four-hour cycle (figure 3.11B).

So Van Voris did the same for the cycles in carbon dioxide, which were of course not as simple as just annual and daily temperature cycles (figure 3.10B). The peaks in the graph indicate periodic cycles in the carbon dioxide measurements. Each peak in the power spectrum indicates at least one loop of connection, which in turn indicates one of the working components contributing significantly to total system behavior. Remember that the microcosms were too incomplete and capricious in species compositions to be replicate sample communities. The fact that they did appear to be replicate ecosystems was indicated by the way several peaks were exactly replicated across the microcosms. This would suggest that the systems shared a basic set of processes that made them replicates in ecosystem terms.

Despite the replication of peaks across all the microcosms, some microcosms had significantly more big peaks. They were the ones that showed greater resilience and resistance in calcium loss. Van Voris and colleagues performed one of the few empirical tests of complexity and its relationship to stability in an ocean of speculation. The little empirical information we have indicates that complexity, at least up to a point, is positively correlated with stability in material biological systems under experimentation.

BIG SCIENCE DIVERSITY EXPERIMENTS

We began this chapter insisting that species and organisms for the most part melt into ecosystem pathways. We have given two exceptions—western salmon and canvasback ducks. Certainly, the majority of species do not map onto ecosystem

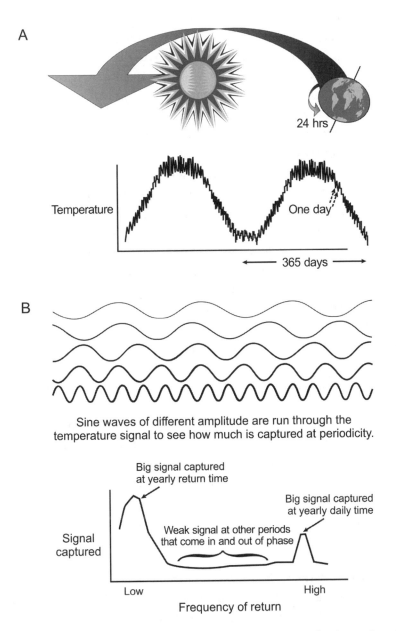

FIGURE 3.11. *A.* The Earth goes around the sun once a year, and spins on its axis every twenty-four hours. This is reflected in annual and diurnal cycles in temperature. *B.* We can pick up these phenomena by running sine waves of different length through the temperature time line. Sine waves of a period of one year will pick up the annual fluctuation, while sine waves one day in length will pick up the diurnal change in temperature. Sine waves in between will go in and out of phase with the temperature and thus will pick up little of the temperature signal. This is a Fourier analysis that takes information in the time domain and expresses it in a graph of periods of different length. Major cycles appear as peaks of variance extracted. For the temperature on the temperate regions of the Earth, the peaks are one year and one day.

processes as species or organisms. Part of the problem with diversity studies is that they are species focused, while at the same time trying to explain ecosystem function. We would not expect a match, but there are so many diversity studies we cannot simply ignore them in this book. Some are worthy, but many are not; the big studies appear to be particularly problematic because of a scale mismatch. Diversity is an issue for ecosystem and community studies, so we could put them in either or both chapters. Those concerned with ecosystem function need to be considered here in the ecosystem chapter.

Diversity is a slippery issue. As Ricotta points out, it is redefined by each user for the local purposes of the argument at hand.[34] With all that said, it does appear a helpful surrogate when all else is in fact equal. In deeply comparable situations, diversity may mean the same thing across a comparison. The complete equivalence of Grime's experimental units is what sets his work apart and makes it successful.[35] One needs tight and rational replicates and controls, which are lacking in many diversity experiments. The questionable work on diversity by Hector and his thirty-three coauthors implies a more general meaning than it has. There is a strategic error of that school addressing diversity.[36] There are statistical problems too that easily swamp modest asserted relationships. The valid work uses diversity in tightly prescribed ways in local situations close to nature. In very equivalent field situations, diversity does appear to be regularly negatively correlated with disturbance. It relates to the extent to which ecological systems have been insulted in some way, perhaps with pollution. This issue goes back to the work of Ruth Patrick, who showed a shift in a lognormal distribution so that common species became commoner and rare species declined in polluted waters.[37] Thus, in some conservation work, diversity can be a useful indicator of trouble, but it is more a surrogate that is not readily tied to any sort of mechanism. We would caution against overgeneralization of that pattern of correlation. We advise mistrust of narratives about how diversity works in general. It seems unlikely to us that diversity is the same phenomenon across ecological places that are significantly different. There are too many causes for differences in diversity: coincidental overlapping of ancient floras; degree of heterogeneity of climate; topographic variation; local peculiar microsites, such as cracks called *grikes* in limestone pavements; histories of epidemics or fires; and so many more causes. Diversity is what ends up happening. With so many causes as options, it is hard to tell which of them are constraints. Medium-number systems are untestable because they are invoked by questions where the causes and constraints keep changing. The many causes invite what Weinberg called a medium-number specification.[38]

We have said elsewhere that diversity is studied largely because it is easy to measure and has an intuitive appeal.[39] But intuition here is misleading. The only time species diversity must go down is when the whole planet loses a species, not a case that can be addressed. All less complete situations might send species diversity up because of unevenness. So the conservation argument for its use is blunted.

Diversity is a "granfaloon" (an incidental collection). Kurt Vonnegut coined the word in his 1963 *Cat's Cradle*. In that novel, a granfaloon is a group of people who profess a shared purpose and claim an identity, but who in fact have nothing significant that brings them together. Diversity is a measurement looking for concept. Diversity is a syndrome, that is to say, it is a pattern without a satisfactory narrative. The effort in the questionable work is a search for a grand narrative, but one with insufficient justification. We are reminded of Gertrude Stein's observation on Oakland, California: "When you get there, there is no there, there." A collection of the number of species just happens, and measuring how many of them there are tells of little that is significant and nothing that is general. The critical work on biogeography did successfully predict diversity on islands based on a balance of invasion and extinction, but it is not clear how those principles apply to places more open than islands.[40]

It seems that diversity of function does relate to ecosystem properties, but that is not a matter of species number per se. Becky Brown performed experiments in a tropical garden where she looked at insect damage.[41] The conventional wisdom she tested was that the stability of tropical gardens comes from their diversity. The accepted view is that when plants of different life forms (e.g., corn, beans, and squash) are planted together, the vegetation does not seem to have pest problems of tropical monocultures. She compared insect damage to leaves in the local uncultivated vegetation to damage to leaves in monocultures and in systems with planted increasing diversity. Her final conclusion was that it was not diversity that reduced leaf damage; it was functional equivalence that mattered. Higher diversity in itself did not help. As long as the physiognomy of the natural vegetation was mimicked, the damage was the same as in nature, even with much lower diversity. Yes, it takes some degree of diversity to achieve that, but it is the mimicry of the life forms, not the diversity, that does the trick. If the addition of species is aimed at mimicking form, diversity soon becomes irrelevant. Of course, experiments on diversity could easily stumble into patterns like Brown found and mistake it for a diversity effect, and we suspect that happens commonly.

The most impressive experimental work is by Philip Grime, who has a long-term experiment using relatively small, seed box–sized plots (figure 3.12). Grime is more convincing than others because he uses species that do indeed occur together on the local moorlands on the Pennines (a line of hilly uplands running up the backbone of England). It is this community that gives the critical all-else-equal. His communities are dominated by fescue (*Festuca ovina*) with sedges and forbs. Becky Brown's work would suggest that diversity only has a role to play up to a relatively small number of species. Grime generally works with only eight species, allowing his work to resonate with Brown's results. Some of Grime's

FIGURE 3.12.   One version of Grime's experimental communities. His seed box communities that Allen visited were in Sheffield, England, in a university garden. (Photo courtesy of P. Grime.)

recent results have linked genetic diversity to diversity in general. In setting up his experiments originally, he took plants from the field and transplanted them into his boxes. He had presumed that most of the plants he collected would be genetically distinctive, coming from an original seed with its own genome. But as he turned to explicit interest in genetics, he found that separate plants were not always distinctive.[42] As he put it colorfully in a personal communication to Allen, who was visiting Sheffield, "There is this one rascal who seems to have got everywhere." Apparently, one particularly successful plant had reproduced vegetatively to come to occupy a significant part of the whole area from which Grime took his transplants. The upshot of the experiment that followed was experimental boxes with higher genetic diversity within which the species present showed more stability and vigor. The success of the "rascal" does not appear to have stabilized diversity.

Some very interesting work has come out since our last edition. It relates to diversity of genetics in crop plants. Again, it is low-diversity situations that give useful results. In traditional Philippine rice cropping, various varieties were grown in the same field. Similarly, Andean potato fields traditionally grow different species, many with different levels of polyploidy. When one species does poorly, the others take up the slack.[43]

A big problem in rice farming is blast fungus. It is possible to breed new resistant strains, but the fungus works its way around the resistance in two to three years. If the susceptible and the resistant strains of rice plants are planted together as a fifty-fifty mix, the yield reflects 80 percent of the advantage of planting the resistant strain alone.[44] The exciting finding is that there is not the usual evolution of fungus to overcome the resistance in the plants. The resistant strains of rice stay resistant year after year. The reason is that the pressure of the regime of natural selection is not to overcome the resistant strain, but is rather to get at the susceptible strain. Thus distracted, the fungus leaves the resistance in place. Given the move to a narrower genetic base than in traditional Philippine farming, preserving some genetic diversity can make a big difference in ecosystem function.

At an even lower level of diversity, we report a beneficial ecosystem function if a single genome of soybean is grown in different wind regimes, and then the plants are mixed.[45] The diversity in our work is two, and it is only two of physiological adaptations, not even two genetic varieties, let alone two species. The message here is that only very low levels of change in diversity appear to change ecosystem function. Michael Huston, a protagonist of diversity studies, says of a much richer diversity study, "There's no evidence from this experiment that 200 species is any better than 50 species." Even the limited mathematical ability of birds could do the counting for most aspects of species diversity: birds count eggs as one, two, many. Two makes a difference, three less so, and any more makes little difference.

The soybean experiment measured ecosystem function with the thermodynamic temperature of the top of vegetation in a wind tunnel (figure 3.13). Photosynthesis captures very little energy from the sun, 4 percent of incident light, and then inefficiencies on top of that for burning the fuel so made. This is not enough to drive ecosystem function, such as moving nutrients from the soil to the growing points of plants. So plants use heat and in this way harness more than 60 percent of sunlight energy. The latent heat of vaporization of water can lift water at pressures over fifteen atmospheres. That is work, and can be detected by the relative cooling of the top of the canopy. The colder the canopy, the more work is being done.[46]

Plants grown in slow air were coolest when tested in wind at that slow speed. Faster air warmed the leaves because of sensible heat from the air. Plants grown in midfast wind were warmer in slow wind because there was not enough wind to effect full evaporation rates. These plants got cooler (that is, they worked harder) as the air speed was increased to the speed in which they were raised. Even faster wind warmed those plants too. The message is that plants less stressed by being in their habitual environment work harder.

The increase in diversity was achieved by mixing plants from different growing treatments. Increased diversity lowered the amount of work done; the two-treatment vegetation was warmer. The reason was the difference in height of the plants, the slow-wind plants being taller because as they were grown in low wind, they did not

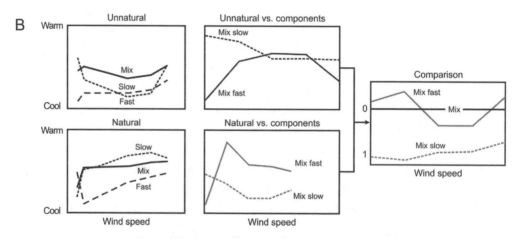

FIGURE 3.13. *A.* The wind tunnel with soybeans used in the experiments of Allen, Havlicek, and Norman (2001). (Photo courtesy of T. Allen.) *B.* Results of Allen, Havlicek, and Norman's wind tunnel experiments. Lines labeled fast or slow represent the temperatures in different wind speeds of plants grown and tested in pure culture. The mix lines represent the temperature of plants grown in the two treatment wind speeds, but for the tests, mixed in a checkerboard pattern. Unnatural mixes were even-aged for both wind treatments. Slow-wind plants grew faster, and so were taller. Natural mixes allowed time for the overstory plants in the wind to grow taller than the faster-growing plants in slow wind. That is, after all, what happens as a stand grows. Slow-wind plants in pure stand warmed as wind was increased. Fast-wind plants got cooler up to the wind speed in which they are grown. The middle pair of graphs shows the respective mixes (natural and unnatural) relative to the temperatures of the respective two treatments grown alone. The right-hand graph is a subtraction of the two middle graphs to see which plants behaved differently when mixed naturally or unnaturally. The fast-wind line there is close to zero, because the fast-wind plants gave the same response in each mixture. The slow-wind plants were, however, much cooler if they were in the understory of the natural mix.

have to divert as much energy away from leaves into stems. So the mixed vegetation had a ragged aerodynamic profile with tall next to short plants. The more ragged profile of the mixture let air into the mixed canopy, and that warmed the vegetation. The problem with diversity as a concept is that it does not address organization. It turns only on complicatedness, not organized complexity.

When, in other experiments, plant treatments allowed more time for faster-wind plants to grow tall, a new version of ragged vegetation profile was possible in the mixed experimental vegetation. This time, the fast-wind plants were tall, where they would be in natural vegetation. The first plants to grow in natural vegetation will be exposed to the wind, just like the contrived new, uneven-aged vegetation. Also, in this new experiment, plants grown in low wind were down in the understory in the contrived mixture, as they too would be in nature. The more natural mixture with fast-wind plants above was cooler than in the even-aged mixtures where fast-wind-reared plants were short and out of the wind. The new regime was contrived not only to be complicated but also to be more complex, that is, more organized.

Clever relativization allowed the researchers to tell how the complexity worked to cool the vegetation and achieve more work. The fast-wind plants had paid with stem thickening to be out there in the wind. Up in the canopy, they protected the understory from the wind. In turn, the protected understory could do a lot more work, and increase total ecosystem functionality. Yes, the complex vegetation was cooler than its only complicated counterpart. And there is the problem with diversity as exposed here; diversity is concerned only with a pile of stuff. Functionality comes from organization, and diversity in mere species number does not explicitly address that.

Note, when a tree that is hundreds of years old falls, a tree that is eighty years old is only a few feet shorter and can take over the role of the top of the canopy. Eighty years of protection allow faster growth. We find in agricultural fields that there are two strategies for weeds where each strategy plays with complexity and organization. Some weeds emerge early and overtop the crop, but others germinate later and allow the crop to overtop them. They then use the crop for support, and reach the light on inexpensive spindly stems. Diversity does not detect or even look for organization that is at the core of much ecosystem function.

There are claims by Tilman and colleagues that diversity increases ecosystem function.[47] That paper in *Science* has been cited 1,270 times, and so has been unfortunately influential. This is part of a more general science political issue, that a fairly small number of ecologists have undue influence over what ecology is published in the two major general science journals, *Nature* and *Science*. We can expect that the review process is fairly circular and closed within the group. The number in the in-group is not large, but big enough for them to get their wagons in a circle. This then generates a positive feedback that reinforces the cycle. Resnick reports data on scientists' dismay that the review process in general is pathological.[48] We have no reason to suspect collusion, although in the United Kingdom, the ecological granting system has too few overly influential players, which can make it look like a conspiracy when it is most likely a single person's point of view. It is encouraging that there was a backlash that promoted Grime to the Royal Society. It does not much need conspiracy for the loops to work, but we do notice a pattern. The papers are widely cited because of the status of the journals in which they occur, not because of the merit of what they say. Evaluation in academe does not much use the information actually written in the papers because that would take too much time and effort on the part of busy evaluators. The original symbols for scientific information have been sucked into a sea of metainformation about number of citations and journal status ratings. We should get back to actually reading the literature, but that is not where the incentives lie, and therefore will not happen.

The Tilman paper has come into serious criticism, generally not in *Science* and *Nature*. The error is the well-known statistical flaw called sampling error. Productivity might be expected to be higher in more diverse experimental plots created with stochastic components because of the greater chance of including a productive species at random.[49] The many authors of Hector et al. (1999) are in the same camp as Tilman. They made such grand claims that finally there was another paper, multiauthored by a collection of leading scientists from outside the big experiment school.[50] Huston et al. (2000) took Hector et al. to task, showing that the claims for the big science diversity studies were deeply questionable for reasons of statistical flaws. The challengers performed their own regressions, and the effects found by Hector et al. statistically disappeared. There is sufficient investment in the experimental protocol such that the political weight of diversity experiment protagonists will not let them back down on diversity increasing function. The politics of science appear notably worse in the United Kingdom compared to North America, which has its own problems.

As we argue here, there are many reasons to question the logic and narrative of the push to ecosystem function in diversity. There is little effort to address organization because of the very conception of diversity, a static notion. By contrast, the literature on complexity and organization is much more promising, some of it coming from business management.[51] It appears that the nonrandom patterns of connection are the bottom line, something ignored in the very idea of diversity. Some of the complexity literature does not play to rate-independent constraints and controls, being all too impressed with rate dependence and emergence. In chapter 4 on community, we expand on Grime's organizing principles as he looks to strategies that plants use. His scheme dissects the types of relationships, so that he can address the functional interrelationships, a complexity issue.

CONCLUSION

The ecosystem criterion is very distinct from all others. Like communities, it bears a complex relationship to landscapes. For the most part, to think of an ecosystem as a place on a landscape is unworkable and certainly makes the concept depauperate. Although it is more inclusive than the community, in that ecosystems contain meteorological and geological components, the ecosystem is not a higher level than community. They are merely alternative specifications. The reduction of an ecosystem to its functional parts does not usually lead to organisms or populations. Only by reducing to functional parts on very specific phenomena does one sometimes find plants and animals as discrete entities inside ecosystems.

The attributes of communities are the end products of biological evolution. Ecosystems do depend on evolved entities for some of their functions like primary production, but any one of a large number of separately evolved plants can do the job. Therefore, evolution is only tenuously connected to ecosystems. Ecosystems are not evolved in any conventional meaning of that term. Nevertheless, some of the same principles that apply to evolution hold for ecosystem development. This is valuable, but can make for a muddled comparison. The particular ecosystems we find take the form they do because those patterns are stable and therefore hold the material of the world in those configurations long enough for us to observe them. We find stable configurations regularly. Given a world where there is almost universal scarcity of biologically available nitrogen, ecosystems develop organization that cycles nutrients. It is stretching Darwinian evolution too far to suggest that there is a group-selection explanation for the sharing and cycling of nutrients. We observe ecosystem functionality, not because there is selection in an evolutionary sense, but because stable patterns persist. Darwinian selection is only one way of achieving stability. Stability can arise without competition and differential survival based on genetics. The ecosystem is therefore a parallel development that is highly structured like strictly living systems, but with a different cause than natural selection in the conventional sense.

4

THE COMMUNITY CRITERION

LIKE LANDSCAPE and ecosystem, the community criterion is not a level in and of itself. Rather, it is a way of describing ecological systems by their floristic and faunal similarities. In the figure of the cone diagram (see figure 1.15A), the community is but one of the columns. There are various conceptions of community that cut into the community column at different scales, offering movement up and down the community column. In the manner prescribed in chapter 1, we compare the community criterion with other related ordering principles caught in cross-section at a given scale. After contrasting communities with ecosystems and landscapes expressed at comparable scales, we look at differently scaled community conceptions inside the community criterion for observation (see figure 1.15B).

We can think of communities as collections of types of organisms on a landscape, but that depauperates the concept of community. The community as a place refers to a happenstance collection of plants or animals, rather than a functional whole with interrelated and interdependent parts. The relationships embodied in the community transcend the places where particular individual plants grow and animals live. The difference between a community and a mere collection of organisms is the accommodation that the different species make for each other. Hugh Iltis knew and spoke to John Curtis for more than a decade. He says Curtis's community conception was of a web of species connected by rubber bands (figure 4.1). The idea is that if you move the status of one species, it changes the tension of all the other species.

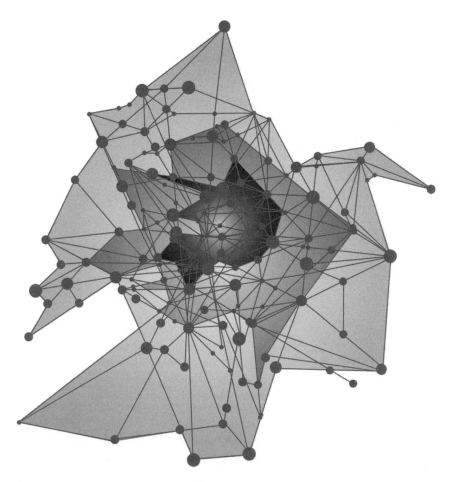

FIGURE 4.1. John Curtis's conception of the plant community as told to his colleague, Hugh Iltis. Curtis saw the community as a set of species connected by something like rubber bands. Remove one species, and the tension among the others shifts. We come from only the verbal description of Iltis, not any actual figure, and have presented our conception of what Curtis meant.

Our equivalent conception is similar and certainly in the same spirit. Species come and go from locations and change abundance. We see the community as wave interference patterns of site occupancy (figure 4.2). Wave interference patterns are common in everyday life, as when the shadow or reflection of a window screen is seen through the screen itself. The screen is slightly displaced from its shadow, so the two grids come in and out of phase. Change your angle and the pattern moves. Another interference pattern occurs in old western movies, where the spokes of the stagecoach wheel get just out of synchronization with the twenty-four frames per second of the film. The wheel sometimes appears to move backward. Bank notes are variously pixelated so that digital copying of them shows interference patterns that give away the forgery.

FIGURE 4.2. An interference pattern painted on a clay egg. The Acoma of New Mexico call this the sunburst pattern. Note that the lines do not relate simply to the emergent pattern. (Photo courtesy of C. Lipke.)

Change the periodicity of major species and the whole community pattern changes because of a shift in the interference pattern (figure 4.3A). A case in point is the keystone species of the intertidal communities of Paine.[1] The *Pisaster* starfish trundles across the rocks devouring all the shellfish in its path, leaving behind a trail of bare rock. The major competitors among the shellfish are mussels. Left alone over time, they crowd out the barnacles, limpets, whelks, and other species. But the starfish sets the mussels back, leaving a space for the other shellfish species to insert themselves for a while. In experiments, Paine removed the starfish and the diversity plummeted as the mussels took over. Paine noticed that there were no starfish on headlands because of the waves. But there was still diversity on headlands because the mussels were ripped off regularly in storms.

An example from plant communities is found in the work of Robin Kimmerer. She studied communities of mosses and liverworts on the cliffs beside the Kickapoo River.[2] That is the same river shown as mapped watersheds in chapter 3 on the ecosystem. The river is in an old landscape that did not have glaciation when the rest of Wisconsin was under ice. As a result, the hills are well carved up by erosion, and the water sheds off the landscape quickly. So the Kickapoo is subject to flooding, particularly since the trees on the hills have been significantly cleared. The top competitor for space on cliffs beside the river is *Conocephalum*, a lobed

A

B

FIGURE 4.3. *A.* The intertidal zone where Paine studied starfish in Washington State. The starfish are evident some feet above low tide (photo courtesy of M. K. Wilcox). *B.* Allen (right) teaching in front of *Chonocephalum*-dominated rocks at Hemlock Draw by a stream in the Baraboo Hills. The liverwort dominates in the region between his hands. Here at least once a year the small stream turns into a raging flood. Kimmerer's sites were closer to a significant river and were therefore flooded more often and showed a clear banding just above the low waterline (photo courtesy of S. Thomforde).

liverwort that moves like a tank over the rocks, shading out all other species. But the sheets of the liverwort are only held down by rhizoid threads, and so floods that drag logs across the cliffs at high water remove liverworts. This releases rock substrate for some half a dozen other mosses and liverworts. At the places generally above the highest water, *Conocephalum* dominates (figure 4.3 *B*). At mid-height on the cliffs, the community is at its richest. Floods break up the liverwort in much the same way the starfish removed the mussels. But down close to the waterline, a moss (*Fissidens*) is the only plant that can take the punishment. It is so small that it can survive scraping and being underwater. Whereas population biologists are largely interested in how the winner wins, the community ecologist asks the opposite question, which is why the winner does not take all. The bands on the cliffs show where *Conocephalum* cannot "take it all," its superior competitiveness notwithstanding.

Both Kimmerer's and Paine's work offers support for an important idea in community ecology, the intermediate disturbance hypothesis.[3] Diversity is highest when competition is intermediate. The idea is one of the few valid uses of diversity, which requires strict all-else-equal. The same community compared across different degrees of disturbance gives an unequivocal signal. Many diversity studies do not meet the *ceteris paribus* criterion. Moderate disturbance takes out the major competitor, so that many species that are not strong competitors can fit into the periods between disturbances. Moderate disturbance thus increases diversity. With infrequent and mild disturbance, the dominant competitor runs off all other species. Under frequent and strong disturbance, only the few species that are tough enough to persist are present. That is what *Fissidens* does.

While a definition of a community as a wave interference might appear esoteric, it is only a formal expression of some commonly held views of what constitutes a community. A community without accommodation or at least a response between members would only be a collection and not worth studying. A community as a thing in a place is a static conception that loses the dynamics that makes the whole. Hutchinson probed the niche in these terms.[4] If a habitat is a place, the niche of a species is the role it plays. But the environment mediates that role, so there is some influence of place put on top of resource space. While a three-dimensional volume, such as a cube, is tangible, it is possible to extend the space to higher dimensions, which are of course intangible. These intangible spaces are called hypervolumes. Distances in such a space can be calculated with Pythagoras theorem. Vegetation described by two species sits at a point on a species A–B plane (figure 4.4*A*). Two points on a plane with different species composition will have different coordinates in the two dimensions of the A–B plane. When you set up the coordinates as the sides of a right triangle, the lengths of the sides are the differences of the coordinates, difference in species values, at the points. The distance between the two points on the plane can then be calculated as the square root of the sum of

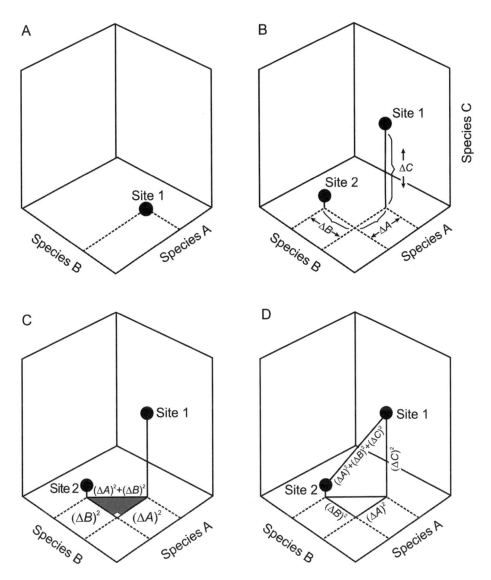

FIGURE 4.4.   A. A site may be defined by the quantity of its various species. In this case, it is a point on a two-dimensional species plane. B. As in a two-dimensional species space, sites may be separated from one another in three dimensions according to differences of occurrence of three species at two sites. C. The distance between two sites in a species space is the hypotenuse of a right triangle, where the base and the side are the difference on the first two species' axes. D. The generalized Euclidian distance is the square root of the sum of the squared differences on each dimension of the space.

the square of the other two sides of the right triangle. Put that plane on the bottom of a volume and the hypotenuse you just calculated on the plane can be used as the base of a new right triangle sitting on its edge in the volume (figure 4.4B and C). The hypotenuse of that new triangle measures the distance of a three-dimensional space, in a volume (figure 4.4D). We are still working in a tangible space, but not for long. A distance in a hypervolume would merely take the new 3-D hypotenuse and use it as the base of a right triangle in four dimensions. Once you have made the fourth dimension tangible, higher dimensions arise from the same operation. And you can do it again and again to measure distance in a volume of any size, building one dimension at a time. Distance in the hypervolume is measured by taking the square root of the sum of the squared different coordinates for as many dimensions as you like. Hutchinson considered the niche to be the hypervolume of environmental resource factors wherein the species can live.

Hutchinson made the distinction between the fundamental and realized niche. The fundamental niche is the volume that a species can occupy without interference from other species. The smaller niche that is available to the species in a community, Hutchinson calls the realizable niche. For instance, the black spruce tree can grow in very wet conditions of a bog, as well as on a sandy dry ledge and all hydrologies in between: that is its fundamental niche. But black spruce does not grow in mesic conditions.[5] It could, but it does not because it cannot take the intense competition for light in the mesic sites: the extreme environmental conditions are its realizable niche. Community studies focus on a set of integrated realizable niches (figure 4.5).

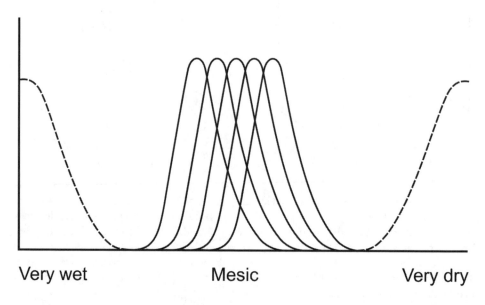

Very wet　　　　　　　　**Mesic**　　　　　　　　**Very dry**

FIGURE 4.5.   In the middle of a gradient from wet to dry, mesic species perform variously well from wet to dry toward the middle of the gradient (solid lines for the species' performance in the mesic zone). But at extreme conditions of very wet and very dry, black spruce appears as part of the vegetation (dotted line at ends of the gradient). It would do well in the mesic zone, but competition for light excludes it. Its fundamental niche is in the middle of the gradient, but it is excluded. The realizable niche is restricted to the ends of the long gradient. Note that the ends of the gradient share the absence of mesic species. The geometry of all this tends to cause curvature in environmental gradients plotted in species space.

A community at an instant is the embodiment of prior processes of accommo-dation and tension. The history of give-and-take between species allows the partic-ular organisms at a site to coexist as community members. Some accommodations may be to site conditions ameliorated by former occupants. Thus, adjustments may be made between community members that do not even occupy sites at the same time. Therefore, there is a distinctly temporal component to communities that extends beyond just the place itself at a moment in time. Anthony Bradshaw showed that plants living together on the spoil of lead mines in the Snowdonia Mountains of Wales were distinct ecotypes. The mine spoil was only decades and not centuries old, indicating that building a community of short-lived species may not take long, but it still takes time.[6]

This raises an interesting difference between plant and animal communities. Space is the separator of plant communities. Patches of similar vegetation are sep-arated in space but are bound together over time during which the same general assemblage persists. It is possible to return to a place and find fairly much the same plants as individuals as one found before. In plant communities the plants anchor the place long enough for time to play out the progress of succession (figure 4.6). However, time tears the experience of animal communities apart. During the nor-mal time period for field observations, members of the animal community come and go. For example, Bell observed the grazers of the Serengeti Plain pass through a given area in a given growing season in a recognizable sequence (figure 4.7).[7] Water buffalo come off the hills onto the lush new vegetation at the beginning of the rainy season. They take the grass to waist height. Horses, without a rumen, eat poor material in huge quantities. Cloven-hoofed animals get more out of their food, and therefore can use higher-quality fodder to full effect. After the buffalo, the zebras come off the hills and mow the grass to a height that is workable for a series of cloven-hoofed ungulates that appear in a sequence of decreasing body size. Animal communities can be bound together over something of the same order of time frame that holds the identity of plant communities in succession. Even so, the capacity of animals to move during the period of data collection does give plant and animal communities contrasting textures.

THE DEVELOPMENT OF THE PLANT
COMMUNITY CONCEPT

The developmental history of the community concept has been a tug-of-war between landscape and organism as the points of reference: the community as a collection of organisms versus the landscape with certain communities scattered across it. The new idea of community as an association was pressed forward by Cle-ments in the Botanical Seminar in Nebraska at the end of the nineteenth century.

Animal ecologist data collection

Populations stream past

Plant ecologist data collection

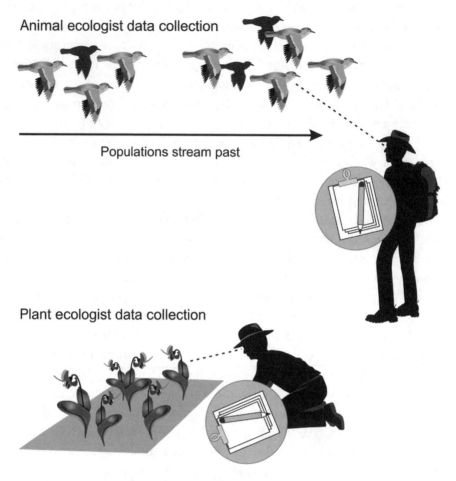

Individual plant organisms are counted

FIGURE 4.6.   Animal community data tend to be collected as populations streaming past. Places are sites for sampling. By contrast, plants do not move, so the samples in plant communities are generally organisms in one place. The plants here are being sampled using a quadrat.

The intellectual context in which the young Nebraskans built the notion of community was the tension between early work of a biogeographical sort as opposed to the young German plant physiologists of the late nineteenth century.[8] The community is also a point of tension between the population and the biome criterion (figure 4.8). The modern view of community denies victory to any side and requires complementarity between landscape, organism, biome environmental determinism, and population biotic determinism. One tension arises from different species of organisms failing to read the landscape in the same terms because each species occupies the landscape at its own scale.[9] The community requires more than one level for adequate description. The organismal prong of communities has its origin

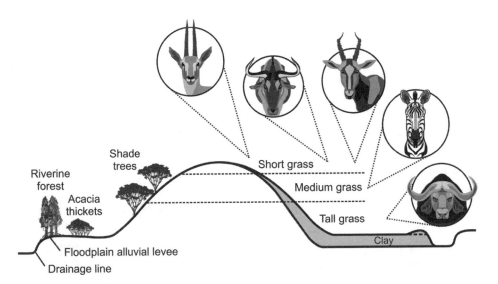

FIGURE 4.7. Bell (1971) considered animal communities on the Serengeti Plain. Just before the rains, all animals are on the only grass left, on top of the hills. As the rains come, the first off the hills are water buffalo; second come the zebras. They are followed by the topi and wildebeest. Finally, the diminutive Thomson's gazelle gleans seeds from the stubble. The order is large-to-smaller species.

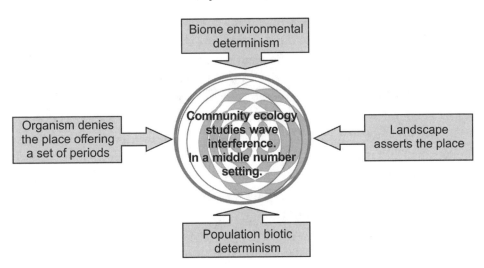

FIGURE 4.8. There are two sets of tension pulling on the community concept. The landscape is where we find the concrete community on the ground. But organisms in the community are of different sizes and occupy the space at different scales. So where and how extensive the community is on the ground depends on the species you ask. The other tension is between population that turns on biological determinism and the biome that turns on environmental determinism. We are unable to see which constraint is in play and when, making prediction of communities difficult. It has some properties of a medium-number system (a term coined by Weinberg [1975]) attached to a wave interference pattern.

in evolution. Important evolutionary notions precede Darwin by a full half century. Furthermore, one of the founders of ecology as a discrete discipline was Frederic Clements, who died a Lamarckian in the middle of the twentieth century and was devoted to that earlier notion of evolution.[10] The organismal pole of community concepts originated with German efforts in plant physiology applied to organismal adaptations.[11] In the late nineteenth century, physiological and evolutionary adaptations were not so much kept apart, which is not all bad, even though it might be dismissed now by evolutionary biologists. The landscape half of the community concept arose a little earlier, through biogeographic exploration in the first half of the nineteenth century. The impetus behind these biogeographic expeditions was a need to record the patterns of life form found in the different zones of the globe. Vegetation physiognomy was an important part of the descriptions brought back to Europe. Young Germans went to the tropics with the notion of adaptation and its physiological basis in plants. Schimper's text covers much of this work; it represents the emergence of botany as a self-conscious discipline.

Having stolen a part of the world from some native people, the imperial powers, including Germany, sent out scientists to codify what they had got for their trouble.[12] In America, it took the form of various natural history surveys. Particularly significant for us were the graduate students in the 1890s in Lincoln, Nebraska, the decade of the Battle of Wounded Knee in South Dakota, and some twenty years after the Battle of the Little Bighorn on the Wyoming border. It was the American version of taking stock after colonial expansion. The link to the German intellectual heritage was though Bessey, the plant physiologist who was the major adviser to the young men doing the Nebraska Natural History Survey. The group was unified under the Botanical Seminar at the University of Nebraska.[13]

The Botanical Seminar produced the vegetation "formation" as a new concept in the 1890s. What we now call communities were then called formations and associations, but that is really only a shift in parlance. The term "formation" already had been in use with plants for some fifty years, but at the turn of the century it took on an independence from landscapes, with the works of the young Nebraskan Clements, and Cowles from Chicago.

Clements took the idea of the association of plants and gave it a set of distinctly ecological explanations. Until the Botanical Seminar did this, vegetation ecology only extended plant physiology into the inconvenient setting of the field. Under Clements and Cowles, plant ecology emerged as a transdiscipline, with its own set of explanatory principles. Transdisciplines can arise, as in the case of ecology, as a failure to achieve a link sought at the outset. The young Americans' agenda was to link biogeography to plant physiology. MacMillan's 1893 thesis on vegetation formations of the Minnesota Valley was the first with a distinctly community ring to it. It is interesting that MacMillan himself retreated in later studies to the physiological origins of the community. He failed to follow through and fully establish intellectual autonomy for the "formation." Perhaps his being before Clements kept him held back, tied to the original agenda, not realizing how radical was the new concept.

The new tool invented by Roscoe Pound was the quadrat (Pound and Clements 1898) (figures 4.6 and 4.14). The quadrat lays down an area that is used as a sampling unit to get numbers on plants in association. All technology rescales the observer relative to the problem. The quadrat does not much appear to be technological, but it critically rescaled the activities of botanists.[14] It is telling that Henry Gleason wrote a paper on the use of quadrats, summarizing their use. Gleason saw the community in similar terms to Clements. Their mutual commitment to quantifying vegetation put them in the same arena.[15]

The principles of the parent disciplines of biogeography and physiology became overextended when applied to species accommodations in the field as measured by quadrat sampling. Inability to make the link gave autonomy to the new entities the Nebraskans found with their quadrats: plant communities. This new conception comes with its own new explanatory principles of succession, invasion, ecesis, and facilitation, none of which are physiological principles.[16] Invasion speaks for itself. Ecesis is the process of surviving long enough to reproduce. Facilitation is plants early in succession making the site favorable for species coming later in the maturation of the association in succession. At the time, there was resistance to Clements's ideas, when a leading British light, Blackman, complained that Clements's big book (Clements 1905) was insufficiently physiological. Blackman was stuck with the original tension between biogeography and physiology. Associations were recognized as having integrity over time such that, in a given association, species composition could change as the processes of environmental amelioration and competition allowed succession. This is nicely stated in the extended quotation from Tansley (1926) in chapter 1. The processes of invasion and ecesis were recognized and investigated. Note how these explanations of community behavior are neither physiological nor spatial but are identifiably ecological.

The community of Frederic Clements was different from the landscape notions that preceded his work. The distinctions can be enumerated: he saw the community in floristic not physiognomic terms; he emphasized sites with vegetational homogeneity, but was not a landscape ecologist; he was cognizant of the dynamic nature of the community at a site; and he was aware that each species requires particular conditions to become established in the community. Clements's conception of the community is really very close to our notion of the community as a tussle between organismal and landscape-referenced views.

In this chapter, we develop the idea that the community is the accommodation between species with different periodicities, as each species occupies the landscape at different scales. The landscape asserts the place, but it is a non-Euclidean space because the organisms occupy it using different metrics for each species. How far is far depends on the organism in question.

Clements's view was very much a reflection of the landscape in which he was raised. The open plains of Nebraska had vast tracts of similar vegetation, with much of the variation attributable to differences in stages of recovery from disturbance. Accordingly, Clements proposed the monoclimax that suggests the vegetation can be characterized by a single climate-determined mature phase over large tracts of land.[17] When Tansley says that the climax community is simply an assemblage that persists long enough to invite a name, he is reining in the excesses of Clements, who reified the community, making the abstraction a material thing growing to maturity. Individual organisms were subsumed in the "superorganism" of the Clementsian community. In chapter 8, we show that the things we see in biology are only realizations of some larger organizing entity, like the species. It is tempting to overstate the importance of the realization, as when we see humans as structures as the central issue instead of mere realizations of humanity. Clements is really only making that mistake with regard to the realization of community principles, the community, so he is not alone.

Beyond the first half of the last century, academic wars were fought over the nature of the community; is it an association on a landscape or is it a collection of organisms? We now see that it can be profitably considered as both simultaneously. We can see here some of the duality associated with Levins's notion of dialectical materialism.[18] Ecologists then were realists to the point that they sought true definitions, without realizing all definitions are arbitrary, chosen and simply asserted. It is a common enough philosophical error in ecology today. Again, this makes Tansley's (1926) position, quoted near the beginning of chapter 1 here, so very impressive in its sophistication.

Henry Gleason championed the organism as the basis for communities in his "Individualistic Concept of the Plant Association." Gleason was raised in the prairie-forest border region of northern Illinois. There, the vegetation on the landscape is dynamic with prairie and forest waging war on each other with fire and shade, respectively. The monolithic single climax for a whole region does not fit the observed facts in Illinois.

Actually, there was remarkably little difference between Clements's and Gleason's considerations of the critical community processes. As to migration, Gleason says that Clements's 1904 account of the literature was "so well summarized" that Gleason need not discuss it further. As to migration of individual species, Gleason (1922:42) explicitly says, "Clements (1904) exposition on the whole subject is without doubt the best general treatment extant."[19] Both recognized the importance of invasion from local seed sources. Clements names ten sorts of invasion. Both ecologists relied heavily on the difference between invasion, arrival, and ecesis. In 1922, Gleason explicitly thanks Clements for the distinction between invasion and ecesis: "The distinction between the two processes [invasion and ecesis] is fundamental, and Clements' clear analysis has done much to systematize the knowledge of the subject" (Gleason 1922:42). Clements's extended consideration of ecesis shows him to be aware of the importance of the performance of individual organisms at a site, a cornerstone of the Gleasonian conception of communities. Nevertheless, the difference in emphasis on tracts of land for Clements, as opposed to the individual organisms of Gleason, drew the battle lines for a quarter of a century of academic acrimony. A close reading of Gleason (1922:42) shows that the only difference between Gleason and Clements was that Gleason wanted ecesis to be acknowledged as being an ongoing process, "continued regularly through successive generations, [that] may lead to broad results far more inclusive than those chiefly considered by Clements," which shaped the community and its dynamics. It is not clear that Clements disagreed on that point.

Gleason's thesis was that the vegetation at a place at an instant in time was the product of the happenstance of the local flora available for invasion, and the selection of individuals by the site-specific environment. Gleason's original statements show that he did not deny the integrity of the larger-scaled system implied by the Clementsian community concept. Apostles of Clements made more than was warranted of the differences between Clements and Gleason. Gleason's supporters in later decades thought that he was completely at odds with Clements, even though a careful reading of both ecologists shows differences that are not competitive but are instead complementary. Gleason preferred to focus on the underlying processes that select individual organisms for a given site; Clements emphasized community integrity across vegetation. Gleason sees small-scale relationships and mechanisms, while Clements focuses on the coarse-grain constraints imposed by the whole community.

Cooper's analogy of a braided stream in the development of communities over eons attempts to reach a compromise between the distinctness of individual patches of vegetation and the obvious continuity over time across tracts of landscape (figure 4.9). The continuity of the processes leads Gleason to emphasize continuous change in vegetation; he has a point. But Clements's view on the large coherent whole causes him to notice that there may be continuous processes, but there are certain forbidden configurations of species. Gleason favors rate-dependent continuity, while Clements notices rate-independent constraints. In animal ecology, Holling notes that forbidden sizes of animal are commonplace and regular. Clements might not have said it this way, but what he means is that certain species combinations are unstable, so we do not see them. And he has a point, too. Greig-Smith reports Webb (1954) as saying: "The fact is that the pattern of variation shown by the distribution of species among quadrats of the earth's surface chosen at random hovers in a tantalizing manner between the continuous and the discontinuous."[20] And that allows both Gleason's and Clements's views to avoid being competitive.

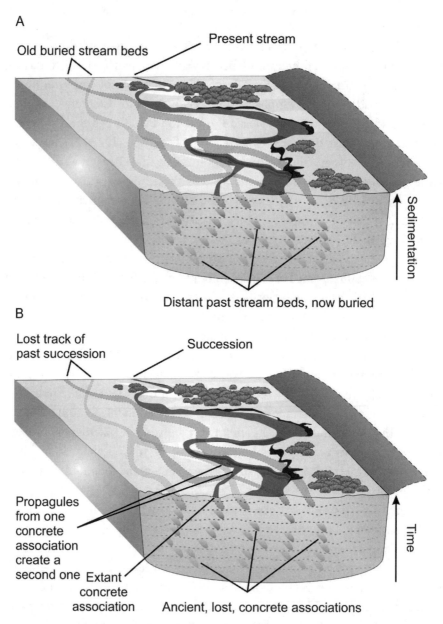

A

Old buried stream beds — Present stream

Sedimentation

Distant past stream beds, now buried

B

Lost track of past succession — Succession

Propagules from one concrete association create a second one — Extant concrete association — Ancient, lost, concrete associations

Time

FIGURE 4.9. A depiction of Cooper's notion of community succession as a braided stream (Cooper 1926). Braided streams flood and overrun the banks of the patterns of flow. When the flood subsides, often new tracks have been cut, and old tracks are buried by new sandbars. The braiding is the coalescence and separation of strands of the flow. Comparing the two representations, we see succession as the flow of the river downstream in the analogy. A cross-section of the path of the river shows various extant channels that represent concrete patches of vegetation caught at a point in succession. The flooding that removes old strands is the equivalent of the demise of concrete examples of the community. The new channels after the flood are equivalent to new concrete examples of a community. The deep cross-section of the river shows the history of past community examples. The whole river system is the abstract notion of community, the category into which concrete example are placed. Cooper's analogy captures continuous and discrete change, as well as succession and the dynamics of vegetation history at a higher level than succession (Allen in Dodson et al. 1998).

We have mentioned another point of tension in the community, biotic versus abiotic predictors. Clements anchors his scheme to the monoclimactic climax: one stable climax repeatedly achieved in a given climate. There is a reason for Clements's point of purchase. Clements coined the term "biome." We have a whole chapter on the biome point of view, but we need to introduce some of those ideas here. Environment does not determine communities because each species has its own tolerances. Many communities occur in one type of environment, and one community can occur in many environments, in a many-to-many mapping. But biomes, by contrast, are environmentally determined. This is only possible because we do not insist on species for biomes. Biomes consist of vegetation that can be recognized by what it looks like. Physiognomy, not species, pins the identity. The life form is a reflection of a physiological accommodation to environment. So biomes can be environmentally determined.

The process of succession left long enough reaches a climax that reflects the final environmental limits. Species replace each other and change the environment until the plants present are the only ones that can reproduce there. Vegetation stops changing, no matter which new species invade. In moist temperate environments, the vegetation is deciduous trees. In southern Wisconsin, that is mature maple/basswood forest. In Tennessee, it is tulip poplar and beech forest. Clements hitches his wagon to the climax, because it is the only time when communities appear to be determined by environment. Earlier in succession, *invasion* and *ecesis* keep changing the vegetation, as Gleason emphasizes. At climax, succession is done. True, much vegetation was not maple/basswood in presettlement southern Wisconsin because much of the vegetation would still have been in succession. Now, with Euro-Americans occupying the landscape, maple/basswood is the rare exception. Although his anchor is an exceptional vegetation, there is a consistency in Clements's view, and it is not at odds with Gleason because the latter does not want to choose climax as his benchmark.

So if biome is environmentally determined, is there a class of study that is biotically determined? The answer is the population. It is not that populations are never environmentally determined, but in population experiments, the environment is often held constant or set to a very small number of conditions. Flour beetle competition can be fixed by certain environmental conditions. But there is a wide range of environments within which the species that is the winner could go either way. One will win, but the ecologist just does not know which. So in those midrange environments, it is the biology of the species and it is population processes that determine the outcome. Biotic and environmental determinism mixed is the other dilemma in community ecology, besides organism and landscape.

Despite almost a century of a search for environmental determinism of communities, the effort has failed. Despite an agenda of population biologists to explain communities in terms of collections of species represented as populations, population biological accounts of community processes are wanting. Thus, communities can represent a medium-number system with switching constraints between biotic and abiotic explanations and causes. The only thing to do with a medium-number system is to change your question to one that can be answered, and one whose answer you still want. So community as a concept is a real monster. It refuses to be defined by place, but communities still give the appearance of being linked to places (figure 4.8). If you want to see a maple/basswood forest, we can take you there. Community also refuses to be environmentally determined, except sometimes. We later discuss Roberts's clique analysis that turns not on environmental determinism but on environmental relations. Roberts's methods represent a critical advance in philosophy.

Given these contradictions and difficulties, the plant ecologists of the 1920s were real heroes. In 1923, Nichols reports having sent out two questionnaires in 1921, the first to a short list of eminent ecologists to set the terms of the second questionnaire, eventually sent to a wider group. In that first eminent group was Cooper, the later author of the braided stream analogy in 1926. He, among others, suggested the distinction between the concrete and the abstract community (Figure 4.9). The concrete community was the example to which we might take you if you wanted to see a maple/basswood forest. It is the example on the ground. The abstract community was to be the pigeonhole in which you put all past, present, and future concrete communities of a certain type.

Cooper's image of the braided stream notes that there are temporary strands of water between sandbars. Each year they are different. For the last two decades, Allen has taken students to the Wisconsin River by Avoca Prairie. Some years we could wade out to the islands, and some years we could not. In the analogy to the braided stream, the flow of the river is succession. The extant concrete communities are cross-sections of the strands of the river. Strands disappear while others come into existence, as do concrete communities. Cutting down into the sandy sediment reveals past concrete communities. The identity of the whole river over ages is the abstract community (figure 4.9).

ANALYSIS OF VEGETATION

In the end, the debates in the first half of the twentieth century were carried forward with a set of statistical techniques that arranged accounts of vegetation on gradients ordered mathematically. Immediately before development of the explicit gradient analyses, soil scientists contributed by talking of changes down hillsides of soil moisture, texture, and nutrient content. These gradients, called "catenas," provide continuous change of substrate for the plants. Catenas give gradients of local conditions that amount to gradients in the factors controlling ecesis.[21] What was needed by the 1950s was a way of giving the patches floristic autonomy from where they happened to be in space.

Much in the way that the quadrat helped Clements, the gradient analysis ordination techniques were a critical improvement in technology. The methods were developed by John Curtis and his students in Wisconsin and Robert Whittaker in the Great Smoky Mountains. Both gave the distinctly modern view of community, with echoes forward to complexity theory. Ordination is a family of data manipulations that place sites with similar vegetation close to one another on a gradient. Dissimilar vegetation places sites at opposite ends of vegetational and environmental gradients. At the same time, David Goodall developed classification devices that place sites with similar vegetation in clusters. Classification is just a discrete version of gradient analysis ordination. Goodall's classifications worked alongside his own and other ordination techniques, all emerging at the same time. These advances allowed the description of community structure to be relatively positioned on flora and fauna, independent of the landscape position. These multivariate methods made it possible to rearrange descriptions of patches of vegetation so that they would be displayed next to their closest relatives on floristic or faunal characteristics. Thus, species representation in a stand of vegetation could assert itself as the organizing principle, unfettered by landscape questions of contiguity.

The gradient analysis ordination techniques fall into two major categories. One type arrayed vegetation along continuous gradients according to its vegetational composition. These are called indirect gradient analyses. The other type ordered sites according to the physical environments at each place, wet to dry, warm to cool. They are called direct analysis, exposing the primacy of environmental determinism as an idea. On gradients of species composition, the individual patches of vegetation are free to order themselves on strictly community criteria, namely biotic composition.

VEGETATION AND ENVIRONMENTAL SPACE

Implicit in both the classification and ordination approaches is the notion of species space. We have already developed the notion of hypervolume space with Hutchinson's multidimensional niche (figure 4.4D). If we make the axes of the space represent species, then we can cast shadows down to a smaller space. The shadow or silhouette should show in a tangible 2-D or 3-D space much of what is going on in the larger hypervolume. An aspen leaf is 3-D, while its shadow is only 2-D (figure 4.10). You can cast a 3-D shadow from a 100-D space with the same sort of projection.

These statistical summary techniques are now out of fashion. The rejection of them is to an extent from ignorance and laziness in getting up to speed on the

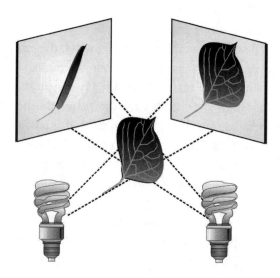

FIGURE 4.10. A 3-D aspen leaf may cast a shadow in 2-D. If the light is directed at the edge of the leaf, the shadow does not make it transparent that a leaf is involved. Only a sliver of leaf is projected. But if the light is directed at the back of the leaf, the shadow is clearly of a leaf. The projection looks very leaflike. We can project from a high-dimension space (say, one hundred species) to a 2-D or 3-D shadow or projection. Ordination techniques find more informative ways to project to a smaller space that informs the ecologist about the main trends in changes in vegetation over time and space.

statistics, justified by the fact that they do not generally test hypotheses. The value of these techniques is that they describe complicated situations in summary form so the scientist can develop hypotheses (figure 4.11). Once the scientists have a clue as to what might be going on, they might have an idea as to what is worth testing. A big problem in contemporary ecological research is that too much of it is trivial. Without the situation well described, the ecologist does not know what to test; this is a common flaw in reductionist ecology where low level, trivial considerations are tested. The emerging field of complexity science is creating new ways to display complexity in summary terms. The need for powerful summary has not gone away, and is being reerected in new display devices working in very much the same way as the neglected gradient analysis of the mid-twentieth century. Reductionists looking for mechanisms (sometimes so naively as to seek the "real mechanism") are not interested, but are they sufficiently intimidated by the new complexity science that their criticisms are muted?

The exceptional ecologist asks important questions. They all champion well-described situations as a must. Paine got a wonderful, compelling description and narrative about diversity and disturbance in the intertidal. Van Voris did that with ecosystem theory and then clearly showed the role of complexity while rejecting diversity as an explanation. Dan Simberloff is the champion of one of the sound

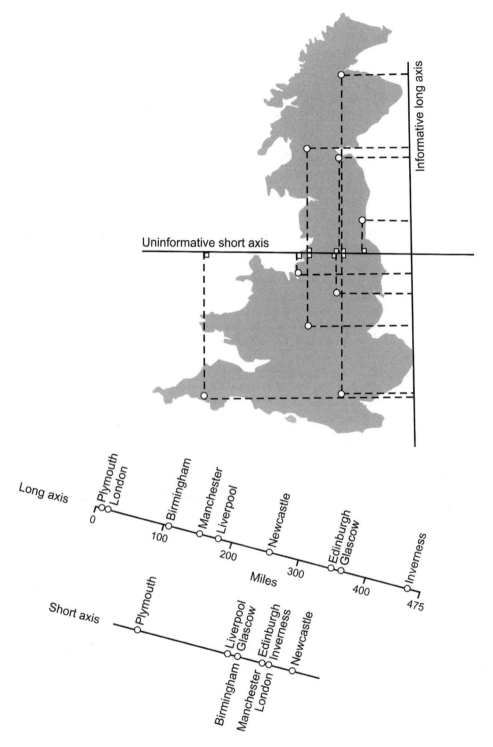

FIGURE 4.11. Two projections of England, Wales, and Scotland. The east/southwest projection only separates the West Country (Plymouth) from the rest of the urban centers. The north to southeast projection captures a separation of the financial South, from the light industry of the Midlands. The Industrial North is further separated, as is the shipbuilding of Tyneside, and both of them are separate from Scotland. Ordination finds better projections. Projections to lines is like casting a shadow.

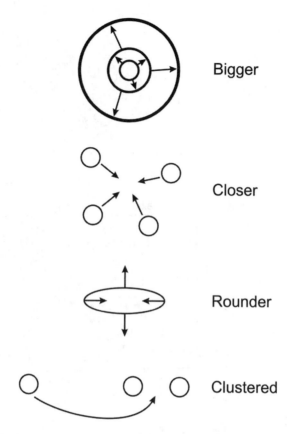

Bigger

Closer

Rounder

Clustered

FIGURE 4.12.   From the principles of island biogeography, Simberloff made recommendations as to the form for better functioning of biotic reserves. But he is still insistent on doing descriptive fieldwork at the specific place to see if theory holds up. Theory is useful, but mostly as a starting point (Simberloff et al. 1992).

theories in ecology, island biogeography. Even so, he has a very healthy attitude to theory that one might not expect. What to do about nature preserves (seen as islands) is clear from the theory at first glance. But Simberloff is greatly disturbed at the lack of tests of the theory in well-described situations in which conservation protocols are developed or rejected (figure 4.12).[22] Interesting questions often arise from a really good description. Communities are hard to test, so description is particularly well developed there. The rest of ecology might take note. The descriptive techniques of community ecology are a good example to follow so as to get ecology asking important questions again.

Given a quantification of the vegetation by species, we can set up the first species as an axis and place the site on that axis according to the quantity of that species at the site. We have shown what we mean here in our earlier discussion of Hutchinson's niches. We can set up the next species as another axis at a right angle to the first. The site can then be positioned on that second axis in the same way. The point on the axes is a stand of vegetation. The site is thus placed on a plane defined by the first two species (figure 4.4A). When we introduce a third species with its axis at a right angle to the first two, then the site can be positioned in a volume that is the space of the first three species. While Hutchinson, in his hypervolumes, generally considered just one species niche at a time, in gradient analysis, there will be a second and third point in the hypervolume, and we want to measure distance between points inside the space. A second site will have different quantities for our first three species and will be positioned elsewhere by three different coordinates in the species space. The distance between the first and second site in the space is a reflection of their differences in composition in terms of the first three species (figure 4.4B). That distance is the hypotenuse of a right triangle, as we have shown in our discussion of Hutchinson's niches. We have been at pains to be explicit about the species space because it is central to many of the ideas and controversies in modern community theory.

The clustering techniques for community analysis work implicitly in species space. Sites with similar vegetation are put into a class because they are close together in the species space; indeed, they are the closest according to the clustering criterion employed by the particular method at hand. Gradient analyses, also called ordinations, use the proximity of sites on axes as approximations of site similarity.

The data that underlie such analyses are matrices of species scores across sites. The data come in as vectors, columns, or rows of datum values, one for each species across all sites. The row vectors of the matrix are sites, while the column vectors describe species. In "normal" analyses, the site entities are clustered or arranged on the axes of gradient analyses. Since a matrix can be transposed so that the species columns become rows and the site rows become columns, there is a converse set of analyses. After matrix transposition, the species rather than the sites appear as points on the axes of gradient analyses. These alternative analyses of species relationships are called inverse analyses.

DATA ANALYSIS BY POINT PROJECTION

We have argued that a silhouette is a projection of a three-dimensional object onto a two-dimensional area. A rotation of the object in the light gives a different silhouette or shadow. Some silhouettes give more information about the original object than others. Similarly, the gradient analysis techniques of community ordination rotate the point cluster in the hypervolume so as to cast an informative shadow of the whole. The criteria for rotation of the point cluster are different between specific ordination methods. Some techniques explicitly project sites from the full dimensional space of the full data matrix to a smaller dimensional space that

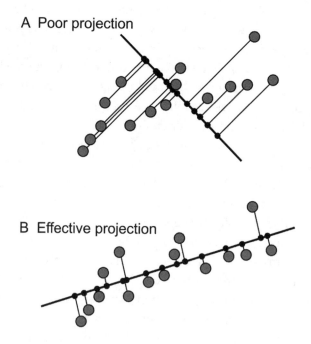

FIGURE 4.13. In an effective projection to the smaller space (*B*), the lengths of the projections are short. Euclidean space involves summing squares and taking the square root. If the sum of squares of all the projections is small, the points projected on the axis are well spread out. Note the clustering of points in the poor projection (*A*) and the spreading of points in the effective projection (*B*) with its minimal sum of squares of the projection.

is tangible. Other methods are not strictly geometric but can still be profitably seen as approximations to projections of points to a smaller dimensional space. Poor projections foreshorten distances that are long in the original large species space (figure 4.13). Effective projections minimize such foreshortening; that is, the points in the lower dimension space are well scattered with high variance. This is what is going on when the sum of squares of the lengths of the projections is minimized in an effective projection. In both normal site analyses and inverse species analyses, and in either cluster or gradient analysis, association of species or underlying species correlation structure orders the outcome.

A great contribution of Curtis and his students was to explicitly recognize the importance of site selection. In retrospect, we can see significant geometric implications to the randomization of sites characteristic of the Wisconsin school of vegetation analysis. Independent samples are collected as vectors (rows of species scores). If the sites to be selected are to be independent of each other, their respective vectors are all at right angles in the vegetation space. The vectors of sites not randomly chosen will not be at right angles to each other because of some sort of bias in the sampling scheme. Underlying the familiar correlation coefficient are geometric patterns. Correlation is the cosine of the angle between the vectors of what is being correlated. Perfect correlation indicates that the two vectors point in the same direction, with one completely informing the other. The cosine of 0° difference is +1. Perfect negative correlation says the vectors are pointing in the exact opposite direction from each other, at 180°, the cosine of which is –1. Independent vectors (say, vegetation samples) are at 90°, with a correlation of zero. By sampling at random, the investigator sets up an orthogonal (perfect right angle) framework. If the sites are not all at right angles to each other, then there is no reliable reference frame against which we can assess correlations of the vectors coming out of the analysis. Without appropriate randomization, the ecologist cannot tell artifactual co-occurrence from that based on a pattern in nature.

Randomization can be time-consuming. If each site must be independent, then the next sampling point may be far away. It is possible to divide the area into subplots and sample them at random in what is called a stratified random sampling. In a stratified random sampling, most points sampled will be fairly close on the ground, and so can be considered faster. There is some loss of independence, but more points are sampled. More points sampled increases statistical accuracy. The Wisconsin school was at pains to get good data (figure 4.14).

By sampling with stratified random patterns, Curtis and his students could see whether it was justifiable to divide communities into discrete types or to accept an underlying continuum of change in vegetation structure. They found the continuum of vegetation change that one might expect if the community did indeed consist of species all with different tolerances, with each site being invaded by a distinct local flora.

GRADIENT ANALYSES AND THE GLEASONIAN COMMUNITY

We have noticed that Clements's argument would suggest disallowed vegetation, empty parts of the species hypervolume. He says less about the continuity of variation of vegetation where it is allowed. Gleason would argue for continuous variation of vegetation. He comments less on empty parts of species space where actual vegetation is disallowed. When Curtis found that his ordination techniques gave continuous variation, he threw his hand in with Gleason. McIntosh was the champion of the continuum hypothesis.[23] But decades ago in conversation at an Ecological Society of America meeting, he said to Allen, "Clements is looking better and better these days." McIntosh saw the compromise. Although continuum of species composition was found in the ordination work, it is worth noting that discontinuity is a relative thing. It is also a matter of how you look at it. Something discrete can appear blurred if it is out of focus. There are techniques that

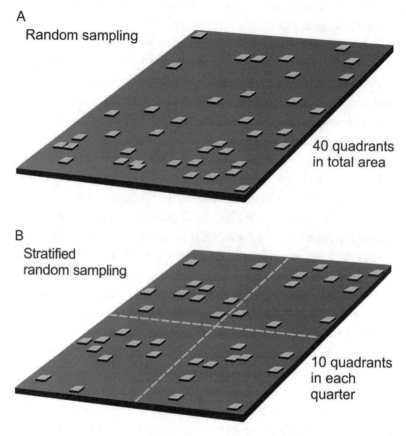

FIGURE 4.14. *A.* All points across the whole area are sampled at once. *B.* The same area has been sampled, but in one-quarter segments of the area at a time. Thus, all consecutive sampling points will be close to each other, thus saving time. This is called stratified random sampling, which loses little of the power of full random sampling, and even makes sure that all parts of the area have been sampled equally with no underrepresented quadrants.

announce groups a priori and then rotate the space to get the best discrimination. Those devices would display Clements's disallowed vegetation as open in species space. Regular ordination techniques do not emphasize that discrimination. Ordination is a sort of lens, which is capable of putting discontinuity out of focus. Furthermore, intermediates between occupied spaces might occur in the middle of a divide because of recent disturbance. As a result, intermediates might well be unstable, and disappear soon enough. Sparsely covered places in species space could easily be discontinuities. There is plenty of room for Clements's basic thesis to be viable. The compromise was found in the work of David Roberts, to which we turn in a later discussion.

There is a large literature that tests the various techniques on data that have been constructed to have certain known properties. Often, known straight gradients in environmental space appear curved and apparently distorted under analysis. We have already argued that black spruce is present at opposite ends of long gradients because it cannot compete for light. It is excluded from the mesic zone (Figure 4.5). We pointed out that the ends of the gradient are made closer by neither having mesic species, a source of gradient distortion. The fact that black spruce occupies both ends means that curvature of gradients is not distortion. Sometimes curvature comes from the interaction of two levels in a hierarchy. Black spruce relates to other species and its environment in a way that is only found at a higher level of analysis. Competition for light is a parameter set high for mesic species in the middle of the gradient. For black spruce, light is a variable that only allows the species at the extreme ends of the gradient. Parameters at low levels (constant strong competition for light) become variable at higher levels (where light is competitive in some regimes, while in other places there is little completion for light). Black spruce occurs in bogs and dry sandy ledges, but only rarely in the mesic zone. Figure 4.15 shows how the gradient in Figure 4.5 would be curved in the smaller-dimension projection. Greig-Smith gives a particularly thorough review of curvature in ordination, showing the geometry of collapsed spaces.[24]

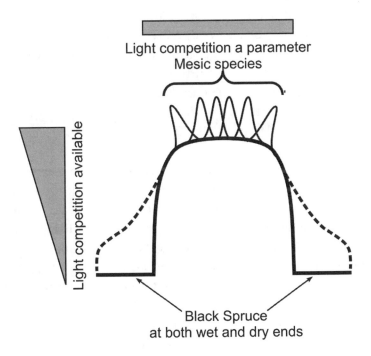

FIGURE 4.15.   If Figure 4.5 were projected into an ordination space, the environmental axis would make a horseshoe. Note that competition for light is high in the mesic zone. Note also that along the full horseshoe, competition for light is a variable, low at both ends but high in the middle. In hierarchies, higher levels turn the parameters of lower levels into variables in that wider universe. The tendency of certain ordination methods to offer curvature of spaces will sometimes be an ability to express two levels of analysis at the same time.

PCA

Bray and Curtis

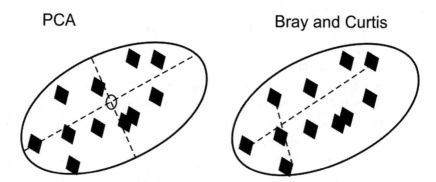

FIGURE 4.16. The same point cluster treated to principal component analysis (PCA) versus Bray and Curtis (1957). In PCA, the axes all must pass through the mythical centroid vegetation that would never occur in nature. Meanwhile, Bray and Curtis has some distortions, but at least its gradients go from one ecological place in nature to another.

All methods have their assumptions, which do not come free. For instance, the gradients in principal component analysis (PCA) have one point through which its gradients must all pass.[25] It is the center of the cluster of points. That mathematical centroid never occurs in nature because it would be too species rich and all species would be too evenly represented. The technique of Bray and Curtis (1957) may not have the elegance of the eigenvectors of PCA with its rich set of influences, but the Bray and Curtis gradients do go from one real ecological site to another (Figure 4.16). The gradients are anchored in actuality, where intuition has room to work. Edward Beals (1973) played a critical moderating role in a paper whose title says it all: "Ordination: Mathematical elegance and ecological naiveté."

The preferred techniques these days are nonmetric scaling devices that minimize stress, not variance. Stress is non-Euclidean because it does not involve squaring differences that is part of the operation of Euclidean space. Neural nets are powerfully nonlinear and those methods can be used in gradient analysis (treated in chapter 1, figures 1.3 and 1.4).[26] The instruction to the neural net ordination is to reproduce the input vector of species occurrences in the output at the bottom of the net. The adjustments in the weights in the net are modified to achieve that. There is a constricted layer that has only two or three nodes. All of the information passing from the top to the bottom of the net must be able to get through that constriction. Once the net is optimized, the values at those nodes are the scores on the ordination axes. That is, the coordinates for a given stand on the ordination axes are the values at the constriction when that stand's vector is fed in at the top of the net. Neural nets are powerful, but their use as just described gives a PCA, so the method is not that strange. The message here is to choose your method with an idea in mind, and respect other methods because they all offer something.

THE DYNAMIC SYNTHESIS THEORY OF VEGETATION

Community ecology appears less experimental and predictive than other ecological fields such as population ecology. John Harper once joked that there seems to be a touch of voodoo magic to community ecology.[27] Indeed, predicting community structure and behavior is difficult because there exists no one-to-one mapping of community composition onto environment. From the environment, it is not possible to predict community composition, or vice versa.

Recognizing this problem, David Roberts drew on the writing of Sukachev, a Russian ecologist whose work became accessible to English-speaking ecologists in 1964.[28] The cornerstone of the Russian work is the recognition that plants interact with each other only through their environment. Therefore, an adequate description of vegetation must equally involve both vegetation and environmental space. Curtis thought that plants are better indicators of the pertinent aspects of the environment than any measure that an ecologist could devise. That may be true, but it does not allow for the two-way interaction between vegetation and its environment. Conversely, the direct environmental gradients of Whittaker miss the vegetational side of the interaction, and therefore are no better. The most naïve environmental determinists say that direct gradient analysis is superior because it uses environments directly. For predictive community ecology, we need a description of the interaction of happenings between vegetational and environmental space. This is in the spirit of Richard Levins's dialectical biology.[29]

Accordingly, Roberts developed a relation between the environmental and species spaces. He maps events and continuous changes reciprocally between the spaces. While some environmental factors may respond to shifts in vegetation, others remain unchanged in the short term even with dramatic shifts in vegetation. Roberts uses the term "elastic" to describe site factors that respond immediately to changes in vegetation. In a forest, light could be one such factor. Tall trees and thick understory shade the ground. However, clear-cutting the forest or a crown fire causes light levels to snap back to full insolation immediately. Light levels snap back to unvegetated brightness. Other factors, like soil pH, change slowly and even then only if there is unremitting pressure from the vegetation. It takes several decades for spruce trees to acidify the soil, but acid soil persists long after the spruce trees have been harvested. Roberts calls factors of this type plastic. They deform slowly, but hold the deformation afterwards.

Elastic environmental factors exhibit short memory from the influence of vegetation, while plastic factors can reflect changes wrought by vegetation long since removed from the site. This is exactly the sort of situation that generates complex behavior. The state of site factors at any time is the result of the interference between processes with very different reaction times. As a result, a myriad of ways can lead to any given state of affairs, and a small change can modify everything. That explains the difficulty we experience in mapping vegetation to environment space with any reliability.

Unidirectional changes in one space may be related to tortuous change in the other space. Consider a two-dimensional species space, the plane of species A and B related environmental space with two factors, say, light and soil moisture. The unvegetated site is high in light and soil water at the origin in the vegetation space, while the same site in environmental space is far from the origin, showing high values for both factors (figure 4.17A and B). It is possible to identify separate places in the environmental space where species A and B find their respective optimal conditions. Some of the work of Ellenberg in Europe involved experimental gradients on which the pH optima were found for the species in his community. If A is a pioneer species and B is a mature forest species, then A is likely to be the first to establish and grow (figure 4.17C). As A grows in stature and numbers, the trajectory of the vegetation in species space is from the origin along the A axis. As species A grows, the ground receives less light. Furthermore, the increasing transpiration draws down soil water (figure 4.17C). In environmental space, the site begins to move down both axes together as it moves toward the origin. If species A does best in conditions between the starting point and the origin, then its growth generates a positive feedback where the plants ameliorate conditions and enhance the growth of their own species. Species A has modified conditions so that they are optimal for the species. During this period, species B finds conditions less than optimal, but better than the conditions of the open site (figure 4.17D). Species B has an optimum closer to the origin in the environmental space than species A. Accordingly in the vegetational space, species B enters the stand, at first in small numbers. The trajectory of the site in vegetation space starts to turn up the B axis in species space, while continuing down the A axis.

Species A early in the process may improve the site for itself but may well continue to change site factors beyond the optimum for A. The increasing stature of species A further darkens and dries the site. Not only does this take the environment closer to the origin than the optimum of A but also it moves the site in environmental space further toward the optimum for species B. At this point, species A is in negative feedback, but B is in positive feedback and takes the site factors further beyond the optimum of A, toward its own optimum (figure 4.17E and F). In vegetation space, the trajectory turns toward high values of B and away from high values of species A. Aspen is a transient forest species and behaves like species A. Its shade suppresses its own offspring but also makes conditions conducive to growth of pine, hemlock, and sugar maple, which in turn behave like species B.

Think of the environmental space in Roberts's model as an undulating surface (Figure 4.18). Hollows on the surface are domains of positive feedback where success of a species leads to further success. For transient species, the surface slopes

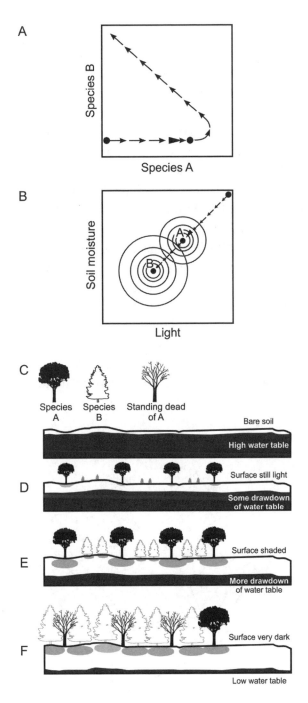

FIGURE 4.17. *A* and *B*. A site at a given time occupies a position in space. Here are the environmental optima of species A and species B. *C* and *D*. Early in succession, there is high light intensity and a high water table. *E* and *F*. As the stand development proceeds, the system moves closer to the optimum of species B as the water table drops and the light intensity at the ground diminishes.

in one direction or another in the vicinity of their optima. If the site is a ball on this surface, it will roll into optimal regions for transient species' optima, but then roll on past into deeper hollows in the region of the optimum of the species that replace the transients. Some species have optima at the bottom of local hollows such that the vegetation cannot escape without application of some external force. For example, heaths in positive feedback change the soil to an acid condition in which they do well. They also shade deeply. Other species then find it difficult to colonize and establish on such poor soil. The heaths thus create a condition and then perpetuate it, thereby creating tracts of very similar vegetation. Similarly, yew in England can form canopies so dense that the shrub holds the site against all invasions of trees for a millennium.[30] The same may be true for rhododendron thickets in the Smoky Mountains of Tennessee, called heath balds, although the Appalachians have a shorter historical record than England and we are therefore less sure of stasis for balds (figure 4.19). Thus, the environmental surface in the vicinity of the optimum of, say, aspen would be gently sloping (figure 4.18, point A), which is conducive to continuous and relatively unguided change in vegetation and site factors. Alternatively, the surface could have local deep hollows that hold the vegetation and environment in stasis, as in heathland (figure 4.18, point B).

There is no firm rule as to which type of pattern will hold sway because the reading of the environment is species specific. Furthermore, there is no reason why the environmental space may not be dissected by strong positive feedback in some parts of the space, while in other parts of the space the surface could be relatively

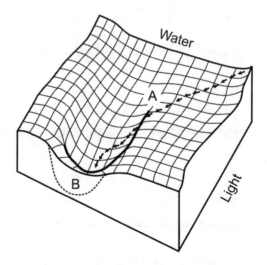

FIGURE 4.18. The environmental space of the example in the text expressed as an undulating surface, where the site changes behavior as it moves along the path shown by the ball rolling from high ground through the continuum of space for species A to the hollow of the optimum of space for species B.

FIGURE 4.19.   Heath balds in the Smoky Mountains appear to be a terminal vegetation type that creates its own optimal environment. High on the ridge, the bald looks deceptively open, but it is an impassable thicket about five meters tall. (Photo courtesy of T. Allen.)

flat with little to retard or direct vegetational change. Different degrees of sculpturing on the environmental surface would clearly lead to different patterns of vegetation typing. A relatively flat surface would lead to continuous change giving the continuum that Gleason and the devotees of gradient analysis expect. Alternatively, an environmental surface dominated by a few deep hollows holding the vegetation in certain states would move to certain states, as Clements might note for his climaxes. Hollows would give a landscape that yields reliable maps of discrete types of vegetation. Such maps were the aim of the European phytosociologists who study discrete vegetation types.[31]

PREDICTIONS RELATING VEGETATION
AND ENVIRONMENT

We see how straightforward movement in environment space is related to complex trajectories in species space. Add to this the complications of differences between elastic and plastic site factors. It then becomes clear how the same vegetation can occur in different physical environments, and how different vegetation can occur in separate sites that replicate site factors. Therefore, vegetation is not a function of the environment; rather, it has a relation to its environment. Mathematicians are familiar with the difference between a function and a relation.

A function is a special case of a relation. If X is a monotonic function of Y, then only one value of X corresponds to each value of Y. A relation does not require only one value of X for every value of Y. For a mathematician or statistician to say that environment is linked to vegetation as a relation rather than a function is unremarkable. Nevertheless, the difference between a function and a relation for mapping vegetation onto the environment is important for ecologists. All this pertains to the model of Curtis and the environmental space model of Whittaker. Both assume that a function links vegetation to environment. Roberts insists that the relation is not a function and so opens the door for prediction on more workable grounds. What the conventional view might call noise in the relationship between environment and vegetation can be recognized as slack within constraints. Thus, Roberts is asking a different question that does not depend on environmental determinism, where both Curtis and Whittaker are environmental determinists, along with almost all ecologists then and many now.[32] Theory in ecology often reflects pressing social issues, and climate change may be the reason for continuing belief in environmental determinism, even in the face of Roberts's insights.[33]

As we have said several times before, prediction is posed against a constraint. Roberts has devised a method that discovers the ranking of constraints. The argument behind them is hugely important. He starts with vegetation data from sites, and groups them according to vegetational similarity. The technical name of the groups is "cliques," groups where all members are connected to all others by a particular level of vegetational similarity.[34] The cliques were maximal cliques, a special case where only the largest groups that meet the condition are considered, with subsets disallowed (figure 4.20A). Thus, it is possible for a given site to belong to several cliques. The method proceeds to work out the sets of maximal cliques in the data for various thresholds of vegetational similarity (figure 4.20A).

Remember that very similar vegetation can occur on sites with different environmental conditions. We want to know how much variation in environment could be tolerated in cliques of a given level of vegetational similarity. That is, how much variation in vegetation can be attached to a certain range of environments?

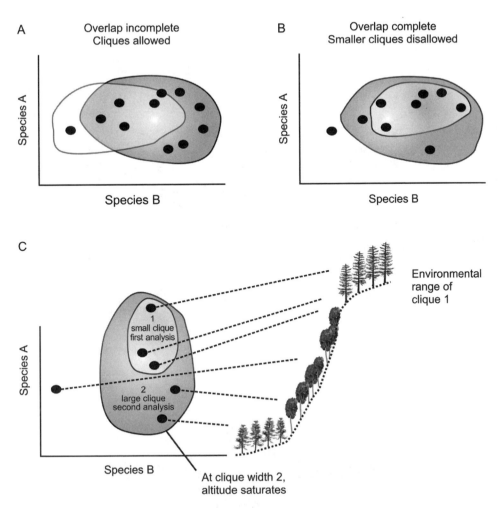

FIGURE 4.20. *A*. In two-dimensional space, the allowed cliques are the largest that still meet the maximum distance criterion. Allowed cliques may overlap, but not completely, so that no clique becomes a subset of another. *B*. Any clique for a given clique width that is a strict subset of another is disallowed. *C*. Two analyses of different clique widths may use small cliques in the narrow clique analysis that are indeed subsets of larger cliques in the wide clique analysis. That is acceptable because the subset clique belongs to a separate analysis. Thus, in diagram C, the smaller clique is allowed because it only occurs in the narrow clique analysis. A valid clique in vegetation space has a certain range for each site factor (e.g., altitude). In this figure, the new wider clique is saturated for altitude; wide cliques can be at any altitude (Roberts 1984).

What is the greatest environmental difference that can still map onto a given amount of vegetational variability? That is a straight many-to-many mapping, but it carries within it the answer to the question, "How many to how many?" Maximal cliques can do that for us.

In a plot of vegetational clique width against widest environmental range within those cliques, of course wider vegetational cliques have more variable environments (figure 4.21*A*). Roberts got one number for environmental variation by determining

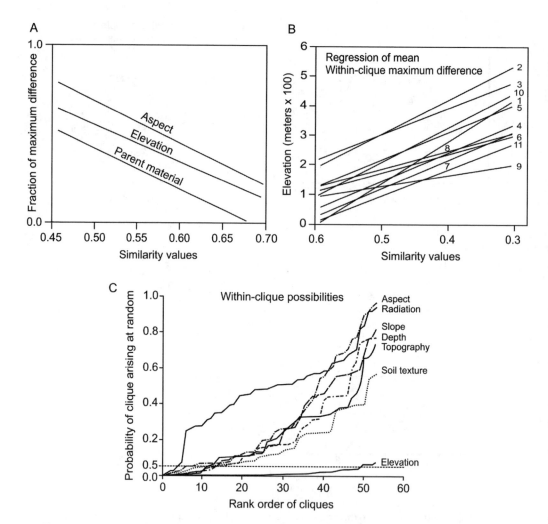

FIGURE 4.21. *A.* The regression lines plotted refer to average width with respect to the factor in question. As clique width increases in the Little Rockies, aspect as an environmental factor is quickly unable to distinguish between wider cliques. However, altitude and particularly parent material can distinguish between wider cliques than aspects, indicating that they are primary constraints. Note that at only moderate similarity of vegetation within cliques (0.45) almost any aspect could appear within a clique, giving such cliques poor predictive power for aspect. However, those same cliques still possess some predictive power as to altitude and parent material. For narrow cliques (0.70) with very similar vegetation inside the cliques, all three environmental factors could be predicted from knowledge of vegetation. *B.* As might be expected, wider cliques show more latitudinal difference within cliques. However the eleven forests plotted here with clique average showed altitude as a constraint much tighter in some forests than others. Forests on mountain 9 were more tightly constrained by altitude than were the forests on mountain 2. *C.* In Bryce Canyon, the vegetationally defined clique that allows fifty-five cliques in the data is consolidated here with regard to seven environmental factors. For altitude, fifty of the fifty-five cliques have less than 0.05 probability of arising at random. The other factors forming the lines above refer to factors where the cliques stand a much greater chance of occurring at random with respect to the environmental factor question. Clearly, elevation is the overriding tightest constraint. Here, individual clique values are plotted, whereas average clique values are considered in diagrams *A* and *B* (Roberts 1984).

the average maximum environmental difference observed between clique members. There are many cliques of a certain size, so if we calculate the maximum environmental variation for all of one size of clique, we can average all those maxima. For any given clique width, there is an average maximum environmental range for every environmental factor.

Very particular vegetation could be generally related to narrow ranges of environment. Less-particular vegetation will be related to wider environmental ranges. For any given environmental factor there comes a degree of vegetational slack (clique width), where clique members can come from any environment (e.g., aspect in figure 4.21A). In a way, the environment of the wider cliques had become saturated as a predictor of vegetational similarity. Even slacker vegetational cliques cannot have a more different environment than the most different seen in the study. The environment's capacity to exert constraint was exceeded by the variation of vegetation in the wider cliques.

The tighter the environmental constraint, the tighter the relation between environment and vegetation. The tighter the environment constraint, the less slack there is in the system. Various site factors were tested against cliques with successively larger vegetational differences. The critical finding was that not all factors became useless at the same clique width. In the Little Rocky Mountains of Montana, parent material from which the soil was derived and altitude both persisted as predictors of clique environment for wider cliques than did site aspect. This means that relatively similar sites can have very different aspect, with some facing north and some south, but will have a relatively narrow range of altitude across clique members (figure 4.21A). Although aspect makes a difference, the more important constraints that will generate firm predictions in the forested mountains of the Little Rockies are altitude and parent material. Roberts applied clique analysis in the mountains of Montana and in Bryce Canyon. In Montana, he studied several mountain ranges, and compared them. He found that altitude constrained to different degrees depending on the particular mountain range (figure 4.21B). In Bryce Canyon, he found that altitude was far and away the most significant constraint (figure 4.21C).

With that knowledge of the ordering of constraints, predictions can be made. Knowing which are the tightest constraints, we can devise experiments to test on limiting factors. With such well-studied experiments, we can at least hope to find the critical explanations underlying community pattern and structure. Knowing the constraints allows the vegetation ecologist to ask questions that avoid implying a medium-number specification of the vegetation.

Some of the best work attempting to prove environmental determinism was performed by Orie Loucks. He developed a series of gradient analyses using environmental factors directly. The scaling on the gradients was nonlinear since, for example, when there is not much water, even less water makes a big difference (figure 4.22).[35] Loucks was influenced by Bakuzis in Minnesota.[36] Bakuzis's critical

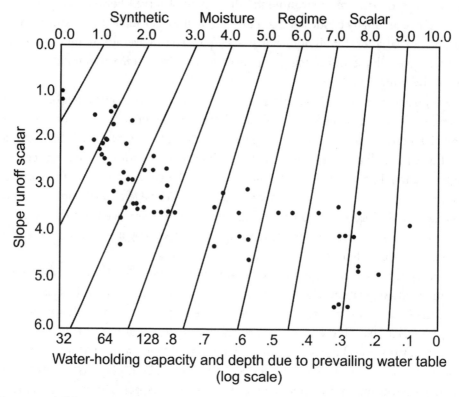

FIGURE 4.22. In his scaling paper in *Ecological Monographs*, Loucks (1962) plotted moisture relations that were relative to the water table in a nonlinear fashion. This is because high water level differences are not influential, but below a critical depth, the small differences in water table make a big difference to plants.

gradients were water, heat, nutrients, and light. The effort was to scale the environment to the way the plants would see it, and that would be a non-Euclidean space, where standard difference is different depending on the environment. How far is far depends on where you are; thus, the units of difference themselves change. Loucks was attempting to move beyond hypothesis generation (most gradient analysis is to generate hypotheses) toward the beginnings of testing hypotheses.

Whittaker's direct gradient analysis in the Smoky Mountains did not use carefully scaled environmental factors, and was hypothesis generating.[37] Whittaker used general gradients that were easy to measure and likely summaries of a complex of factors. He used altitude for temperature and a ridge to hollow gradient for water. Confidence in both approaches comes from how Whittaker's vegetation description coincides with the results of Curtis in Wisconsin on Curtis's first and third axis plane (Bray and Curtis 1957) (figure 4.23). Despite their misconceptions about environmental determinism, comparison between Curtis and Whittaker is a worthy precursor to Roberts's clearer view.

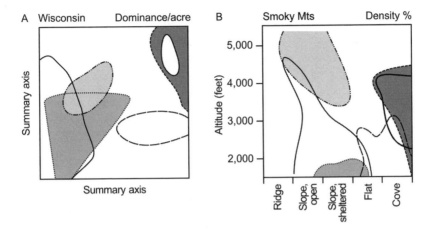

FIGURE 4.23. *A.* Bray and Curtis (1957) compared their indirect analysis (in vegetation space) to *B.* Whittaker's direct gradient analysis of altitude (heat, in Bakuzis's terms) with moisture from dry ridge to wet cove. The altitude of Whittaker maps onto temperature by latitude in Wisconsin (which is flat relative to the Great Smoky Mountains). Tree species are plotted with the same style of line between analyses.

COMPETITION, RUDERAL, AND STRESS STRATEGIES

We have given Phil Grime's work on diversity some laudatory attention, but there is more. We have suggested that his work stands out because his systems are very well described. He knows what he is investigating before he starts his experiments. Now it is time to lay out his descriptive work (figure 4.24).

Grime goes one better than other direct gradient analyses in that he starts from three fundamental strategies of the plants. He boils it all down to how plants deal with their circumstances through evolving in a space defined on fundamental principles. Grime's three gradients are each anchored at one end by a particular strategy: competition, stress, or disturbance (ruderal). It is accordingly dubbed the CSR classification. The ordination is on a plane that takes a triangular form.[38] The respective other end of the gradient for a particular strategy is the middle of the line, across the triangle, that connects the other two strategies.

Grime has two sorts of data to back him up. In general, he sticks to the British flora because it is limited enough for a more or less complete treatment. Greig-Smith actively chose to teach the monocotyledons for the same reason; he could cover them all in one term. Only the high points of the dicotyledons were covered, but all monocotyledon families were in Greig-Smith's course. For Grime, there are two editions of a grand work by Grime, Hodgson, and Hunt (2007) that summarizes data on all the common species. Allen was privileged to write the foreword for the second edition. All the common plants are organized on many factors, for

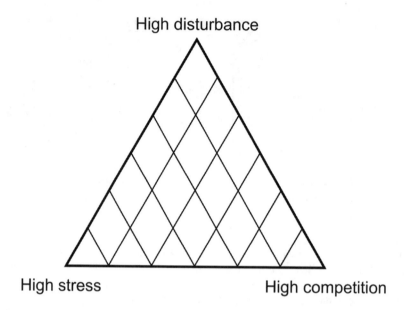

High disturbance

High stress   High competition

FIGURE 4.24.   Grime's triangular ordination of the major types of environment applied to plants. The corners are singular as to type of environment. The space in between can apply to particular plant species with compromised adaptations to all three types of environment. There is another section theoretically possible, but disallowed in nature. High stress and high disturbance does not allow accumulation of enough capital as a response to disturbance, so we never see it.

instance the incline of the ground on which that species occurs. The second source of Grime's information is his relentless experimentation.

Tilman has suggested that competition for mineral nutrients is the key to species representation in a community. We know where he is coming from because like Tilman, Allen's thesis work also studied algae, in which nutrients explain a lot. Tilman's central idea is that the competitors consume nutrients, driving levels down below where other species can only break even on effort to capture nutrients, considering the return on that effort. There are many factors and the best competing species drive nutrients down on their respective best factor, says Tilman.[39] This mistaken position comes from the focus on individual species in agricultural settings, where lower nutrient status in uninteresting.[40] To quote M. Caccianiga and colleagues:

> The contemporary ecological mindset borrows heavily from agriculture, in which periods unsuitable for growth are disregarded and thus resource acquisition and plant growth are seen as continuous processes . . . Life forms in chronically unproductive habitats are primarily defensive adaptations against these perilous episodes, not aggressive adaptations to subtle variations in resource availability and competition during milder periods.[41]

Grime's experimental work shows that Tilman's arguments are simply not valid in nature. If the nutrient status at a site is chronically very low, the species there are stress tolerant, the S species. Put those species in a more moderate environment, and they do not grow any faster. They have sold the store when it comes to using nutrients. To survive in low nutrient status, such as on serpentine soil, apparently they have had to give up the ability to take advantage of better nutrient regimes.

In a competitive C environment, there is also stress on resource availability because the competitors drive the nutrient levels down until they are more or less unworkable, as in the stress environments. In that phase of competition there is more or less stasis. The difference from the stress-adapted plants is that competition-adapted species can respond to release from stress.[42] Grime has shown in the laboratory and the field that plants from stressed environments cannot respond to resource. Data on plant traits from around the world show that competitive strategies decline in more stressed environments.[43] When a light fleck gives a transient patch of sun, or an animal urinates in the forest to give a jolt of nitrogen, the competitor species leap to, and soak up the source quickly. Thus, the competitive strategy of C-adapted species is not to drive nutrients down so as to win. Winning in competition is not about the average level of resources, the first moment of the distribution. Rather, it turns on the second moment, an ability to take advantage of pulses.

Grime is accepted on the stress and disturbance gradients. The reason is that stress and disturbance are both driven by the environment exerting control over the vegetation. There is little feedback from the plants onto the environment. Stress and disturbance can therefore be modeled as relatively simple situations. But Grime has been criticized for his ideas on competition. The reason is that competition is a reciprocal relation between plants and their environment. The stress on plants in competition is of their own making. There are two causes operating at different levels. Such a move between levels of analysis introduces contradiction as to causality, as it always does. The rate dependence of nutrient flow butts up against the rate-independent coded information in the plant strategies. One cannot strictly model situations like that. However, we do have narratives available, and they do not have to be internally consistent. We can use models to improve the quality of the story, as we discuss in chapter 8 here, and that is what Grime does. That takes more work and flexibility on the part of the listener. Grime's narratives are compelling.

Grime has used his ordination space to follow trajectories of succession. He has plotted out areas in his space occupied by different life forms. And he has been able to characterize species as to strategy, and even follow it through the life history of the plants in question.

INDIVIDUAL-BASED MODELS

Most of the models we have presented so far use sampling and statistics. Statistics work on distributions and give essentially one summary number. Often, the mean is adequate for summarizing a collection of information (Figure 4.25). But sometimes we need to see how much variability is in the population. For that there is another single number, the variance. That takes the deviations from the mean and sums the square of all the deviations. The square root of the variance is the standard deviation, which is in the same units as the mean. That emphasizes the effect of big deviations, as big numbers multiplied by themselves again yield big numbers. The variance is called the second moment or the mean square deviation because of the 2 in squaring the deviations. Other times, there may be deviation more to one side of the mean. To capture that asymmetry, we use the skew. How much big variation is piled up to one side is found in the third statistical moment. And indeed, the skew is calculated like the variance, but it involves 3, the cube of the deviations. Statistics give you one number.

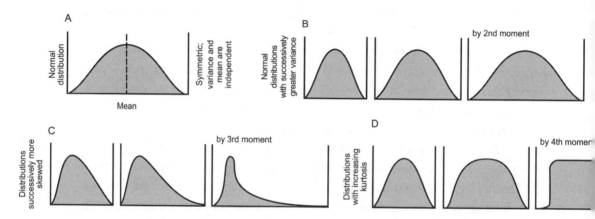

FIGURE 4.25. Frequency distributions show how many times a measurement of a certain quantity appears in a set of measurements. *A*. The most common measurement found appears as a peak on the distribution. *B*. If the situation of items is independent and additive in their relation, the normal distribution arises, the classic bell-shaped curve. It shows a mean in the middle at the peak and a spread that is independent of the mean. The mean is the sum of all measurements divided by the number in the population: straight values are summed and divided by straight numbers in the population, so the mean is the first moment. But variance measures the dispersion on either side of the mean. It squares deviations from the mean, and averages the squared value. The squaring is the reason why the variance is the second moment, with the variance called the mean-squared deviation. If the distribution has a longer tail on one side than the other, we can measure that tendency with the skew. It is called the third moment because deviations are cubed. The fourth moment uses the number four, and is called the kurtosis, a measure of how square the shoulders of the distribution are (Allen, in Dodson et al 1998.).

A bell-shaped curve, perhaps for seed size, is symmetric. But as the seeds grow, the bigger seeds have more with which to grow, and so they grow faster. The bell-shaped curve becomes skewed to the right. The longer the plants grow, the bigger gets the skew. The skew is one number, but it captures ever-bigger differences from small to big individuals. The reason we can capture all that difference in the one number of the skew is that all the seeds are on the same growth curve. It is just that the bigger seeds start farther up the growth curve. Fundamentally, there is only one situation, but with growth in it.

Now we get to the nub of this section. Sometimes bigger is not simply quantitatively bigger, it can be qualitatively different. Take two populations of fish where one simply contains larger individuals than the other.[44] Over time, the means of both populations would get larger. So would the respective variances and skews, more so for the population of fish that started with larger-sized individuals. But if big fish are predators, then differences would be qualitative over quantitative. The small individuals become prey. First, the small fish in the population of smaller fish would disappear. Next, as the predators got bigger, the whole small population and the small fish in the bigger species would disappear through cannibalism. In the end, there would be one small population of very big fish, a very different outcome to the one that can be described with means variances and skews (figure 4.26).

The same can happen in a forest. Bigger trees do not just have more biomass; they take light away from the understory and therefore will not only grow faster but will also live longer than most shaded individuals. In such situations, statistical descriptions fail. Because bigger is different, we have to model individuals separately. That is the basis of individual-based modeling.

Shugart built the FORET model by simulating as many as seven hundred trees individually.[45] The growth and survival of each tree was modeled by species-specific values as responses to shade. The parameters for the species were fixed, but they were applied as probabilities, to capture the happenstance of trees dying in a forest. The trees were very simple, rather like cocktail umbrellas: stem and disc canopy. The total biomass on the plot was limited. This caused the simulation to be of a given area, whose size was derived from maximum biomass possible. The biomass maximum was for an area of 0.1 hectares, although area itself was not specified. Another peculiarity was that there was no horizontal information in the program. As far as the model was concerned, the cocktail umbrellas were all sitting on top of each other. Shade was calculated by summing the shade of all taller trees relative to the one whose shade was being calculating at the time.

When Shugart first grows a new type of forest, he has to calibrate the model so it works with the new species. He has to set species parameters as to shade tolerance and growth in the given environment. But in the end, he gets a very convincing simulation of forests from Tennessee to Australia. His adjustments of species is not tinkering to get the model to work, as some criticisms have suggested, but are derived from independent data about the new tree species in the

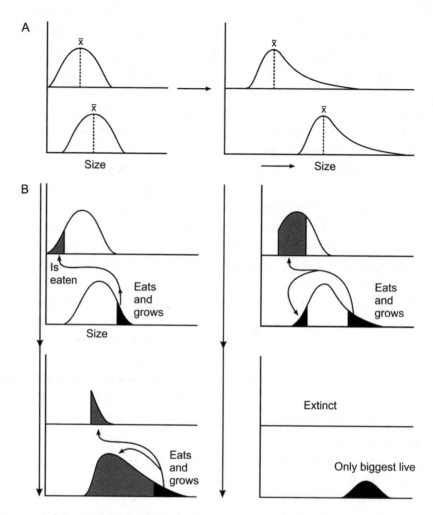

FIGURE 4.26.   A. Two populations of fish, one of larger individuals, the other smaller, are plotted on the left side. The mean of the bigger species, x^{bar}, is further to the right. Over time both populations grow and therefore increase their means, variances, and skews. Basic statistics applies because there are only quantitative differences between individuals. B. The same population differences as in diagram A apply between two populations of fish. The difference here is a qualitative distinction where bigger fish do not simply get higher on the growth curve but also become big enough to eat smaller fish, which is a qualitative change. At first, big fish eat smaller fish and are even cannibalistic. Instead of an increase in the mean, variance, and skew in both populations over time, the end result is one species of only big fish. Qualitative differences cannot be modeled at the level of populations, and must be modeled one individual at a time.

new environment. Those data are: (1) conditions from species range maps; and (2) local knowledge of foresters in the region. Those simulations were good enough to use as generators of data for looking at the effect of climate change (figure 4.27A). True, FORET output is not perfect, but what is remarkable is how very good it is. The issue is not perfect simulation, but how few parameters the modeler can use and still be impressively realistic in the results.

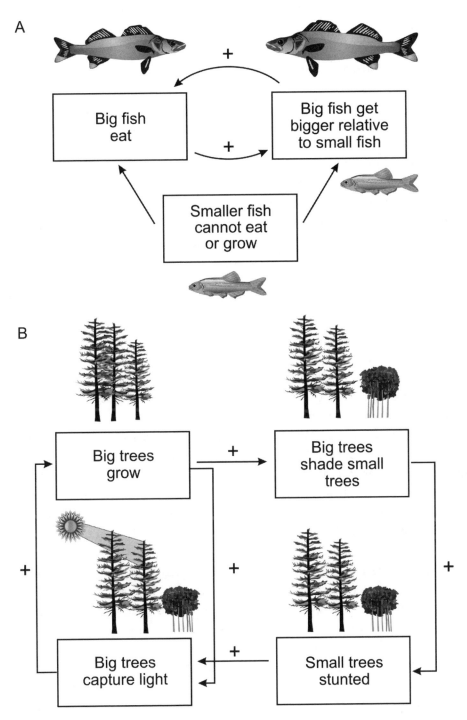

FIGURE 4.27. *A.* Feedback diagrams showing big fish affecting small fish. The difference is not size but a qualitative difference between prey and predator. *B.* Like the big fish, big trees have an effect on small trees, stunting them. A positive-feedback diagram showing that being big is more than having more with which to grow. Big means resource capture and that small trees get shaded and therefore stay small and tend to die.

Failure of a model to predict from reasonable assumptions is normal science refutation, the normal way to find things out.[46] C. S. Holling once complained to Allen that he could not get his budworm model to fail, so it was teaching him nothing. All assumptions are false, but you simply cannot get away with some. Science is not about truth; it is about finding which lies you can afford to tell. There is a case to be made for outrageous assumptions like those Shugart made (cocktail umbrella trees all on top of each other). With such assumptions, if the model fails, well what did you expect? But if it succeeds anyway, that tells you that all that reasoning you put into reasonable assumptions does not matter. Shugart's unreal assumptions give deep insights. His finding is that, although there is horizontal placement of trees in a real forest, for the most part those assumptions do not matter because almost everything in a forest is determined by verticality. Simulations of the roots in prairies show that same dominance of vertical influences in the soil.

CYCLING INSIDE COMMUNITIES

Individual-based models have been useful in calibrating the dangers of habitat loss in fragmentation conservation issues. It is important to put a stake in the heart of attacks on individual-based models. In a remarkably restrained rebuke given how egregious was the mistake, Mooij and DeAngelis (1999) report that the errors fed into simulations by Ruckleshouse, Hartway, and Karieva (1997) were not the 16 percent claimed in the publication, but were more like 16,000 percent.[47] It makes sense to put in error to see if the model is robust to some noise. It is called sensitivity analysis. But an error that is three orders of magnitude greater than what was planned will, of course, generate failure. It could do nothing else, so the dismissal of individual-based models by Karieva, Skelley, and Ruckleshouse (1997) is simply ill founded. If you can get your mathematics right, you will find that individual-based models are powerful and useful.

Shugart's experience with forests growing in simulated stands led him to have a keen interest in cycling processes in communities. In forests, these are called gap-phase cycles. If you run FORET for seven hundred simulated years with it set for species in Tennessee, trees grow until usually there is one big tulip poplar left, shading out a beech tree. Then the big tree comes down. We have used something of a narrative style in this book. In a wonderful slim volume, *How the Earthquake Bird Got Its Name, and Other Tales of an Unbalanced Nature* (2004), Shugart also tells stories. If you are to read just one other ecology book, let it be *Earthquake Bird*.

Shugart's chapter on gap-phase analysis in that book uses the ivory-billed woodpecker, saying that it was always rare because of relying on four-hundred-year-old dead trees that are good for only a few years of foraging. This leads Shugart into a discussion of trees that come down destructively creating a gap, as opposed to trees that do not create gaps. Tulip poplar comes down hard, which suits its seedlings. It needs gaps and it creates gaps. Other trees do not make gaps. The beech tree does not make gaps because it moves between generations by stump sprouting. We see here strategies of trees that map onto Grime's species strategies. The tulip poplar creates its own disturbance, to which it is adapted. In that chapter in *Earthquake Bird*, Shugart goes on to give the best account we know of Watt's midcentury paper on cyclical processes in communities.[48] Even climax forests cycle trees in the over-story. Watt was the first to lay out gap-phase cycles.

Watt also described several other cyclical systems in communities. On the hillsides of the Cairngorm Mountains in Scotland are bands of vegetation that move over the years across the landscape. Under the windbreak of heather and bearberry already established, bearberry moves out over bare ground (figure 4.28).

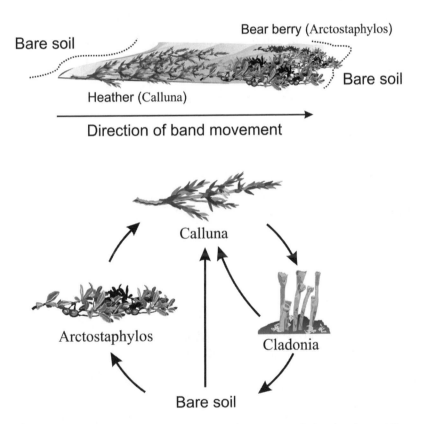

FIGURE 4.28. Watt refers to the cycle of bearberry (*Actostaphylos*), heather (*Calluna*), and lichen (*Cladonia*). Protected from the arctic wind, bearberry colonizes open ground. Bearberry gives cover to heather, which cannot colonize open ground but grows well in the windward side of the bearberry. The lichen has a similar relationship to the heather. But as the band moves away from the wind, the lichen cannot take the wind and dies to leave open soil. The next band to windward eventually invades the open land. A set of striped vegetation bands moves away from the wind in unison (Watt 1947).

That moves the whole complex multispecies band of vegetation forward. It keeps moving forward. But in a given spot, the bearberry grows old and declines. Heather moves into that place on the ground, with the older, declining heather providing a windbreak and the declining bearberry making room. That moves the heather band forward behind the bearberry. As the heather declines, it offers lichens enough of a windbreak to move the lichen band forward into what was the heather band. As the windward side of the lichens takes a buffeting from the wind, it breaks up and leaves bare ground. After a period of bare ground, bearberry moves the band onto open ground, but can do so only because of the windbreak of the band behind it. In his experience in Australia, Shugart encountered shrub vegetation that appears striped like tiger's skin, when seen from the air. That cyclical system works on colonizing areas with more water, and then using it up.

Watt speaks of tussock fescue grasslands (figure 4.29). The grass grows and captures soil in its leaves. That builds the soil mound. This makes the soil on top of the mound drier, which causes the decline of the grass. The declining grass leaves room for lichens that do well in those protected conditions. But the lichens tend to lose soil to the wind, not gather soil from it. In the end, the lichens disappear leaving a new hollow for the cycle to start again.

A fourth cycle Watt reports is sphagnum mosses growing up to make tussocks. The hollows, being close to the water table, are moist, an ideal situation for growth of the sphagnum moss. The tops of the hummocks become dry as the sphagnum moss grows and that stops growth of the sphagnum. Then the moss grows in the hollow, causing it to overtop the tussock, as it makes a new tussock. A cross-section through a bog reveals lens shape growths of the history of tussock and hollow, and back to tussock. Bogs grow higher through this process, with the moss pulling water ever upward. Watt's work was clearly talking of landscape patches, while at

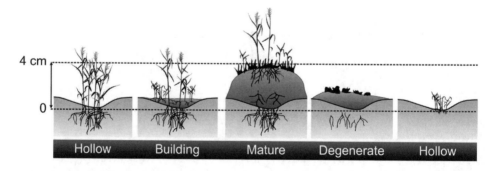

FIGURE 4.29.   Watt (1947) also speaks of fescue (*Festuca*) establishing in hollows with less wind and drying conditions. The grass catches particles that build up a hummock, which is invaded by lichen species (*Cladonia*). But the hummock becomes too dry and windy for the grass, which dies. Without the grass, lichens die as the hummock blows away, leaving a hollow, wherein the cycle starts again.

the same time addressing community processes. His paper is a standout contribution, as a harbinger of the revival of landscape ecology that would have to wait at least thirty years.

THE POPULATION BIOLOGY AGENDA

The ability to express their systems in equations has led to great confidence on the part of population biologists. The more they quantify, the more kudos they receive. This has engendered a confidence in population biologists who imagine that once we get the population biology down, we can explain communities. That view does not understand that community ecologists work to answer an opposite question—not "How do the winners win?" but rather, "Why do the losers not lose it all?"

But, of course, it took one of the best population biologists to show that communities are more than a set of population parameters. Joan Roughgarden with Steve Pacala conceived a critical question.[49] The pattern of Roughgarden's work is to drive a problem through until it is well described and properly understood. Roughgarden's thought was, let us go to the simplest communities to see what makes them communities. They chose what they thought were the simplest communities, lizards on the Lesser Antilles in the southern Caribbean, where there are either one or two species per island. Over thousands of years, there have been favorite spots for lizards, and the fossil remains of their teeth tell of the size of the owners of the teeth. Roughgarden and Pacala had a means to probe the past (figure 4.30).

On islands where there is only one lizard, the species appears at what we presume is an optimal climatological size, fifty millimeters. The lizards on the big Caribbean islands, say Cuba or Hispaniola, are larger at about one hundred millimeters. There are many species on the large islands, and environments are varied, so it is one long competitive mud-wrestling match, where size helps, and there is no conclusion. But on the islands where there are two lizards, the smaller one, presumably the one that was there first on its own, is smaller, about forty-five millimeters. One can reasonably suggest character displacement. The smaller one has been naturally selected to be smaller to get out of competitive confrontation. Remember the ghost of competition past that we discussed earlier.

Another critical observation is that the smaller lizard is generally losing the battle and is retreating up the volcano. When it is forced up to the top of the volcano, that is the end of the species because there is nowhere else to go on the small island. The history of tooth size at the favorite spots tracks a sequence that indicates Roughgarden and Pacala were not dealing with communities, but rather multiple populations. The bigger species had been ripped from its communities on the larger islands like Hispaniola. Traveling probably on logs, the big species arrive on the islands and the lucky ones establish new populations once in a few millennia.

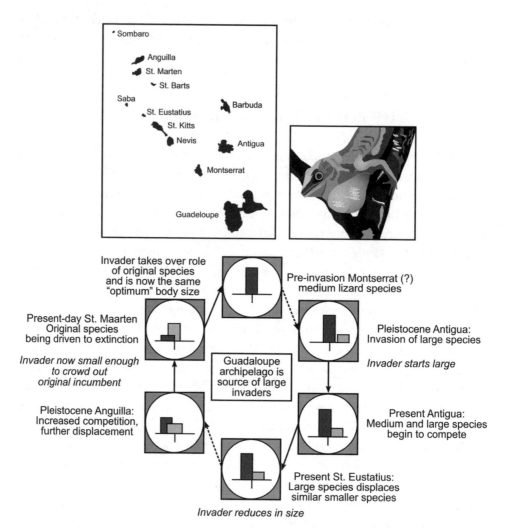

FIGURE 4.30. The Lesser Antilles are variously occupied by one or two species of Anolis lizard. From several lines of data, Roughgarden and Pacala triangulated on the pattern, and concluded that on islands with one lizard species, the species was about the climatic optimum at 50 mm Apparently, invaders (light gray bar) from big islands like Hispaniola are bodily bigger (100 mm), their size reflecting their battles in real communities. But the bigger species is at first uncommon. At first competition, the invaded species gets smaller to avoid competition in what is called character displacement. Then the bigger invader evolves in the face of little competition from the incumbent, which begins to decline. Eventually, the bigger invader species is reduced in bodily size close to climatic optimum size. This ultimately drives the old incumbent (darker bar) extinct, and we are back to a one-species island again with a new darker bar (Roughgarden and Pacala 1989).

The process identifies a sound principle in population biology, but it does not apply to communities. That principle is competitive exclusion. It has appeal because it introduces certain finality, which is rare in ecology. In biology, there is rarely a workable time zero, and almost never a final conclusion. If two species are in the same niche, one will be driven extinct by the other, but that is largely a theoretical or contrived situation. Experimental contrivances with *Tribolium* flour beetles show that one species always wins.[50] The experimental conditions force the species into the same niche. In the same niche, differences in humidity and temperature change the species of the winner. In an intermediate range, one species will also always win, but which one it is becomes uncertain. The principle does not apply in communities because the environments have a wider range in communities, and there are more species involved and more environmental heterogeneity. In communities, change in environmental factors is fast enough across space and time for the loser not to lose.

In the upshot for Roughgarden and Pacala, when all the evidence is marshaled, there is a cycle. On islands with one species of lizard, the constraint on size is environmental. Then a larger, superior competitor arrives and forces character displacement on the smaller, weaker competitor. As the winner begins to take over, it is selected by the climatic environment to become smaller. Presumably, this allows the still marginally larger species to put pressure on the character-displaced smaller species. In the end, there is competitive exclusion, and the invader is selected to become the new species at the climatic optimum of fifty millimeters. Then the situation is primed again to go into a new cycle with the eventual arrival of another big species from Hispaniola. Thus, the two species islands are not communities, but are instead two populations going through competitive exclusion. Rather than a community, the islands are more like traffic crosswalks, not places in themselves. The islands are places of transit.

CONCLUSION

The community is an important criterion for organizing ecological ideas. However, it is itself a complex notion, meaning different things for each major taxonomic and resource-sharing group. As the scope of a study is expanded in time, space, and heterogeneity, the nature of the community changes radically. The only universal that applies to all communities is the definition that we used at the outset. A community is a complex whose parts are organisms assigned to some taxon. The types of organism are the key to a given community. To be worth studying, the community parts must be accommodated to each other in some way; otherwise, the community is only an arbitrary collection. However, different types of communities are variously integrated by spatial or temporal contiguity. Animals can come and go over time, and succession can alter the flora.

The community is something other than a place in space, for that is a landscape consideration. It is also other than a set of taxonomically undefined producers and consumers in mass balance, for those are the attributes of ecosystems. Taxonomic identity is required to be a member of the community. There is certainly some evolution that underpins community structure, but evolution is far from a determinant of community structure. That reservation expressed, the example that indicates at least microevolution inside communities is the work by West, who looked at the fossil record to compare British vegetation from different interglacials.[51] If the ice comes off Britain early, sea level is low and the English Channel is dry. Hazel (*Corylus*) with its big nuts can colonize quickly, and it becomes a major part of the flora for the next seventy thousand years. But if the ice comes off Britain late, the polar caps are melted and sea level is high. The English Channel, full of water in those times, is a barrier to hazel, which arrives late and is a minor component for that interglacial. Apparently, there is a founder effect that sets the stage for tens of thousands of years. Those species that are already there in Britain in that interglacial maintain themselves with a genetic founder effect. The hazel has to squeeze in later, and its populations will become adapted to that and apparently remain so. One would expect that seventy thousand years is plenty of time for hazel to recover, so a genetic barrier is the most likely cause of its lack of success in some interglacials.

That being said, communities come and go after a few thousand years, depending on how long it has been since the last ice age, how far were the refuges of the various species, and how fast the species can move across the landscape. For instance, the American beech only made it just across Lake Michigan, and occurs only in the eastern counties of Wisconsin, where it is successfully a major component of the maple-beech forests. Its migration depends on exquisite timing, in that at the end of the last ice age, Lake Michigan was a stream.[52] The rest of the state is free of beech.

In another example, hemlock (*Tsuga canadensis*) is a relative newcomer to the northern parts of Wisconsin, only in the last few thousand years. However, in the unglaciated southwest of Wisconsin, it appears to have survived as relics on north-facing cliffs made of porous sedimentary rocks that let cold air through to cool the face. The picture of Allen teaching (figure 4.3*B*) was taken at one of those cool gully refuge sites. The conifer branch on Allen's left is hemlock. In the last ice age, the area where Allen is sitting was not glaciated. It was cold just south of the glaciers, just the climate for hemlock. But as the site warmed and glaciers retreated, the hemlock died out everywhere, except in those southern refuge sites. The northern forests of hemlock had to trek all the way back from the southeast, the Appalachians. Potter and colleagues have looked at the genetic diversity of southern refugia, but not the northern hemlock, so the genetics to clinch the deal have not yet been done.[53] But the story is supported by physiological differences between northern and southern populations.[54]

If communities are so ad hoc, it is not reasonable to expect that much deep accommodation exists between species. So what is the usual organizing factor for the regularity of communities that we see? It will be stability of certain configurations. Like a child's LEGO set, the community is made of ready-made pieces. You can make a lot of things out of a LEGO set, but only those whose pieces (species) fit together to make something stable. We see what is stable, and might mistake it for something more dynamically organized.

5

THE ORGANISM CRITERION

O F ALL the criteria, the organism is the most tangible at a human scale. That tangibility gives powerful insights, but is a double-edged sword. This chapter observes the blade cutting both ways. The conventional view of organisms surrenders itself to organismal tangibility, emphasizing the most readily apparent aspects of organisms. Behind this openness to an organism as a self-evident thing is a veiled anthropomorphism. There are good biological reasons why humans can recognize each other; natural selection of our own species has put a premium on cooperation within our species. An intelligent caution when dealing with human strangers seems a helpful character for us all to have.[1] Our own species resonates with our personal selves, which makes us vulnerable to being probed by other humans. We are open to other humans, at the very least as sources of human diseases. Members of our own species are among the most tangible parts of our experience. In emphasizing tangibility of organisms, we take full advantage of our facility to identify our own kind. The perfect example of an organism is embodied in ourselves.

This chapter follows the same plan that has been used in describing other ecological criteria. The organism needs to be considered as just one criterion among many for looking at ecological systems. First, we lay out the primary characteristics that make the organism distinctive (see figure 1.15A). Then we put the organism in its environment (see figure 1.15B). There follows a strictly structural account of organisms as mechanical systems. Plants and animals represent two different ways of being an organism, and we therefore include a section contrasting them in terms of scale. The genetic basis of organisms makes them a most distinctive type of entity. The organism concept links ecology to genetics; accordingly, this chapter ends with a discussion on genetics and evolution.

One of the best examples we know of for putting the organism in its environment occurs at the International Crane Foundation (ICF) near Baraboo, Wisconsin.[2] It is a conservation program founded in 1973 by Ron Sauey and George Archibald. Much as Grime sticks largely to the flora of Britain because it is manageable, the Crane Foundation covers cranes because the genus is globally distributed but is represented by a fairly small number of species.[3] There are just fifteen species worldwide and the ICF has them all. The ICF website displays a map where a click on an image of the bird shows where that species is found. The species occur in tropical and temperate climes, with North American, European, Siberian, Chinese, Indian, and Australian species. The ICF has a captive breeding program. Although situated on 225 acres, the program is necessarily limited. It is highly organized. In order to conserve all crane species, some of the rarest species are not bred in some years so that the resources can support all species in programmatic fashion (figure 5.1).

Cranes are a wonderful device for conservation because they are very charismatic. All the species have a magnificent presence. Beyond conserving the species, the ICF uses the crane as an instrument for restoring ecosystem function. Each

FIGURE 5.1A.   The International Crane Foundation has a remarkable collection of rare cranes; it includes all the species in the world. The whooping cranes are as rare as any, displayed here to appear in a natural setting for visitors. (Photograph by the International Crane Foundation, reproduced for publication with permission.)

FIGURE 5.1 *B*.   Many cranes are bred in the International Crane Foundation's crane village, from which the public is excluded. Their breeding program follows a particular plan, wherein only some species are bred in a given year. The species that breed change over time, as the plan prescribes. This might even involve sacrificing the eggs of rare species to have the adults raise the eggs of another equally or more rare species, the species that is part of the breeding plan for that year. The plan involves many aspects of maintaining the population, such that all species reproducing all the time has been identified as a suboptimal use of resources. Saving the rare species involves counterintuitive decisions for a complex management system wherein full-time maximization of reproduction is not the best plan. (Photograph by the International Crane Foundation, reproduced for publication with permission.)

crane species has its specific relationship to its environment. The ICF uses cranes in a sort of reverse engineering of ecosystem function. If the cranes are doing well, that is an indication that the system in which they live is properly functional. For instance, the wattled crane is found mainly in Zambia and is one of the species most dependent on wetland quality. The ICF works with governments and nongovernment agencies. They are remarkably effective, building pride in the indigenous people for the conservation effort. Southern Africa is densely populated, but the wattled crane is coming back, and is an indication that the crane is working as a lever to improve human husbandry of the river system and its estuary.

We take a process-oriented view that defines the organism as a suite of parts responding to external stimuli. An interesting point of tension arises here, for the interplay between the organism and its context is an excellent way of defining the organism, quite apart from the role it plays in the larger system. In a sense, the organism is not captured when we see it as a collection of parts or internal processes, but it is found when the organism becomes a whole that is the interface between the parts and the context. The way the organism acts on its environment in response to stimuli is the sum of its internal functioning. Arthur Koestler coined the term "Holon" and recognized it to be "Janus-faced" to emphasize the part/whole duality raised here. We take care to avoid the naive stimulus-response of the behaviorists in psychology. Koestler uses the difficulties with the Skinner box as a way of unpacking the notion of Holon. It is not a linear input/output device.

As a counterpoint to a process-oriented description of the organism, we go on to look at the organism in strictly physical structural terms. This leads very comfortably to a consideration of the organism criterion at different scales. Many researchers have looked up and down the organism column in our grand conceptual scheme, and we end with reports of their findings.

REIFYING THE ORGANISM

The archetypal organism is human, and other organisms variously represent departures from ourselves, roughly in the following order: cuddly and childlike; warm but big; scaly and cold; immobile and green; and finally, microscopic. The further away from being human is the organism in question, the less the more formal attributes of organisms apply. To abstract the general characteristics of what makes an organism an organism is the following set of three attributes: (1) a genetic integrity that reliably coincides with (2) a physical discreteness in space. An attribute that makes organisms whole is (3) physiological coherence. Organisms need to have at least one of those characteristics, and preferably more, to qualify as organisms. Plants reproducing by vegetative means often display strong interconnections between parent and offspring, so the discreteness characteristic breaks down in plants. These problems are not restricted to plants, for the branching colonies of the simplest multicellular animals, bryozoans, and coelenterates, (figure 5.2) present a similar ambiguous condition. Some sponges can be put through a sieve and still reaggregate to form a whole sponge again. Is one dealing with an organism throughout that process? Clearly there is cause for some equivocation no matter what the answer.

If we are correct in our surmise that the organism is an anthropomorphic construct, it should be no surprise that goal seeking is very much part of the organism concept. This is not intrinsically an unscientific view for, as Erwin Schrödinger suggested, biology without purposiveness is meaningless.[4] For example, the notion of pathology depends on a role and purpose that is not met in the diseased organism. Evolution by natural selection has amplified an explicit functional directedness in organisms. Organisms do things for a reason. However, focus on purpose and consciousness unduly segregates the organism from the other critical ecological entities.

FIGURE 5.2.   The parts (A) of and the whole colony (B) of *Membranipora membranacea*, a bryozoan (photos courtesy of D. Padilla). Primitive animals bud and branch in a manner similar to plants; the branches can break off and become unambiguous separate organisms. The branching colonies are altogether more ambiguous, straddling the line between organism and population.

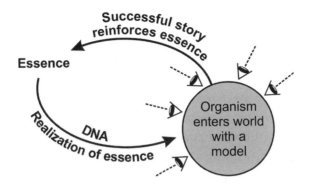

FIGURE 5.3.   The relationship of the essence to the realization. The essence might be a species. The essence gives the realization information about things the organism has never experienced.

Landscapes, ecosystems, communities, and populations do not seem to manifest centers within them for active planning and preemption, but that should not preclude us from using analogy to compare the organism with other ecological entities, as long as we remember it is an analogy and take responsibility for taking it right through.

The purposiveness of organisms comes from the flow of information between levels, and that makes organisms in particular complex systems. Purpose introduces a duality between coded information and matter/energy flux. A move to rescale in a system described in terms of thermodynamics will change the situation. J. B. S. Haldane made this point well when he spoke of a wonder horse that was enormously powerful because it was twice the size of a normal horse.[5] He then painted a picture of the beast lying down, panting and sweating, and unable to exert itself because it was too large to dissipate heat. It has a surface-to-volume ratio that does not work. In contrast to rescaling a thermodynamic system, if you rescale the delivery and application of information, it moves easily and largely unchanged. A mainframe computer can communicate effortlessly with the smallest handheld device. The diffuse Internet informs your telephone easily, once you have bought the "app." And so it is with organisms. They receive information across vast scale differences.

Because information can be rescaled without change or compromise, a whole species that has existed at least several millennia can give the benefit of experience to its realizations of organisms that are small and relatively short-lived. The species would be the essence of its realized examples, organisms (figure 5.3). If the reader is uneasy with the notion of essence, we promise ample treatment in chapter 8 on ecological narratives and models. The notion of purpose in an organism has no mystery. It comes from its species simply learning from experience. Since that information can be embodied in some small temporary thing, like an organism, then purpose suddenly looms large. A property of organisms is that they anticipate and meet expectations. With evolution by natural selection, there is a taboo among biologists that we can never say evolution aims purposefully into the future. Evolution comes from the past: it is not going anywhere in particular. What makes the organism an appropriate object for endowment with purpose and anticipation is that it know things of which it has no experience borrowed from its species. The organism is coded in an adaptive way that uses the experience of its species as a predictor for what happens in the organism's life.

The organism has no business knowing how to predict its environment because it, in itself, has had insufficient experience. This creates a sort of wrinkle in time, because the organism embodies things that were learned before it existed. The change of scale at which information is applied forces the strange concept of purpose to the fore as something that cannot be avoided. Teleology cannot be dismissed as only a story because stories are matters of substantial belief and consequence. The molecules in your body do not know their purpose or intention, but you do, and so does your dog. Looking at organisms as objects of investigation, they cannot be understood without the notion of purpose and significance. That is easy to see in human sentience, but the same applies to the adaptations manifest in the lowliest of organisms. Organisms embody purpose.

All sorts of modern findings challenge the status of the organism as an absolute. Certainly, there are abundant data that indicate the eukaryotic organism is an arbitrary assemblage, a collection of other organisms of the prokaryotic type. Thus, even the archetypal organism, the human creature, is an aggregate of other organisms at the cellular level. The mitochondrion and chloroplast each has its own genome distinct from that of the nucleus. This puts some tension in one of the key characteristics of organisms, the eukaryotic organism's genetic integrity. In his *Lives of a Cell*, Lewis Thomas said he felt as if he was tingling when he remembered the mitochondria in his cells.

Bacteria are prokaryotes, very different at a cellular level from eukaryote plants, protists, animals and fungi. As to information in eukaryotic organism systems, genes move between host cells and prokaryotes captured long ago by endosymbiosis. Mitochondria are organelles that perform respiration inside all eukaryotes. They derive from free-living aerobic bacteria. Having been co-opted by endosymbiosis into eukaryotic cells, most of the captured bacteria's genetic material has been sequestered by the nucleus of the host cell. That is a big difference between mitochondria and the prokaryotes from which they derived. Eukaryotes are large at both cellular and organismal levels. Bacteria are of limited size because of the essential problem of moving hydrogen ions in the energetics of the bacterial cell. Bacteria do not have a bounded nucleus, but they do have a mass of genetic material, just one in the middle of the cell. If the prokaryotic cell gets too big, the sites in the cell where the mechanics of respiration and other functions are performed are too far from the genetic material with instructions on how to do it. It is a thermodynamic issue of information carriers not being able to move fast enough across a larger cell. But eukaryotic cells have no problem there. Most of the information in the mitochondrion has gone to the nucleus where it is part of executive function, but some genes stay dispersed in the population of organelles. The mitochondrion keeps the genetic instructions for actually doing its job, respiration. There are bacteria that are large by bacterial standards, and they have dispersed multiple genomes, but it is many copies of the full genome. The boost that eukaryotes get is multiple copies of a greatly reduced genome. So the genetics is economically distributed through the cell at the sites where respiration is done, no matter how big are the eukaryotic cells.[6] Codes have no trouble meaning the same thing at different scales, but the rate-dependent mechanics of respiration will fail if the prokaryote is too big.

The same happens in business when firms are taken over. Warren Buffet's Berkshire Hathaway took over Dairy Queen in 1997. Whether or not Dairy Queen expands in Brazil will be decided by the central administration at Berkshire Hathaway. But for the most part, the day-to-day decisions are made by Dairy Queen, which is still a quasi-autonomous organization. In biology there is sometimes a double endosymbiosis. It appears that some red algae (eukaryotic cells with organelles) were themselves engulfed by green plants, giving multiple membranes. Normally there are just two membranes around a mitochondrion, one for the bacterial origin and another for the vacuole that engulfed it. With multiple endosymbioses, there are more membranes left over, two for each time something was engulfed. The same double endosymbiosis occurred in Dairy Queen, which was one of the first firms to franchise in the first half of the twentieth century. Dairy Queen is itself a mass of franchises; the whole Dairy Queen entity had engulfed individual franchisees before it was engulfed by Berkshire Hathaway. The resulting hierarchical organization is particularly efficient.

Lichens present a similar problem of multiple origins, but since they are clear symbionts, one could assert that they do not count as organisms proper. Lichens are algae held within a fungal matrix. However, the discovery of promiscuous DNA calls into question the genetic integrity of many living things that had been good organisms heretofore. It has even been suggested that land plants are better seen as reverse-phase lichens, with the formerly algal matrix supporting a secondary fungal partner. The nucleus of the vascular plant cell is suggested, in this modern speculation, to be an amalgam of algal and fungal genetic material with fungal cellular morphology emerging in certain specialized cells, like the germinating pollen grain.[7] The genetic integrity of the organism slips away from us, and organisms begin to look more like happenstance collections (figure 5.4).

To avoid confusion, we treat the organism as a heuristic. Certainly, the organism is a special entity, but the manner in which it is special does not stop us from asserting that the organism is a convenience. For this argument, consider another special but arbitrary entity, the species. As in the organism, there is also a veiled anthropocentrism in the species concept. The perfect species breeds freely among species members, but suffers significant infertility in the parents or the offspring when mating between species. Furthermore, there is significant morphological homogeneity inside species and consistent morphological differences between

FIGURE 5.4.   Lichens on a tombstone. Lichens are an amalgam of fungi and algae. They take on discrete bounded form, with the fungus providing a context in which algae grow. (Photo courtesy of J. Will-Wolf and S. Will-Wolf.)

species. Zoologists, and vertebrate zoologists at that, feel comfortable identifying the species as being the only taxon that is other than arbitrary. Botanists know better: hybrid swarms, clones, and species with inconsistent chromosome numbers all make the species as arbitrary as the genus, family, or order.[8] Nevertheless, the species is the foundation of the whole taxonomic system in that the species name is overwhelmingly the one given to most organisms in most discussions. The name of an organism is its Latin binomial down to species.

The species is special, but just like the organism it is not so special that it deserves to be reified.[9] Rather, it represents the level of variability that humans can readily tolerate in a collection. The distinctiveness of species is in us, not in nature; the same goes for organisms. We have a large vested interest in organisms because we fit the criteria for the most typical of organisms ourselves. If the organism is based in nature, it resides in human nature.

EMERGENT PROPERTIES IN ORGANISMS

Although the organism concept is not helped by reification, organisms are robust observables with emergent properties. The coincident mapping of distinct characteristics onto a single structure makes those characteristics emergent properties. Accordingly, emergent properties of organisms are: (1) a certain physical discreteness; (2) genetic homogeneity; (3) recognizable physiological subsystems that perform various service functions like circulating resources; (4) coordination of parts, even in the most simple organisms; (5) irritability or response to outside stimuli; and (6) reproduction with a certain genetic consistency.

The last in the list is what make individuals the favorites of evolutionists as the default level of selection. The first two characteristics map onto most organisms, particularly mammals. In many species, an individual is apparently very capable of recognizing self from not self. This is particularly true for placentals, like us, where the act of reproduction nibbles very close to overt, deleterious parasitism. An entopic pregnancy in the fallopian tube causes fatal parasitism. The control systems of internal housekeeping have to be unambiguously separated from those of defense against invasion in placental reproduction. Even so, in the case of rhesus-positive blood groups, confusion arises. If the rhesus-negative mother recognizes the blood of a rhesus-positive as alien, her immune system attacks the trace of that blood in her blood stream. The antibodies enter the baby and attack its blood. This is worse for second rhesus-positive babies because the mother's system says, "I've seen this infection before."

The self/not self edge in vertebrate organisms is set up to be sensitive in the ecology of early fish with no jaws, the Agnatha. Lampreys represent them today. There is much danger in that ecology of lamprey parasitizing other lamprey instead of prey species. Parasites often have to deceive the host that the invader is "self." In the movie *The Empire Strikes Back* of the *Star Wars* series, the reader may remember that Han Solo landed his spaceship on the larger ship so as to escape into the self of the pursuing enemy ship. The British spy ring, headed by Kim Philby in the early 1960s, was all from establishment private schools (called public schools in Britain) and Cambridge University men.[10] In that way, they appeared as "self" to British intelligence; it took the Americans to spot them. Cancer is difficult to treat because it is of the very self it is killing. Not all the things we call organisms have all the properties of the ideal organism; for example, some creatures lack a convincing structural separateness from their fellows. Nevertheless, things that meet more than one of the standards for being considered an organism can be helpful if not perfect examples. Not all the entities we consider organisms will be good by all standards. The authors are concerned with the ecological significance of the organism criterion in general terms, rather than a pursuit of the perfect definition of an organism or the perfect entity that fits such a definition. Scientists should get definitions to work for them. Too often the elaborate vocabulary encourages biologists to fixate on definitions, so the scientist ends up working for the terminology.

THE CRITICAL SUBSYSTEMS

To provide some redress for an overemphasis on the structural aspects of the tangible organism, let us proceed with a process-oriented approach. James Miller has suggested, in his *Living Systems*, nineteen subsystems that perform various analogous roles across a range of highly organized, scaled entities from the cell to the international global political system (table 5.1). In Miller's scheme, organisms seem to be the archetypal biotic system, and the nineteen analogies appear to be mostly based on recognizable things that organisms do. Its appeal is that it is a distinctly process-organized approach, which identifies the critical things that need to be done.

The analogy to social systems explains why something as integrated as the bloodstream is split into two subsystems by Miller. One description of the bloodstream subsystem is for a pattern for delivering energy in the form of sugar, and the oxygen to use it. The other description is for blood flow that delivers hormones as part of information transfer. Miller gives those two aspects of the bloodstream each its own name, the "distributor" for moving sugar, as opposed to the "channel and net" for moving the information in hormones. In social systems at Miller's higher levels, the two sorts of job appear on their face to invoke two different systems, and would never be seen as equivalent. The separation would be of railway lines for delivering coal from the telephone system for moving information. In the body, the same physical system—the blood—does both jobs, so Miller gives the blood two names. This lines up with his major categories of living system properties. One list is of structural and material subsystems, whereas the other list processes information. That is how we have tabulated the generalized properties mentioned earlier, exactly as did Miller himself.

Management systems appear particularly well suited to Miller's scheme by virtue of the human component that decides management action. The management hierarchy goes as high as the problems of international treaties addressing acid rain, as well as Amazonian deforestation that sits at the door of the World Bank and its policies. It is less clear where we might find the decision-making centers in communities, ecosystems, landscapes, and biomes. While useful enough, the organism analogy can be pressed beyond advantage in ecological systems at large.

ORGANISMS AS PERCEIVING
AND RESPONDING WHOLES

For all its detailed process- and information-based emphasis, Miller's scheme only tells a part of the story of organisms. It is distinctly part-oriented and barely addresses the way the organism relates as a cohesive whole to its environment. A much older work was originally published in German by Jacob von Uexkull and contained the

TABLE 5.1

These subsystems are of a general type, although they are particularly applicable to organisms. Miller applies them to cells, organs, organisms, and four levels of societal systems up to international politics.

1. *Reproducer* creates new versions of the system to which it belongs.

2. *Boundary* contains the system and protects it from its environment, permitting only some inputs.

Matter-Energy Processors

3. *Ingestor* takes in material: mouth, seaport.

4. *Distributor* moves material around inside the system: blood delivers sugar, or road system.

5. *Converter* changes inputs of matter-energy into more useful forms: teeth, sawmill.

6. *Producer* creates associations of materials used for repair and growth: leaf, industry.

7. *Matter-energy storage:* fat, warehouses.

8. *Extruder* removes waste: urinary tract, sewage works, and trucks dumping in landfills.

9. *Motor* moves the whole or the parts; muscles, trucks.

10. *Supporter* maintains spatial relationships: skeleton, buildings, or the landscape.

Information-Processing Subsystems

11. *Input transducer* senses the environment: skin nerve endings, satellite dishes.

12. *Internal transducer* transfers signals inside the system, often to a different medium: synapses, glands, or telephone exchanges.

13. *Channel and net* moves information around inside the system: blood transporting hormones, nerves, or telephone lines.

14. *Decoder* converts inputs into a private internal code: ganglia, foreign news analysts, newspapers.

15. *Associator* performs first-stage learning: short-term memory, intelligence agencies.

16. *Memory* performs long-term learning: reflexes, habit forming, scientific research institutions, libraries.

17. *Decider* is the executive information system: brain, political administrations.

18. *Encoder* translates internal private code into a public code used in the environment: speech centers, translators, speechwriters.

19. *Output transducer* sends out information: voice, policy statements, embassies, radio transmitters.

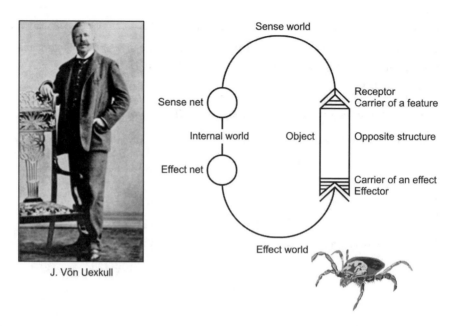

J. Vön Uexkull

FIGURE 5.5. Von Uexkull, a wood tick, and its umwelt, which is very austere. There is a receptor-sensing arc in the organism, starting with the thing in the world that impinges on the organism. The other half of the cycle is the motor action arc that affects the world through the new stimulus.

term *"umwelt,"* a word not easily translated into English.[11] Von Uexkull's translator used the term "self-world." He was trying to convey the notion of the private world in which each type of organism lives, a world determined by what the organism can detect and how it responds to stimulus. The organism itself defines the functional environment. Anything that the organism does not perceive is not part of the environment in the umwelt. We use "self-world" and "umwelt" interchangeably.

In von Uexkull's scheme, all the complicated input pathways and special subsystems of Miller disappear. In fact, von Uexkull's organism becomes a straightforward input/output device. While one is cautious to embrace anything resembling the behaviorists' rude stimulus-response models, there is a certain appeal to von Uexkull's perspective, as a rescaling of the organism so that it functions as a unified whole. The example he gives, the wood tick, is enlightening (figure 5.5). It indicates a much greater sophistication in scaling than psychological stimulus-response models, for which it might be mistaken. The female wood tick is modeled by von Uexkull as being a perceptual arc joined to a motor arc. Connecting these two arcs on the outside, making a circle of response, is the object in the environment on which the tick operates. This is a cleverly simple scheme, for a track once around this circle is the phenomenon of irritability, the same phenomenon that required at least seven of Miller's subsystems. The tick hangs in a tree just above the height of its prey. The first stimulus is butyric acid emanating from the skin glands of the mammal. The tick responds by dropping from the tree. The second stimulus is the shock of hitting the hairs of the mammal. This stimulates a response of running around. If the tick lands on something cold, she has missed her prey and must climb back up the tree. If it runs around and finds a warm membrane, the heat stimulus causes it to start its piercing response. The tick has no sense of taste and will puncture any soft membrane and consume any fluid of the right temperature. It will puncture rubber and suck glycerin, if that is what the scientist offers.

Of the many signals that could be taken to indicate the presence of a mammal prey item, only three are read, and then in a particular sequence. Only three signals stand out like beacons to guide her to her prey. It is the very paucity of signals in the tick's self-world that allows her to make the unerring response. True, there is some luck involved in finding the prey, but another critical factor comes to the tick's aid, time. Von Uexkull reports that ticks have been starved for eighteen years and still survive. The metabolism of the tick has been scaled almost beyond comprehension, so that its self-world works.

In a charming walk through the worlds of animals from insects to humans, von Uexkull repeatedly points out the striking differences in the perceptions important to the organism. He shows how external cues readily noted by us, such as a fertile female insect under a soundproof cover, pass unheeded by males who one might have thought would have an interest. Instead, they flock around a loudspeaker making the noise of a female insect. The speaker is carrying the appropriate sexual signal in their umwelt. Von Uexkull reports a mother hen continuing to peck unbothered when her distressed chick is in sight but cannot be heard. If she can hear but not see the distressed chick, she tries to come to its aid. It depends on whether the right or wrong cue is given. Von Uexkull shows how slow organisms like starfish ignore fast cues. He has a set of three figures of the same room where he shades what is of interest in the respective umwelt. Humans see cutlery and books, while the dog only sees food and a couch to sit on. An insect only sees the lightbulb over the table.

The important point in all this is that organisms are deeply context-dependent. Von Uexkull's organisms extend beyond their tangible bodies to become one with their context. By looking at organisms this way, we see beyond the physical, tangible body and observe a new entity, the scaled self-world that is often only fleetingly detectable by humans. We cannot see the organism's umwelt, although it is probably more important for our understanding than either the tangible creature itself or a full accounting of Miller's subsystems inside it. Von Uexkull presses on us a rescaling of organisms that gives a glimpse of the relationship of the organism criterion to other ecological criteria.

In general, small organisms have a small, spatially defined self-world, although a bird views a wider landscape than does a large mammal. It is the size of the umwelt of the organism, relative to the size of our own, that determines which organisms are used for which type of ecological investigation. That is why birds are used for island biogeographic studies.[12] Their self-world is larger than our own, so we use them disproportionately often for studying the ecology of going to and from places. The organism's occupancy of the landscape is determined by the scale and texture of its umwelt.

When it comes to population ecology, umwelts are important. When a species appears to be in danger of extinction, it is usually the case that its umwelt has been in trouble for some time. The problem may be that the environment has changed. Sometimes the change is for the better. The "earthquake bird" in the title of Shugart's book about unbalanced systems is noteworthy because its numbers greatly increased with the great New Madrid quake of 1812. The event had influence over a million square miles, centered just south of St. Louis. The disturbance changed the landscape, creating scrubland and causing areas of swamp. That fed right into the umwelt of Bachman's warblers. It is called the "earthquake bird" because it became common in the nineteenth century where the earthquake happened. The new landscape was perfect. Since that time, the landscape has recovered through natural processes of swamps filling in, but also because of intense Euro-American settlement. The last recorded finding of its nest was in 1937. Most of Shugart's chapters refer to animals that are either suddenly abundant or recently extinct. In all those cases, it is that the animal's umwelt either started to resonate or not resonate because the environment was different. A threshold is crossed and the population either booms or busts. The ivory-billed woodpecker is the star of Shugart's first chapter called, "The Woodpecker That Was Too Picky." Experts are divided as to whether or not it was rediscovered in the last decade, but it was always rare, as stated in our community chapter's discussion of the gap phase (chapter 4). Its umwelt only recognized a resource that took centuries to build and was available for only four years. Its purported extinction will have been caused by it being a species on the margin already, having its habitat drained and destroyed by human intrusion. In his later chapter on island invasion, Shugart refers to the great moa birds, whose size led them to misperceive the danger of humans. The week of this very writing, Lonesome George, the last of his variety of giant tortoise of the Pinta Island variety, died. All the giant tortoises of the Galapagos Islands have been driven close to extinction because their umwelt of "danger: retreat into shell" simply did not work, as sailors preyed on them for meat.

The California condor also had an umwelt mismatched with humans. Other birds on the brink have had some but not all their last few members taken into hatchery programs, with some left free to teach the umwelt to the hatchlings. But the condor had a wrong umwelt that needed changing; so all twenty-two were captured in 1987 and bred.[13] Unafraid of humans and their devices, condors had trouble with landing on corrugated iron roofs, hitting power lines and encountering traffic. So the breeding program fed them via humans in condor costumes so the chicks would not see humans as friendly. All encounters with humans looking human were engineered to be upsetting. With their new umwelt of "humans are nasty and dangerous," the released birds are now doing well. We started this chapter with the International Crane Foundation. They used ultralight airplanes piloted by crane costumed conservationist to lead juveniles from Wisconsin to southern migration sites. That migration umwelt not only stuck but also has been improved by whooping cranes, who have remembered and then improved upon the self-world humans resurrected for them.

Just because an animal has an ecology that generates huge populations, it is not safe when its umwelt becomes inappropriate. The last passenger pigeon died September 14, 1914, in the Cincinnati Zoo. In the last half of the nineteenth century, flocks of the bird darkened the sky. A paper that is a favorite of Robert May showed in mathematical descriptions why it is distinctly possible that common species may be some of the most vulnerable to extinction.[14] Those authors did not discuss umwelts, but we can see how one out of step can reduce numbers drastically.

If we want to understand the ecology of organisms, we need to know their umwelt as much as their physical form. Some are surprising. Southern burrowing owls have moved into the Eglin Air Force range. They have been there since 1985, and recently reached some fifty-five in number.[15] Apparently, the dangers of low-flying jets and bombs do not enter their umwelt. The bombs are not aimed at the owls and they seem to know that. The good thing is that unexploded ordinance keeps humans out. The absence of humans is apparently the big plus in the owl way of looking at the world.

Beavers influence ecosystems as they build dams. Part of the beaver's umwelt is to recognize places where a dam could be built. They also construct dams with the expectation they will be washed out. That is why blowing up unwanted dams fails to discourage beavers. The animals eagerly put them back. The trick that does work is to put a tube in through the dam and extend it out toward the middle of the pool. Tubes are not in the beaver's umwelt and so they do not know what to do about it.

BASIC STRUCTURAL UNITS IN ORGANISMS

We have worked through a process-oriented approach to organisms, both internally with Miller and externally with von Uexkull. There is, however, another side to organisms concerning their physical structure. We turn to counterintuitive, tension-compression structures of architects like Buckminster Fuller. This class of structures has stimulated Stephen Levin, "a recovering orthopedic surgeon," as he would put it, to see organisms in a different light. Following is a discussion of Levin's model.[16]

Feeling uneasy about the conventional model of his colleagues for neck and back problems, Dr. Levin went to the Natural History Museum in Washington, D.C., to see if looking at dinosaur skeletons could help. He left the exhibit troubled by the unlikeliness of a lever-girder model for a dinosaur's neck and pensively strolled across the mall. Then, minutes after looking in puzzlement at the neck of a great extinct monster, he was in front of the Hirshhorn Museum, where he encountered the breathtaking tension-compression sculpture of Kenneth Snelson. Like Newton's apple, the statue changed Levin's view so he would never see organisms and skeletons the same again. The statue, "The Needle Tower," is just twenty by seventeen feet at its base, but is as high as a four-story building. It is essentially rigid and consists of cables and light girders. The remarkable feature of the design is that the girders do not touch each other, being instead held in a helical tower by a web of cables. The statue is neither suspended from anything rigid above, nor is it a stack of compression members sitting on top of each other.

From this experience, Dr. Levin saw the underlying model for not only backs and necks, but also other biological structures including knees, cell packing, and bacteriophage viruses. Levin suggests the icosahedron as the basic unit of construction in organisms. When he says basic, he means the way that small icosahedra pack to give a larger icosahedron ad infinitum. This nesting allows the form to be basic at many scales and levels of construction. It is fractal. An icosahedron is a mathematically regular solid with twenty triangular faces. Triangles are important because they are the only two-dimensional forms that are not deformed by pressure as long as the sides remain connected, straight, and the same length; a rectangle, for example, deforms to a parallelogram (figure 5.6).

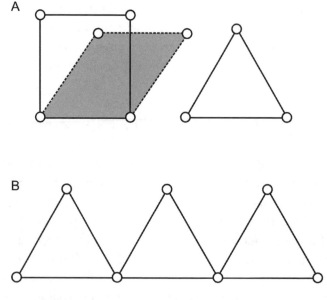

FIGURE 5.6.   *A.* Triangles are the only two-dimensional structures that do not deform under pressure. Rectangles, for example, deform to parallelograms. *B.* Airplane trusses may be readily made from triangles.

The icosahedron is made of triangles put together. It has thirty edges and twelve corners (figure 5.7A–C). If the edges are rigid, then pressure at any point transmits around the thirty rigid surface members, placing some under pressure and others under tension, in a regular pattern. The twelve corners each have five edges that come together at a point, which is the corner. If all edge members are rigid, some of these edges are under compression while others are under tension, depending on the position of the corner relative to the loading of the whole icosahedron. But it is possible to transfer all compression away from the outside of the structure.

The trick is to connect each of the twelve corners to its opposite number with one of six new compression members spanning the middle of the icosahedron. Opposite corners push each other away from the center in opposite directions, working through the new rods in the middle. The "opposite" corners are not strictly opposite, so the connecting rods are slightly eccentric and slip tangentially past the center and each other without touching. These compression members push the corners away from each other. In this way, all edges joining the corners become a tension member; that does not sound astonishing in itself, but it creates counter-intuitive situations.

With the edges universally under tension, remarkably all thirty can be replaced with cord or cable. Together, these tension edges hold apart the six compression members spanning the middle. The tension icosahedron, as it is now called, is a rigid structure, but the compression units float, suspended by the skin of edges, all under tension. Pressure applied to this structure causes an increase in tension around all the edges almost uniformly, and this distributes the compression load around all six of the suspended compression members evenly (figure 5.7D).

The tension icosahedron has counterintuitive deformation characteristics. Pressure applies tension all around the surface. As one presses down on the form against a surface, instead of flattening like a pancake, it grows smaller and becomes more compact. As one tries to pull the tension icosahedron apart from opposite sides, instead of stretching into a sausage, the whole thing just becomes bigger; that is, not only does it become longer between the poles under tension but it also becomes proportionally fatter around the girth and so keeps its shape.

The relationship between stress (pressure) and strain (deformation) in the tension icosahedron can be graphed as a curve (figure 5.8). That curve is radically different from the same relationship in Hookean girder-lever structures (constructions where compression members press against each other, as in office blocks or toy model building kits). Hookean material deforms at a constant rate with increases in stress until there is a short period of slightly faster deformation, immediately preceding system failure. In contrast, the tension icosahedron resists deformation with the first application of stress, but soon deforms considerably. This rapid deformation is not, however, an indication of incipient system failure, for after a short period of significant distortion with increasing stress, deformation

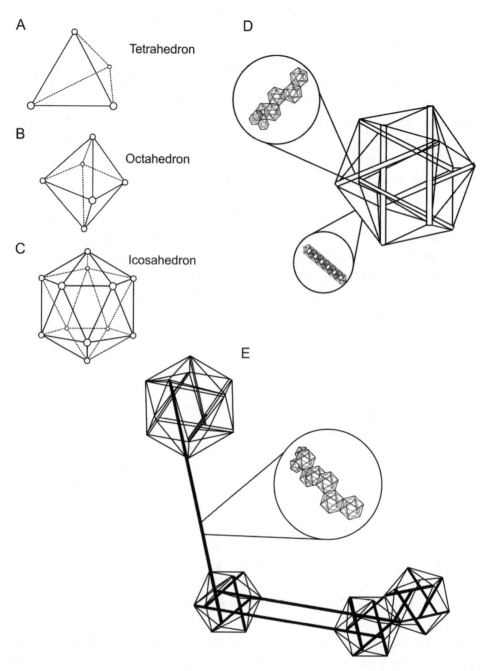

FIGURE 5.7. The first three solid objects (*A, B, C*) are made from triangular faces. Others are larger, and in chemistry, such carbon structures are called fullerenes after Buckminster Fuller. The right icosahedron (*D*) has compression members between opposite points. There are twelve points and therefore six compression members. Since the points are not quite opposite, the compression rods pass each other without touching. In this arrangement, all the sides of the triangles become tension members, which hold compression members floating in a web of tension. Higher-order rod structures can be made from icosahedra or tetrahedra. *E*. Icosahedra can be fitted into spiral twisting columns that become compression tension beams. Strain on any part is transferred evenly to all parts and thus it is not limited by the normal constraints of beam construction. Two levels of icosahedra are used to model the human arm (Levin 1986).

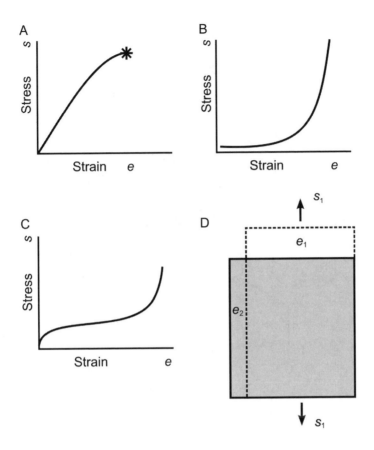

FIGURE 5.8.   A. Typical stress-strain curve of a girder/lever Hookean material with compression members touching. *B.* Stress-strain for typical animal tissue and icosahedron. *C.* Stress-strain curve for typical rubber. *D.* When a solid is stretched by a tensile stress s_1, it extends in a direction of s_1 by primary strain e_1, but also contracts laterally by a secondary strain e_2 (Levin 1986).

essentially stops. A lot more stress causes very little further change in form. The tension icosahedron is very tough at high stress because it distributes the load evenly around the whole system. Rubber and animal tissue have stress-to-strain relationships like that of an icosahedron, not the linear stress-to-strain relationship of a Hookean girder system.

The tension icosahedron can form hierarchically nested structures. First, consider a three-dimensional increase in size. An icosahedron is made from packing spheres around an empty space and connecting their centers with straight lines (figures 5.7 and 5.9). The icosahedron approximates a sphere, and thus icosahedra formed from twelve subunits can themselves be packed in dozens to form yet a higher-order icosahedron ad infinitum. Plant parenchyma cells pack together in this manner to make tissues for larger units like organs.

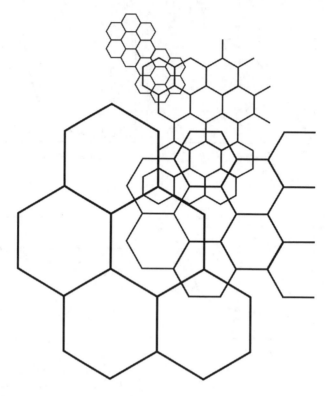

FIGURE 5.9.   Icosahedra can pack to ever-higher levels (Levin 1986).

Now consider a one-dimensional aggregation to form a rod. Beyond packing to make increasingly higher-order solids, it is possible for icosahedra to sit on top of each other to form a column with a spiral twist to it (figure 5.9). An icosahedron connected into the column fits by triangular faces that are not quite opposite. Think of the separate icosahedra as holding these two triangles apart by tension rather than compression. A column of icosahedra is a column of these triangular interfaces held apart by tension. Therefore, the column is not a conventional rod but a tension-compression system that transmits stress throughout the "beam" evenly through the web of tension.

The remarkable effect of building higher-order structures from tension units is to make the strain within the aggregate units independent of gravity. Stand an icosahedron tower on one end and it suffers the same stresses and shows the same strain as if one stood the tower upside down. Snelson's "Needle Tower" could stand just as well on its top; it is gravity independent. This is because the pressure on the base, whichever end it may be, is transmitted throughout the whole tower. That explains how gymnasts can stand on their hands with no ill effect. It also explains how we can stand on our feet, which is also by no means a trivial exercise.

The living bone units in the foot are far too small and soft to bear the load if the structure were the skeleton of Hookean girder construction. Their load lightens because the foot is a set of tension icosahedra.

It seems that joints in vertebrate skeletons are not simple levers, despite the conventional wisdom. It is well known among specialists in the knee joint that it functions like those children's toys where three strands hold a string of tiles together. The tiles apparently form a single ribbon except that, counterintuitively, simply folding the ribbon allows tiles at the fold to flip over and make a branch off the main axis. The analogy is valid, and the knee does flip back and forth in that fashion (figure 5.10). The orthodox model of the joint as a simple lever with

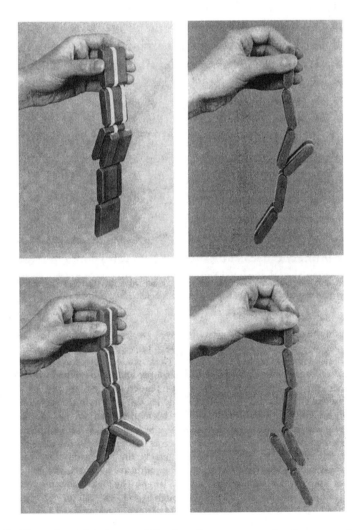

FIGURE 5.10.   The children's toy of flipping tiles operates under the same principles as the knee joint. The sequence shows the stages of the counterintuitive flipping of the tiles. (Photo courtesy of C. Lipke.)

compression across the knee joint has a problem. It is at odds with the other conventional wisdom of the knee as a flip joint. For the knee to work as a flip joint, which it does, it must be under tension, not compression. The knee and the other major joints are tension icosahedra, not simple articulation points between levers. The membranes around the fluid in the knee are not strong enough to act like a cushion air bag in an automobile crash. The synovial fluid cannot be compressed to give support between the shin and the thigh, and yet there is always space between the two bones, even when the knee is loaded. With an icosahedral model, the ligaments take the load. In knee surgery, Dr. Levin would tighten the ligaments, not to pull the joint together, but to pull the bone compression members apart. When he got the tension right, the gap across the knee joint would open, not close.

When orthopedic surgeons have to fuse an ankle damaged beyond repair, they have an unreasonable amount of difficulty getting the bones to knit. The anklebones are soft and removal of the cartilage should result in rapid fusion, but it does not. Since the ankle joint is icosahedral, the bones being pulled apart by the ligaments would explain the problem. Dr. Levin found that cutting the ligaments, which are superfluous after fusion, would collapse the system and cause the desired freezing of the joint.

Back pathology is a matter of weakened tension members in an icosahedral stack. The soft disks between the vertebrae are not compression members, but are shock absorbers for when the icosahedra are exhibiting high-frequency pulses of strain. Weakened tension members slacken the stack, putting the disk under chronic low-grade compression, for which it was not designed. Often, disks rupture when the victim is doing nothing that might lead one to expect strain. The reason is that the primary cause is unfelt chronic compression. From the widespread occurrence of icosahedral structures in nature (figure 5.11), Dr. Levin now knows that the rationale behind even successful surgical procedures is completely wrong. This distracts him from the parts of the surgery requiring focused manual dexterity, and so he has given up surgery. He now treats backs only with less-invasive means. Some of these therapies involve stretching exercises that are focused with precision on the particular tension member responsible for the pathology. The sort of generic stretching prescribed by conventional practice is much more hit-or-miss because it has no rational predictive model.

Ecology has had a mechanistic agenda for some time now. The naïve are looking for the real mechanisms. We can use Levin's ideas to disabuse such hard-line mechanism. We said that joints are not Hookean girder-lever systems, and they are not. So why is it that we look at our arms and see the elbow joint as a lever mechanism? As an exercise, we would ask the reader to sit up in the chair, not using the back to stabilize the torso. Now repeatedly bend and straighten your elbow. Also feel your stomach muscles. You will find that your stomach muscles ripple. The reason is that your arm and body is a web of tension compression. Your elbow

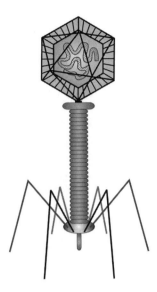

FIGURE 5.11.   A bacteriophage virus with a characteristic icosahedral form, so that form crosses wide ranges of size.

moves changing the tension in the muscles, including your stomach muscles. When a person falls from a height on the heel, it does not break, but the tendons and muscles attached to the fourth lumbar vertebrae are ripped, perhaps to the point of being torn apart. It is that same transmission of tension and compression that moves your stomach.

The elbow cannot be a lever for several reasons. If it was a Hookean system, muscles and tendons would tear, bones would be crushed, and there is not enough energy to allow the body to move. The reason it cannot be a lever is not apparent in conventional discussions of levers because the shoulder is never included in the argument; it is never in the figure. If it were, then the lever argument would be seen on its face as false. First, the bicep does not connect the forearm to the upper arm, it connects to the shoulder. A flexing bicep should lift the whole arm up and forward as if making a rude gesture, but for the triceps that connect the forearm around the elbow to the humerus as well as to the shoulder. There are two friction-less joints between the forearm and the shoulder (they have the friction of an oiled ball bearing). No lever there. Worse still, the shoulder is not one bone, but rather three, also connected with frictionless joints (figure 5.12).

Archimedes said, "Give me a lever, and a place to stand, and I can move the world." In ignoring the shoulder, orthopedic researchers are denying him a place to stand. All muscles in the end are pulling the body down onto the floor. If you have a nonlinear graph, no matter how nonlinear, you can narrow the view to show a part of the graph as linear. It is a local linear approximation. By taking out

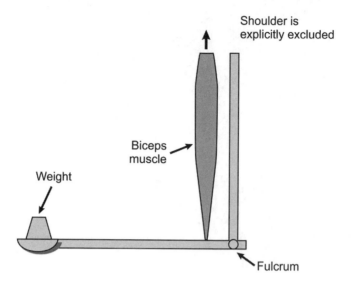

Shoulder is
explicitly excluded

Biceps
muscle

Weight

Fulcrum

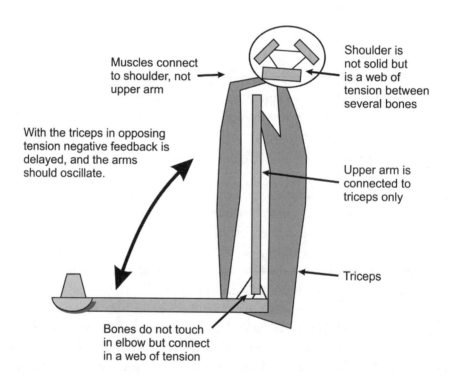

Muscles connect
to shoulder, not
upper arm

Shoulder is
not solid but
is a web of
tension between
several bones

With the triceps in opposing
tension negative feedback is
delayed, and the arms
should oscillate.

Upper arm is
connected to
triceps only

Triceps

Bones do not touch
in elbow but connect
in a web of tension

FIGURE 5.12. The arm as a lever is always represented with the shoulder missing. Levin shows how it is not actually a lever system at all. Between the end insertion points of the biceps muscle, there are two joints with as little friction as lubricated ball bearings. In fact, the arm is a set of related tension/compression icosahedra, as shown in figure 5.7*E*. There is no linear mechanism to it. There is a set of compression members floating in a web of tension.

the shoulder, one is narrowing the view of the elbow so that it appears as a lever when seen superficially. Ecological mechanistic realists are happy to seek "the true mechanism." Mechanisms tend strongly to linearity or at least strong seriality. We would expect less willingness to seek the "true linear approximation," but that is what naïve mechanists are saying.

THE FORM OF PLANTS AND ANIMALS

There is much more to the form that organisms take than just the abstract generalized unit of construction, the icosahedron. Let us turn to some of the principal patterns of difference across the organism criterion.

One of the recurrent ideas in this book is a contrasting of plants and animals. We have done this not to maintain traditional academic demarcation lines between botanists and zoologists, but to allow a proper unified treatment that recognizes the concrete differences in scaling when they occur. Sessile animals do appear plantlike because the environmental experience is similar. Viewed from above, plants possess radial symmetry even as they have a top different from the bottom. With roots at one end and stems at the other, a view from above a tree yields a circular structure. This is in contrast to advanced mobile animals that are bilaterally symmetrical. However, to say that animals are bilaterally symmetrical is to look at the problem the wrong way around. This is because the concept of symmetry recognizes the similarity of the sides rather than the differences between both the front and back and the top and bottom. Those differences are a much more insightful matter. Bilateral symmetry is more interestingly a bipolar departure from radial symmetry.

E. J. H. Corner has made an appealing argument that explains the patterns of symmetry we see.[17] He says that radial symmetry is the default condition for organisms. In radially symmetrical organisms, the same developmental model can be used for development in every direction. This is the form commonly found in phytoplankton that floats in the brightly illuminated surface layer of the ocean. They are so small that the photoic zone is spatially enormous to them and functionally three-dimensional, even though it is only the skin of the ocean. Since the cells rotate as they float, their environment is not polar; light only incidentally comes from above. With no polarity in their environment, there is no stimulus for plankton to be anything other than radially symmetrical (figure 5.13).

The center of the ocean is a nutrient desert, so productivity over most of the ocean is painfully low. This is the answer to Paul Colinvaux's question in the chapter "Why Is the Ocean Blue?" in his book *Why Big Fierce Animals Are Rare*. It should be green, but nutrient status denies that productivity. However, close to land, the waters are not only nutrient-rich, they are saturated with gases. Since

A   B   C

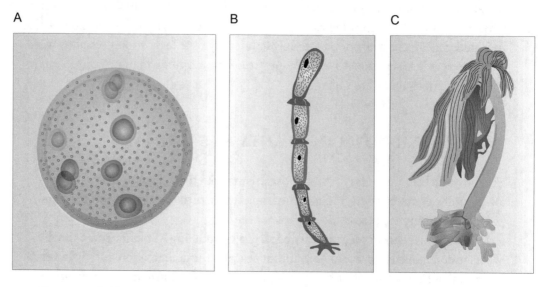

FIGURE 5.13.   Planktonic algae are not anchored and therefore have no commitment to any notion of "up." As soon as a plant attaches to a surface, up and down are polar and the organism becomes polar. As such organisms get bigger, the top is not only farther from the bottom but it also shades the bottom, making the environment more polar. The plant form becomes more polar and the parts more specialized. A. *Volvox* spins. B. *Oedogonium* is simply polar. C. Large brown algae like *Postelsia* have a blade for photosynthesis, a stipe to achieve height, and a holdfast for keeping its spot.

there is a solid, brightly lit bottom close to shore, there is no danger of sinking out of the photoic zone, and every reason for cells to settle on the bottom. Should that happen, the plants find themselves in a polar environment where light comes from above and the substrate is below. Bottom-colonizing plants now live in a two-dimensional environment and thus space becomes a limiting factor. Growth up into the third dimension is the only solution to crowding, so the organisms on a surface become simply polar.

Small organisms interfere with their environment very little. But when they grow bigger, their presence modifies the environment. As they grow bigger, two things happen to make their experience of the environment more polar. First, as they grow further from the plane of the substrate, the top of the plant finds itself in an environment increasingly different from that close to the substrate at the base of the plant, simply by virtue of being increasingly distant from the base. Second, the large upper part of the plant modifies the environment on the plane by shading it. With this increased polar experience, the plants have every reason to differentiate between top and bottom. Increasing size also changes the organisms' surface-to-volume ratio. The way to correct it back again to one where there is enough surface to support the increasing mass of the plant is to change shape by becoming flat. Space being in short supply, the large plant faces three problems that it solves by morphological differentiation. The leaf or frond is the flat structure that increases photosynthetic surface. The stem or stipe lifts the photosynthesizer into the canopy. The root or holdfast keeps the plant in possession of its site.

There are remarkable parallels between algae and land plants across unrelated groups for reasons of the physical structure of the environment. Plants, particularly big ones, are polar because their environment is polar. A tree viewed from above is as radially symmetrical as a plankton cell because there is not significant directional difference across the plane in environmental quality.[18]

Animals also live on the plane and so they too have a top and bottom. The critical difference between plants and animals is the strongly directional movement in pelagic animals. Because the front end arrives in the new place first, there is a temporal polarity to the experience of the moving animal (figure 5.14). The front is the best place to put sense organs, and so they evolved there. The bipolar environment dictates the bipolar animal. A departure from this arrangement is the flounder. It starts life with normal bilateral symmetry, but as it comes to live on the bottom, it lies on its side. The eye that would face down migrates to the other side. It lies on the eyeless side with two eyes pointing up (figure 5.14B). The flounder gives up its last axis of symmetry because its left and right side are now made polar by the axis of up and down. Instead of bilateral symmetry, the flounder has no dimension of symmetry, only three directions of polarity.

The two-dimensional ocean floor is one dimension down from three-dimensional open ocean. That leads to competition for space. Surface-living plants put an enormous amount of effort in reproduction, wasting millions or billions of propagules just to make sure that one of theirs is there when the site opens. An emerging strategy is to go around the whole life cycle having to establish only once (figure 5.15). The brown algae *Fucus* establishes with the fertilized egg. *Laminaria* establishes with the spore. Notice how the embryo in Laminaria (figure 15 B) uses the same place for establishment as the spore-derived female plant. Ulva, with its isomorphic life cycle (sexual and asexual plants look the same) has to establish with zygotes and spores, running the gauntlet twice. The trend is away from the sexual and asexual generations being alike, toward each part of the life cycle being specialized and different.

There is a parallel trend to asymmetry within sexual reproduction. Sexual reproduction is an oxymoron. Reproduction is one goes to two or many, while sexual reproduction takes two and makes them one. A similar trend toward specialization in sex occurs in parallel to the specialization of the several whole plants in the life cycle. The trend is from gametes being identical (isogamy, in figure 5.16A) to oogamy, where the egg and sperm are very different (figure 5.16C). The egg is large and immobile, ensuring that the new plant makes it (reproduction). Meanwhile,

A

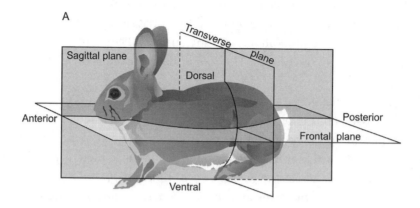

B

FIGURE 5.14.   Animals on a plane differentiate a top and bottom like plants. But because they move, they have a front that is also differentiated. Thus, bilateral symmetry is not so much as its name suggests, but is in fact a bipolar departure from radial symmetry. Flounders start as bilaterally symmetric, but when they move to the bottom to live, they sacrifice the only symmetry that they have, with both eyes on one side. The flounder (*Cleisthenes herzensteini* [Schmidt]),is still considered a right-eyed species, even though some are left-eyed. (Courtesy NOAA Photo Library. 1957)

the sperm is spent freely since they are small. Sperm specialize in taking care of the sexual side of sexual reproduction.

The open water does not make demands on available space, but it does offer the opposite problem. When it comes to finding a mate, the third dimension makes the space too big. Actually, it is worse than that; not only do you have to be in the right place to find a mate but you also have to be there at the right time. So it is a four-dimensional time/space. To deal with this, many sexual cells in many species

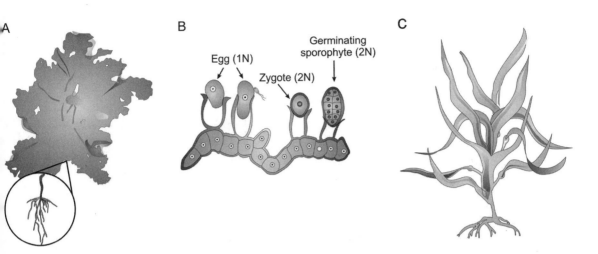

FIGURE 5.15.   *A.* The haploid sexual generation of ulva has the same form as the diploid spore-bearing plant. *B.* The sexual and nonsexual generations in the advanced brown algae are commonly different. This is the small sexual female generation of a large brown kelp. Egg production to fertilized egg, and growing new sporophyte are in sequence, left to right. The multicellular attachment on its right end will become the large plant in diagram *C. C.* The big kelp seaweed you see at low tide are always the spore-bearing generation (mature sporophytes).

of alga are released on a signal, like the next tide after the neap tide. Gametes are released synchronously, and time disappears. So the problem is only a 3-D spatial search. Another strategy is for the egg to give off chemical attractants to the sperm. Thus, the eggs move from existing in a zero-dimensional space of the point that is the egg to the egg and chemical filling a three-dimensional space that is easier to find.

The strategy of changing dimensionality to solve a problem is common in organisms. If a male moth is searching for a female ready to mate, she will help with pheromones. But if she is a mile upwind, the huge space in which the molecules are spread makes finding her a daunting task. The male has huge, comblike antennae. The pheromone lands there. The molecule is now in a 2-D space of the comblike antennae. The pheromone moves to the main stem of the comb. The molecule is now in a one-dimensional space of the spine of the comb. The molecule is passed to the base of the comb, where a counter records it in a zero-dimensional space of the point of measurement. With that pheromone "Geiger counter," the male moth can tell if he is going in the right direction because the frequency of hits goes up.[19] An insoluble problem in a certain dimensionality may have its solution in a space of different dimensionality.

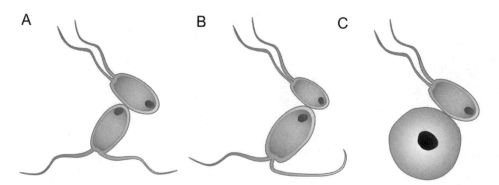

FIGURE 5.16.   *A.* Isogamy means gametes are the same size with both moving. *B.* Anisogamy has one larger and another smaller gamete, but usually both move. *C.* In oogamy, there is motile sperm and a much larger, immobile egg. Algae show all three conditions that have evolved many times, but land plants are all oogamous.

SIZE PRESSES PLANTS AND ANIMALS APART

Size produces another architectural difference between plants and animals. This time, the argument turns not on motility but on energy sources. Because animals depend on a high-energy food source, their cells have to remain soft and naked so that they can take particles into the body. Plants, however, only require small molecules, and they can therefore surround themselves with dead exudate that is the fibrous cell wall. Much of the mass of tissue of the land plant is dead, the living part being encased in a dead cellulose fortress (figure 5.17).

Single cells are of limited size (some unicellular algae can be centimeters long), but multicellularity opens up new opportunities for being big. As an organism grows, its mass increases by a cubed function of its length, while its surface only increases by a squared function. The larger the surface, the more mass can be spread to give a relatively low pressure on the surface. But the cubed function of volume or mass grows faster. Therefore, as the surface-to-volume ratio becomes smaller because of increased size, the organism runs into structural problems. Being made of tough woody material, the plant just keeps growing. It functions on the principle of the building brick. It stacks them up. The animal, being made of soft flexible material, quickly presses against structural limits (figure 5.18).[20]

If the animal toughens up its outer surface to keep from falling apart, it will be unable to absorb food particles. The outside of an animal has two jobs that interfere with each other. First, the skin keeps the animal's innards within. Second, the outer surface must be an interface for food and waste products. The solution to the dilemma is to divide the outer surface into two parts, one structural and the other nutritional. It is a radical resolution; the animal turns itself inside out, or rather

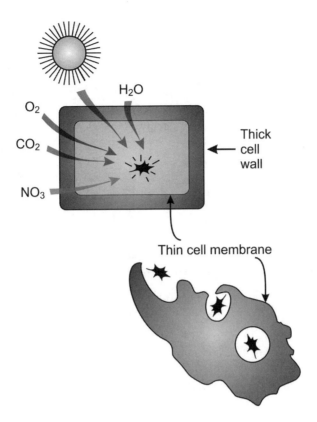

FIGURE 5.17. Because the plant cell uses only small molecules, it can exist in a tough supporting cell wall. Animal cells, however, need to be capable of ingesting large, high-energy food particles.

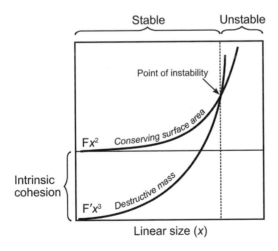

FIGURE 5.18. There are two facets to stability of a structure: (1) intrinsic cohesion that is constant for a given material; and (2) surface area that increases as a squared function of length. The destructive force is weight coming from mass that increases as a cubed function of length. Despite the head start of intrinsic cohesion, the cubed function of weight eventually overtakes the squared function surface and the structure fails.

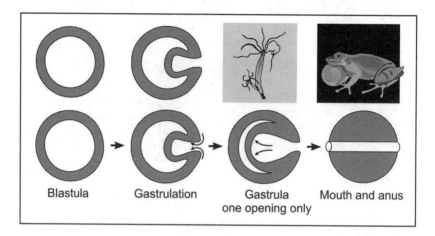

Blastula　　　Gastrulation　　　Gastrula
one opening only　　Mouth and anus

FIGURE 5.19.　In the process of gastrulation, the hollow ball of cells tucks itself to form a cup. The cup becomes the gut in the mature animal.

outside in. Early in the development of large multicellular animals, the process of gastrulation tucks part of the outer surface of the embryo inside to make the gut (figure 5.19).

As a consequence, the animal has an inward-looking circulation system. Being a bag, the animal fills up and therefore manifests determinate growth as a rule. Plants have an open-ended circulation system and indeterminate growth because they are not forced into the animal's inward-looking morphology and physiology.

Plants found many of our clever architectural devices before human designers. First consider the forces applied to a stem as opposed to those of the root. Stems feel sideways-bending stress. Roots feel tension. The vascular system in plants is lignified, and is accordingly available to give strength. Plant stems are sometimes hollow. Big members of the carrot family such as *Heracleum giganteum* have a cavity an inch wide, with a solid outer ring of stem not a quarter of the width of the hole. Bamboo has huge cavities. Most herbs have thin-walled parenchyma in the middle, offering no support. Stems resist bending by making a cylinder of strong tissue. It works on the principle of the I beam girder (figure 5.20). If you bend a horizontal girder by placing a weight in its middle, the upper surface is compressed while the lower is stretched. The middle portion of the beam neither expands nor compresses.

Take a sheet of paper and hold it horizontally by one end; the other end will be bent down because it cannot support its own weight. And yet if you roll the paper into a tube, it becomes strong and rigid. The difference is that the upper side of the cylinder is moved away from the plane of bending. Also, the lower side is moved in the other direction. The further away a surface is from the neutral middle, the

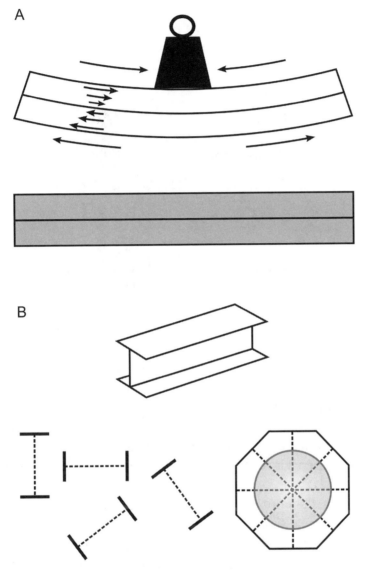

FIGURE 5.20.   *A.* When a horizontal girder bends from downward stress in the middle, the top is compressed and the bottom is stretched. *B.* Plant woody material in stems is arranged as a cylinder, sometimes with a hole in the middle, sometimes with thin-walled cells in the pith. The cylinder is comparable to a set of girders whose edges touch, obviating the need for the middle of the girders.

more it must stretch or compress if the whole is bent. The middle vertical sheet of metal in an I beam offers little strength in itself. All it does is keep the top of the I beam away from the bottom. A cylinder is just a set of I beams connected at their edge. Now the top and bottom plates do not need the middle plate because they keep themselves from the middle.

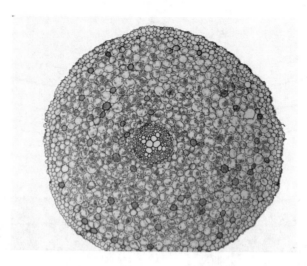

FIGURE 5.21.   Cross-section of buttercup root. The xylem in the root is in the middle, so the root functions like a cable. (Photo courtesy of the University of Wisconsin Plant Teaching Collection; Mike Clayton, photographer.)

But what about roots? They too need to resist strain, but it is tension. Furthermore, roots need to wrap around rocks and soil as an anchor. Hollow cylinders would not work because they would resist bending; roots need to bend. If they did not bend, the cylinder would be crushed. Accordingly, at the transition between the stem and root there is a complicated twisting of vascular tissue so that the strengthening material is in the middle of the root. Instead of a girder, the root is a cable (figure 5.21).

Tree boles use the principle of the cantilever as one side balances the other. The V-shaped vascular tissue in the leaf petiole invokes the principle of the suspension bridge. Buttress roots on tall tropical trees work not as buttresses but as guy wires supporting a tower. Hit one with a hammer and it does not thud, it rings like a guitar string.

THE WORLDS OF SMALL AND LARGE ORGANISMS

The question of size of organisms emerges again, but this time not to separate plant and animal experience. It pertains to the difference in the critical physical forces that press on big as opposed to small organisms in general. Water and gravity have different meanings depending on the size of the creature involved.

The bigger the organism, the more marginal is its existence in the face of gravity. Mastodons used to break limbs regularly, while beached whales suffocate under their own body weight. By contrast, small creatures fly in air without wings. As J. B. S. Haldane observed, in a fall to the bottom of a mine shaft, an ant flies, a mouse is stunned, a man is broken, and a horse would splash.[21] Even under a static load, larger organisms need to be formed in a distinctive way that accommodates the effect of gravity on their bulk. Thick solid columns support both elephants and large plants. Sequoia trees and hippopotami both look stocky because both have heavy loads to bear. Should the break occur, it involves a separation across a cross-sectional area. With a linear increase in height, the thickness of the support structures must increase on a squared function to provide the same strength under static loading.

While trees are stockier if they are large, actually they do not increase the base of the trunk on a squared function. A big tree does not have the same proportions as a sapling, but it also is not thicker at its base by as much as a squared function of its height. In terms of static loading, bigger trees are not as strong as small trees. They increase in thickness not on a squared function of power 2, but with an exponent of 1.5. Thus, smaller trees are overbuilt for the stresses that a structure of their size would experience. This overbuilding of trees emerges as one of the strategies for staying upright. The overbuilt bulk of a tree sits as ballast at the bottom of the tree; it remains upright on the same principle as lead-loaded chess pieces or a plastic, round-bottomed clown doll that pops back up again when a child hits it. A big tree stays up in a storm because of the parallelogram of forces on its center of gravity. Falling has to move the center of gravity of the tree, sideways yes, but more important, it must be lifted a small amount. That is why felling a tree in a certain direction is achieved by cutting a notch into the trunk on the side you want it to fall. The notch moves the pivot point from the base of the tree to the V of the notch. With that pivot point, the falling tree only moves down. The lift component is reversed. The parallelogram of forces is turned upside down, and the weight of the tree now tends to move the treetop sideways (figure 5.22).

The exponent of 1.5 keeps constant, not strength under static loading, but rather a flexibility factor as trees increase in size. This constancy of flexibility indicates two problems for trees that are more important than strength under static loading: first, if a tree keeps the same flexibility as it increases in size; and second, the limbs will bend to the same degree no matter what the size. This means that the form of the canopy remains constant as the tree grows. Keeping a constant outline to the canopy is important for keeping leaves and branches positioned constantly relative to each other.

The singular constancy in canopy form fits with the modular construction of leaves. That modularity requires a conservative plan for growing a particular shape of leaf, as well as for placing leaves to account for self-shading effects. It is easier for the plant to keep the context of the leaves constant by holding canopy architecture constant across size, something that can only be achieved by normalizing on flexibility rather than strength under static loading.

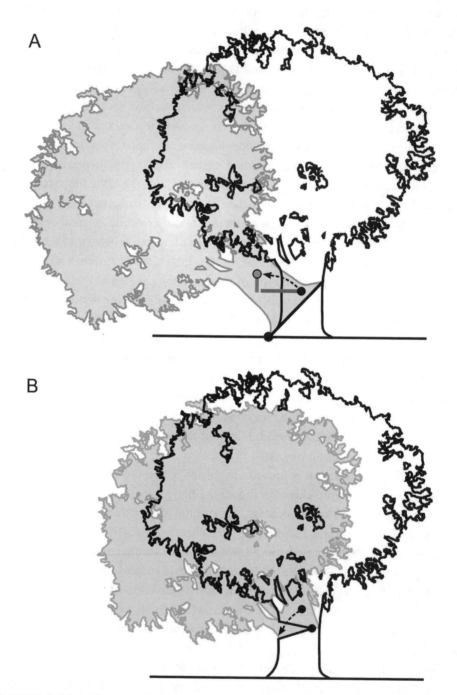

FIGURE 5.22. The mass of a tree is in its bole. *A.* The sideways thrust of the wind has to lift that mass at the center of gravity. See the parallelogram of forces, with its slight vertical component, but enough so the tree sits stable on its bottom. *B.* Cutting a notch out of the base of a tree moves the pivot point from the leeward side of the tree to the base of the notch. Now the lifting component has disappeared, and the tree falls under its own weight.

Only exceptionally, as in climbers like ivy, do plants change leaf form with age. The conventional view is the distinction between leaves at and below the canopy: sun leaves versus shade leaves. The microcosm wind tunnel experiments of Allen, Havlicek, and Norman would indicate that the distinction is in fact "wind" versus "no wind" leaves. Allen made informal observations on the British Dorset south coast. A wall on the landward, leeward side of the tall castle tower was covered with ivy. The tower acted as a windbreak for the sea breeze. The leaves were "sun leaves" in places exposed to wind where the tower was not a windbreak. The sun leaf form did not occur on top of the wall where there would be more light unless it was a part of the wall in the wind. It looked like the mature form leaves were wind leaves. The life form of climbers is focused on getting to the light, but that means moving into the wind too. Roset herbs, where a few inches make all the difference, also have different leaf forms. The leaves in the roset on the ground are bigger and rounder than leaves on the stem. Wind again may be a player here.

In deserts, the area of the leaf for photosynthesis comes at a great cost in water loss. Desert plants that have evolved to minimize their leaves sometimes evolve them into spines or scales. That tends to be a one-way trip, so in a return to a moister environment, something else must become flat. Butchers broom is a favorite of botany professors because its "leaves" have flowers growing out of them. So that must mean that those flat things cannot be leaves because the rules say, "leaves are terminal." They are actually cladophylls, stems that have evolved to look like leaves, so as to solve the surface-to-volume ratio problem (figure 5.23).

TEMPERATURE REGULATION IN ORGANISMS

Surface-to-volume ratio figures large in the evolution of our species. The conventional wisdom was that as humans started to move upright, the hands were freed for tool use. This gets into a positive feedback so the brain gets bigger, encouraging further clever use of the hands, and thus encouraging more upright walking. The expectation was that upright stance would be associated with larger cranial capacity. On November 30, 1974, Don Johnson discovered a more or less whole specimen of *Australopithecus afarensis*, a 3.2-million-year-old hominid. In their celebration, a tape played the Beatles, "Lucy in the Sky with Diamonds." The name stuck. With the discovery of Lucy, the conventional story fell apart.[22] Bones of this sort had been found before, but Lucy was an unusually complete specimen. A cranium with one name could now be linked to a knee and hip of a species thought previously to be from some other animal. When the legs were identified as walking upright, with Lucy's small cranium about half the size of early humans, the old theory collapsed.

FIGURE 5.23.   *Ruscus aculeatus.* The flat leaflike structures are cladophylls that are stems flattened to give leaflike structure. See how the cladophylls are subtended by scales. These scales are actually the leaves, and the cladophylls being stems grow out at leaf nodes (in leaf axils). The bottom right image is a cladophyll (stem) with a leaf on it, subtending a flower. The second stem next to flower is last year's fruit stalk. There is stem in the axil of the leaf on the phylloclade that is too small to be seen here. That stem has small scale leaves on it. In the axils of those scale leaves grow flowers stems. The stalk of last year's fruit comes from that short shoot. Flowers are stems, and petals are leaves. So the sequence is: primary main stem; scale leaf; secondary stem is a leaflike cladophyll; leaf on cladophylls; tertiary stem in axil of leaf; small scale leaves (unseen here) on tertiary stem; in axils of those scale leaves grow a quaternary stem that is the flower (bottom right). (Photos courtesy of University of Wisconsin, Madison, Botany Department.)

The puzzle was solved when Falk (1992) noted that the cranium of *Australo-pithecus* was solid, whereas the cranium of *Homo* is porous; it has blood vessels in it. Her theory is called the radiator theory of brain cooling. The trick is that the human cranium allows blood from the cooled scalp to flow inside the cranium next to the brain, cooling it. Blood leaving the brain is cooler than the body temperature. It was not that the old theory was fundamentally wrong; it was that there was a sleeper across the track, and the large head could not evolve until the problem of cooling limitation was solved. A big brain produces heat on a cubed function of size, while the surface of the head, the cranium, goes up by only a squared function. Finally, evolution solved the problem by putting an air-conditioner on the scalp. If Lucy had had a brain selected for size and smarts, she would have died of heat stroke. Hair does shade heat input to an extent, but heat loss is the deal breaker. Hair is for actively cooling the scalp, not just interfering with the heat coming in. The hair works as a wick, helping the sweat to evaporate faster. And the latent heat of evaporation of water is huge, a massive 540 calories.

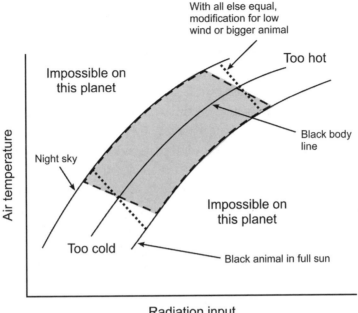

FIGURE 5.24.   Porter's scheme for displaying the climate space of animals. The right-hand long edge of the space is the limit of input of heat to a black animal in full sun. That is a physical limit on the planet with our sun. A white animal's full insolation line would be a little to the right. The night skyline on the left of the box shows how at a given air temperature, an animal at night receives less radiant input. The black body line applies when the environment is neutral to the animal, neither sun input or night failing to return radiation. The range available to the animal is surrounded by the heavy dotted line. Beyond the upper limit is too hot, and below the lower limits is too cold. The diagonal connectors between night sky and full sun are the upper and lower limits. The connectors pivot in the wind to allow an animal to keep cooler in wind at the top of its space, and become too cold at the bottom of the space. The finer dotted lines show the effect of a given wind speed and the limits of larger animals. The steeper the fine dotted line, the greater the effect of radiation is because of only slow wind. A high wind line would be flatter and show greater effect of air temperature. Bigger animals (with all else equal) feel radiation effects more, and their fine dotted line would be steeper, as shown here. Warm-blooded animals pivot their response to wind and size adjustments toward the night skyline, because they have the ability to increase physiological heat production in the cold (Porter and Gates 1969).

Warren Porter has spent his distinguished career applying engineering scaling to temperature control, mostly in animals. His early work modeled animals as cylinders.[23] It is remarkable how far one can go with such models in adept hands. The space in which he modeled his animals was an ordinate of air temperature against an abscissa of radiant input (figure 5.24). Measurements in the space show positive correlation of two axes of air temperature and radiant input. Warmer air warms surfaces of the surroundings that then radiate more. The blackbody line is neutral between sun and night sky. When a lizard goes down its burrow, it is retreating to a blackbody zone.

The animal has a zone, its climate space in which it can survive with regard to body temperature. The space is a curved rhomboid. Wind, body size, and animal color change the size of the climate space.

With all else equal, smaller animals are more influenced by air temperature. Large animals are more influenced by radiation. This can be seen in the angles of the top and bottom limit. Increased fat and fur has more or less the same effects as increasing body mass. Insulation and size give the animal delay in losing heat. Physiological activity generates heat and can be adjusted in warm-blooded homeotherms. Some lizards in the desert venture outside their viable climate space, but they can only do this for a while before they must retreat to the blackbody radiation environment of a burrow. Some animals have special huge surfaces for radiating heat away. The jackrabbit's ears allow it to venture into very hot places. The elephant is so large that it needs its large ears to get rid of the great heat load of such a mass of heat-producing tissue. The big sails on the back of stegosaurs were highly vascularized, presumably for getting rid of heat. Porter did experiments on small metal models of the dinosaur to show that the sails are organized not down the middle of its back, and not as paired protrusions down its back. He found that alternating left- and right-sided sails, a very unusual arrangement, achieved the most efficient heat loss. Stegosaurs had the most efficient arrangement for losing heat. Recent improvements in modeling metabolism indicate that big dinosaurs had a heating problem as they moved. *Tryrannosaurus rex* would not have been able to run as it is shown doing in recent movies.

With the coming of truly massive computational power, Porter is able to simulate animals appropriately proportioned so he can move them in space and calculate the effects of inertia.[24] Gone are the days of "consider a spherical (or cylindrical) cow." We are great advocates of bringing in the humanities and arts to science because they have insights that scientists lack. Mitman tells of how film has influenced biology, but we are now moving on.[25] Porter is liaising with creative artists to produce videos of movement in animals that allows him to collect data from simulations.[26] The results for elephants so far are well inside experimental error, and are so good as to expose some experimental errors. Using these dynamic models, Porter's models for humans walking show that hairlessness appears as important as any adaptation of our genus. Nakedness combined with wearing skins opened up a huge climate space for us, allowing us to move well outside the climate space of our hairy closest ancestors.

FLUID DYNAMICS

The fluid dynamics of air and organism size is only one place where fluid dynamics comes to bear on scale problems of organisms. The Reynolds number takes into account the size and speed of the object and the viscosity of the fluid. All those different units are canceled out to give a dimensionless number, the Reynolds number, which predicts turbulence in the system. For very small organisms, water is functionally very viscous.[27] That is, the Reynolds number is small. Accordingly, for small organisms, swimming actions like a breaststroke kick do not work. Kicking fast on the stroke does no good because the return stroke pulls the organism back to where it began, even if the return stroke is much slower than the original kick. Animals of our size sink struggling in quicksand or mud because wet sand and mud are viscous. Microscopic animals in water are like us in quicksand. Accordingly, they swim by other means than fish or humans; they often screw themselves through the water like a propeller. The world of small organisms is not just a small piece of our own, it is different.

Let us turn to other physical properties of water as it affects organisms of different sizes. One of the most distinctive things about our world is the ubiquitous presence of water. Water is a very peculiar substance. Because of its particular molecular and atomic characteristics, water allows visible light to pass. That is why visible light, as opposed to any other wavelength, is important to biological systems for photosynthesis and sensory perception. The density of the commonest form of solid water is less than liquid water (ice made with the shock of liquid nitrogen has a different configuration and so, unlike regular ice, it sinks). Therefore, normal ice floats and saves water bodies from freezing solid, thus enabling aquatic species to survive cold winters.

For the most part, water is macromolecular, with bonds breaking and forming readily. That is why, despite the small size of the single water molecule, water is liquid at temperatures that prevail on most of the surface of the earth. A critical property of water is its spectacularly high surface tension. If water had a more normal surface tension, the ocean would not have significant waves (compare pouring oil on the water) and erosion would be minimal. Furthermore, raindrops would not form as easily, thereby breaking the water cycle. The ecosystems of the earth and its landscapes would be radically different.

The critical danger for an ant is from the effects of surface tension, a force hardly detectable by large organisms. Although an ant can lift many times its own weight, it is almost completely incapable of escaping from a drop of water, and would almost always die once it was submerged in any body of water. A needle of solid steel can be readily floated on the surface tension of water. This is the principle on which some insects depend when skating across the surface of water as if it were ice.[28]

Pond skaters are the well-known example of insects using the surface tension, but other insects do it too. Allen has observed yellow jacket wasps treading on water when they needed to drink. They were not skating, but they did distinctly move their feet on the water on Allen's just-watered seed box in order to get water on a hot day. Quite unlike the ant, some of the very largest organisms may not be threatened by surface tension, but rather they depend on it for their daily functioning.

Trees are commonly higher than thirty-two feet. Suck water above that height and the tube cavitates, creating a near-perfect vacuum above the water column. The tallest trees can pull water up hundreds of feet only because the woody tissue of the trunk consists of microscopically fine columnar spaces. Because they are so fine and exhibit such large surfaces inside these tubes relative to the volume, the water in these columns is held together by surface tension. Under as much as fifteen atmospheres of suction, water is hauled to the tops of trees as if it were piano wire, not a liquid. Trees suck so hard that it can be detected in tree trunks being slimmer during the day.

Comparing the size of organisms amounts to moving up and down the organism column in the cone diagram (see chapter 1, figure 1.15A). Knut Schmidt-Nielsen finds physiological and behavioral patterns that are remarkably simple across a wide range of organism sizes.[29] If one plots the log of organism size against the log of a series of metabolic or behavioral traits, the line is often straight. These unexpectedly simple effects of size, measured sometimes by mass, other times by linear dimensions, include: speed of running, swimming, or flying; the energetic cost of the same; oxygen demand; number of heartbeats per minute or per lifetime; longevity; and many more (figure 5.25). Sometimes the line is only straight

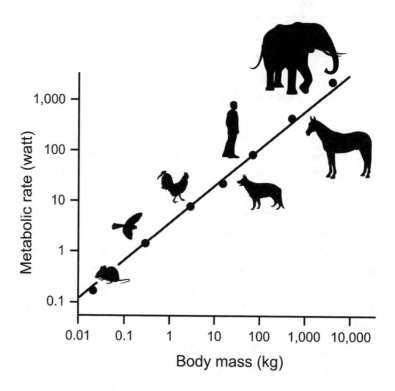

FIGURE 5.25.   The size of an organism relates to the speed of many aspects of the organism's functioning (data from Schmidt-Nielson 1984).

for a given class of organisms because the fundamental differences in strategy for dealing with a given problem may change the slope of the line. Sometimes the equations for the various powers in the relationships make intuitive sense because of squared surface areas and cubed volumes. However, the form of some of the simple relationships is less than intuitive. The tangibility of the organism criterion allows us to see some remarkably deep patterns.

GENETIC VERSUS STRUCTURAL DEFINITIONS

While the organism is usually genetically homogeneous, this does not mean that the genetic aspects of organisms relate simply to their form or life history. Anne Fausto-Sterling and Gregg Mitman show how the course of the study of genetics in relation to organismal form has been enormously influenced by the choice of organism. Their historical analysis shows that at the turn of the century, the flatworm *Planaria* was the organism of choice for investigating heredity and development.

In genetic sequestration, the cells whose descendants will become gametes are separated from the other lines of cells that will become the main body of the organism, the somatic cells. The critical feature of the flatworm is that it does not sequester its germ line. This distinctive, but by no means unique, genetic arrangement greatly complicates the animal as an experimental model. For example, flatworms have remarkable powers of regeneration; slice the head many times and each slice becomes a new head. It can reconstitute a new organism from a very small piece of its body and still develop in roughly the same orientation. The genetic expression of the flatworm is much more environmentally oriented, and is much harder to place unequivocally as Darwinian as opposed to Lamarckian. The model for development in flatworms is forced to be more outward in its orientation; the system demands holism.

By contrast, the fruit fly does sequester its germ line, and from that comes the simple genetic models in favor today. All higher organisms do that. Had the fruit fly not become the experimental organism of choice, the central dogma of the primacy of genetic influence would have emerged in a subtler form than the strident version that is now the conventional wisdom. The fruit fly is much better behaved as a model and leads to reductionist genetics of inheritance and development. Because it allowed an answer to local, focused questions, the fly gained the ascendancy in an acrimonious battle that was basically won for reductionist genetics early in the twentieth century.

Had *Planaria* been the winner, we would be asking broader questions about genetic control and development. It would be a much more ecological environmentally ordered scheme than the one that prevails today. Epigenetics invokes change from genes being turned on or off rather than present or absent. We would have understood it earlier with *Planaria*. Furthermore, the issues raised by the flat-worm model have not gone away, they have merely been neglected by the prevailing paradigm that is powerless to address them.

SCALING IN DEMOGRAPHY

Since our last edition, scaling in demography has blossomed into a branch of the discipline called metabolic ecology. A champion there is Jim Brown of University of New Mexico. Metabolic ecology has used branching patterns, particularly in animals, to look for universal scaling rules for form. The idea is not completely new, being one of the conclusions of Allen and Starr decades ago.[30] It derives from Schmidt-Nielsen's work on scaling animals, looking at the exponents on scaling laws (figure 5.25). In general, metabolic ecology seeks to explain form in terms of challenging flows. The slope that Brown's school prefers is a 3/4 rule. Earlier work on plants used a 2/3 thinning rule, but the 3/4 rule is gaining influence.[31] However, it is very hard to distinguish 2/3 from 3/4 in measurements.

Beyond scaling animals, Brown has looked at scaling laws in cities.[32] His most recent arguments are applied to production in nation-states:

> The exponent for the scaling of energy use as a function of GDP, 0.76, is reminiscent of the three-quarter-power scaling of metabolic rate with body mass in animals [Kleiber 1961; McMahon and Bonner 1983]. This may not be coincidental. In a very real sense both animals and economies have "metabolisms." Both consume, transform, and allocate energy to maintain complex adaptive systems far from thermodynamic equilibrium. The energy and other resources that sustain these systems are supplied by hierarchically branching networks, such as the blood vessels and lungs of mammals and the oil pipelines, power grids, and transportation networks of nations. (Brown 2011:21)

While cautious in his conclusions, Brown calculates that the pressure on the global system is unsustainable because of limits on the resource gradients. He uses the approach of macroecology and robust correlation that appears to have held for the last twenty-five years. The data on economic substitution indicate that Tainter's observations on diminishing returns on effort are not saved by the new technologies.[33] Diminishing returns apply just the same to biotechnology and electronics as they have to conventional technical and engineering progress in the first half of the twentieth century. There was significant objection to Brown's paper from the engineering and economic community, but the work prevailed due to measured action from editors. Sometimes peer review works.

Not from the same school, but in a related way, Giampietro has been able to plot a large number of variables against three sets of human-related variables: (1) eight indicators of nutritional status and physiological well-being (e.g., life expectancy, average body mass index); (2) seven indicators of economic and technological development (e.g., Gross National Product or percentage of GNP is agriculture); and (3) nine indicators of social and technological development (e.g., number of phones per thousand, or pupil-to-teacher ratios).[34] With his graphs, Giampietro can see a kink in historical data that breaks as a nation in question crosses the demographic transition. That transition has happened in almost all industrial nations, Britain years ago, but Italy recently. It is the change into extended life spans and distinctly lowered reproduction rates. In the 1970s, Hugh Iltis would rail against the booming population in Italy, blaming religion and attitudes toward birth control. Now Italy has the lowest population growth in Europe, and presently needs to import workers from Turkey to make up for the shortage of labor. Italy is still Catholic, but it has crossed the demographic threshold. A key pivot point is increased education of women, who then gain authority and can choose to have fewer children. All the industrial nations have turned that corner, and Giampietro can see it happening elsewhere.

A remarkably optimistic but not unrealistic approach to human demography has been coming over the last decade from Hans Rosling, with a series of breathtaking TED talks.[35] He has created dynamic graphs that plot the world's nations with variables similar to those used by Mario Giampietro. But Rosling looked at national demographic variables over time. With his downloadable Gapminder software (http://www.gapminder.org/), he found that since the 1960s, many third world nations now display the same statistics that the United States had in the 1960s. In the 1960s, China, South Korea, and India had patterns close to the United States in 1900. The gap between the third and first world is now closed to make a continuous cluster. Even though fourth world nations in Africa remain well behind, Rosling found that they are on schedule. In 1960 they were not like the United States in 1900, as were Asian third world nations, but much more like the Middle Ages of Europe. He finds it encouraging that Africa, while far from being one place, is now at about the stage of the United States in 1900.

Brown's more pessimistic view is in line with Tainter's (1988) ideas on societal collapse. We have already explained that the problem is becoming canalized in greater specialization and loss of slack. Essentially, resilience is lost. Across a wide range of historical societies, Tainter shows that societies collapse when problem solving fails because of diminishing returns on complexity that comes with problem solving. The environmental determinist view of Diamond, to the extent its scholarship is sound, merely describes the incidental trigger that started off the collapse (fire, earthquake, little Ice Age, invasion). For a firm rebuff of Diamond's scholarship, see Smil (2005) and Tainter (2008).[36] We think that Victorian Britain could have ridden out global climate change because they were not as specialized, nor fuel-dependent, as we are now. We have our doubts for the present first world. Organismal metabolism uses gradients, and evolution moves toward optimization in organisms within physical limits. Tainter and Allen have compared the thermodynamics of ant colonies and Classical Rome.[37] Insect colonies, like human societies, can be helpfully understood in thermodynamic terms and metabolism.

CONCLUSION

The organism being a tangible entity, offers a wealth of signals to human observers. As with other easily read objects of study, there is a problem working out what in the organism is generalizable and what comes from our own humanity. As human observers, we are likely to be overpowered by obvious signals so that we miss more interesting and subtle aspects of organismal ecology. A whole new world emerges if we try to look at the organism in the bubble of its umwelt. The physical organism itself is isolated and meaningless until we watch it reading its self-world. There is much to be said for allowing the organism to be seen not as a tangible, but by means of the closed loops and processes behind the structure. Let it turn into a set of fluxes associated with filters. By struggling against the tangibility of the organism throughout this chapter, we have endeavored to enrich the concept of organism. We tried to make it more than a reification of what the naturalist finds on a walk. If this chapter has achieved anything, we think it has laid the groundwork for integrating organismal biology into a wider view of ecology.

6

THE POPULATION CRITERION

A. D. Bradshaw (1987) had seen it all, and we hope that the complexity science of which we are proponents is not subject to his critique. But what he says does ring true with regard to population biology:

> Once we spent a lot of time looking at the behaviour of species, then ecosystems became important. Now, nearly any ecologist who feels he must be respectable will work on populations. Although an historian may be able to see a proximal cause for what happens—the arguments of a persuasive scientist or a novel and interesting discovery—the ultimate causes may be little other than those which give the world its flourishing fashion industry.[1]

An ecological population is a collection of items, usually organisms that are in some way equivalent. The equivalence may come in the population members being of the same species. Even so, there are many populations where members, while still equivalent, are from different species. For instance, the winter flocks of black birds in Wisconsin are not blackbirds, they are birds that appear black, coming from different species. Sometimes being from the same species is insufficient for population membership. In epidemiology, there are at least three populations within the species: the infected, the susceptibles not yet infected, and the immune who have recovered from the infection. So the equivalence that makes for a population is specific for the purpose of conceiving that population. There is one central assumption in population biology with regard to population members, that is, "when you have seen one, you have seen them all." This is, of course, a relative statement, but if you cannot say it, you simply cannot do population biology. Sufficient equivalence is what allows population equations. It makes the individuals summed to give N in a population sufficiently similar so that they are additive. If you cannot say that when you have seen one you have seen them all, you cannot compute N.

These are conventional ways of defining populations, and some valuable work can be achieved this way. But we would argue that population biologists are looking for the wrong thing for most of the phenomena they investigate. James Kay made a parallel startling claim that particle physicists were looking at the wrong thing, particles.[2] Even though investigations of particles have been the device for ushering in the nuclear age, Kay said that particles are not the issue. He noted that structures, discrete entities, come from constraints. With the right constraint, discontinuity occurs. When particle physicists put material into a large accelerator, the more energy is applied, the faster goes the stuff, and most important, the tighter are the constraints put on the material. The tightest constraints cause new particles to appear. It makes sense; as Kay said, particle physicists should be looking for constraints, not particles, which are only realizations of constraints. Particles come from constraints, so physicists should be looking at constraints. We question conventional population biology as addressing the wrong units. But we will acknowledge that one can still do some useful work, as do particle physicists, while actually looking at the wrong thing.

The wrong focus is on reproduction. Population growth is often cast in terms of female fecundity. Most telling is that before much population biology had been done in ecology, Lotka told them what to do, but somehow ecologists did not notice.[3] If you want the best account of anything, go to the first generation: Bach for fugues, Lotka for populations. In the 1920s, refugees from physics came into population biology bringing with them differential equations. Go to those physicists (Lotka, Volterra, and Lascaux) for the best version of the essence of population biology. This was in contrast to the statistical approaches to ecology that prevailed uncontested to that point. There was a flowering of the statistical methods for ecological description with Curtis in the 1950s through to Roberts in the 1980s with gradient analyses, all coming from those early descriptions before the physicists arrived. It is a cruel irony that description has been eclipsed by a population biology that appears not to have understood its own origins and main tenets. Lotka (1956) is quoted by Giampietro (2003):

> Birth rate does not play so unqualifiedly a dominant role in determining the rate of growth of a species as might appear on cursory reflection. Incautiously construed it might be taken to imply that growth of an aggregate of living organisms takes place by births of new individuals into the aggregate. This, of course, is not the case. The new material enters the aggregate another way, namely in the form of food consumed by the existing organisms. Births and the preliminaries of

procreation do not in themselves add anything to the aggregate, but are merely of directing or catalyzing influences, initiating growth, and guiding material into so many avenues of entrance (mouths) of the aggregate, *provided that the requisite food supplies are presented*.[4] (emphasis in original)

Giampietro goes on to say that same idea was expressed by Lascaux. Giampietro (2003:343) translated Lascaux: "Both for humans and other biological species, the density is proportional to the flow of needed resources that the species has available."[5]

Giampietro's point here is that populations grow not so much by sex and procreation, to which Lotka refers as "preliminaries" to birth. Birth does not increase the population size; it only provides the vessels into which growth may be put, as long as there is enough food available. Birth is more catalytic than substantive. When Hoekstra and Michael Wolfe of Utah State were a graduate student and post doctoral fellow, respectively, at Purdue in the 1960s, they performed statistical analyses on populations of North American species of deer.[6] Deer are primary consumers that commonly get to be an adult weight of at least seventy-five pounds in its southern range and more than two hundred pounds in its northern range. Deer populations, regardless of population or body size, consistently have some individuals that live to be around twelve years of age. Hoekstra and Wolfe's interest was in how small populations that were subject to various environmental pressures (heavy or light predation, etc.) were still able to have individuals that lived to about twelve years, regardless of the total population size. After trying several different analyses, they found that once a deer population cohort lived past the first year and entered the breeding population, the remaining individuals had a relatively constant mortality rate. What differentiated the individual deer population's size was the mortality rate of sexually immature individuals. So getting past the vulnerable first year was critical.

Translating into what Giampietro and Lotka are saying, a deer population needs enough energy input to get a critical number past their first birthday. A lot of that first year is spent consuming energy and learning how to survive as a deer. After that year, the mortality rate is consistent across populations of almost all deer populations they investigated in North America. Thus, being born does not contribute to the population but only introduces the option to do so. The critical energy need is for getting a cohort up to their breeding age and then they live out the rest of the population cohort's life span at a relatively consistent mortality rate and until a cohort passes into the dust. Thus, food is translated into information.

Our species certainly does that, as it invests a lot in children. They have a lot to learn. In general, mammals make that investment in information more than other animals. Birds are probably in there too. Even fairly opportunistic species like deer, as Hoekstra and Wolfe showed, invest energy in information. Insufficient food limits not the number of children born in the third and fourth world of humans, but health issues and government policies limit survivorship through those early years. After the one-child policy in China, we could expect the population to rise again. However, that policy might well have taken China over the demographic transition. All first world nations are now across that transition, such that countries that had high birthrates in Europe now have low birthrates.[7] The conventional and reasonable cause for the transition is more informed and educated women who then take more control over their reproduction. The one-child policy appears to have taught women, not so much by the normal route of education, but by forcing on them the experience of having small families. Thus, information and energy inputs are far more important than procreation, as Giampietro and Lotka would argue.

There is much condemnation of the consumption of meat in the first world, but there is a bright side to it. Are there any nations that are vegetarian, with roughly the same landmass and range of climates as the United States? The answer is yes, and it is China. There is more meat being consumed in China today than decades earlier, but as China grew its huge human population close to its present size, it was on a largely vegetarian diet. Given ideas of environmental determinism, one might expect the United States to have a much larger population. The reason why it is relatively small is that Americans consume large quantities of meat. That limits the amount of food available because meat is more expensive to produce. With a vegetarian diet, China has been able to grow its population in a way that was unavailable to North America for cultural reasons. Pollution in China is much worse than in the United States simply because of the size of the population there. American predilection for beef has put a limit on pollution, which has saved the North American environment.

A human population will grow to the limits of its food supply; at least it always has heretofore. Chapter 2 introduces the notion of high and low gain, significantly in the section entitled "Energetics of Movement on Landscapes." Now gain is back, once more tied to energetics. Our species is a low-gain species with deep planning. Its investment is in a small number of children, most of whom do make it to maturity. But that steady but sure strategy can still ratchet up the numbers if the food supply is adequate. In other words, give this low-gain species enough food and it will consume that food in a high-gain fashion. Birth control does introduce a certain freedom to choose, but it will not limit our population so long as the food supply is large enough to support greater numbers of humans. Birth control in our species limits births so that all offspring survive, thus actually increasing human population growth. We do not recommend a Malthusian outcome,[8] where "famine stalks in the rear, and with one mighty blow levels the population with the food of the world."[9] However, it does seem important to understand the limitations to population growth in a way that appears to have escaped population biologists in the recent mainstream. Lotka and his fellow pioneers would not have missed it.

We have been at pains to condemn diversity as a device that reflects ease of computation rather than actual utility and intellectual coherence. Diversity is easy to calculate, but may not be something we can understand. Population biologists, having ignored Lotka in his critical statement, are mistaking their units as collections of equivalent organisms. This is much the same error as is made by devotees of diversity. By only counting the vessels into which the size of a population fits, mainstream population biology makes it convenient to mathematize its studies. And population data are easy to collect too. Just count them. The mainstream of population biology is counting the wrong thing. Einstein is variously quoted as saying, "make your model as simple as possible, but not simpler."[10] A population biology that views the effort complete when it has summed the number of organisms in a population is an example of "too simple" in Einstein's terms.

POPULATIONS AS ECOSYSTEMS

Giampietro's insight puts the population criterion unexpectedly close to the ecosystem criterion. He looks to metabolism and makes a distinction not generally made.[11] There is an endometabolism, which works inside the system (perhaps organism or society) in question. That is fed by an environment that has its own metabolism, an exometabolism relative to the system at hand. The exometabolism also receives material and energy dumped into it by the outputs of the endometabolism. That relationship was broken down with devices such as holon, taxon, environ, genon, and creon by Patten and Auble early in the development of ecosystem theory.[12] Except for holon, those terms have slipped away, and with them the conceptualization of biology and ecology in those terms. We all know that there are inputs and outputs in ecological systems, so what has been lost? It is the distinction between the different metabolisms. There is an irony here in the way that ecologists are so critical of economists who externalize and then forget the environment in a studious neglect in less than full accounting. And yet, the critical error of largely ignoring or glossing over external metabolism is almost universal in population biology and most of the rest of ecology too. Both endo- and exometabolisms involve rate-dependent fluxes, as well as rate-independent constraints, and must be respected as separate but linked. Parts of the full situation are from time to time recognized by ecologists, as they are by even errant economists, but linking endometabolism with exometabolism has only been done formally by Giampietro and his team.

In these terms, the population is a means of capturing, using, and seeking energy.[13] The different vessels in the population (individuals) play different roles among them, and have different potentials. The simplest distinction is those who can work and those who do not. Those doing work in first world human societies are all working-age adults. In beaver society, it is all adults. There is a second distinction in nonworkers: juveniles, as opposed to those too old to work (figure 6.1A).

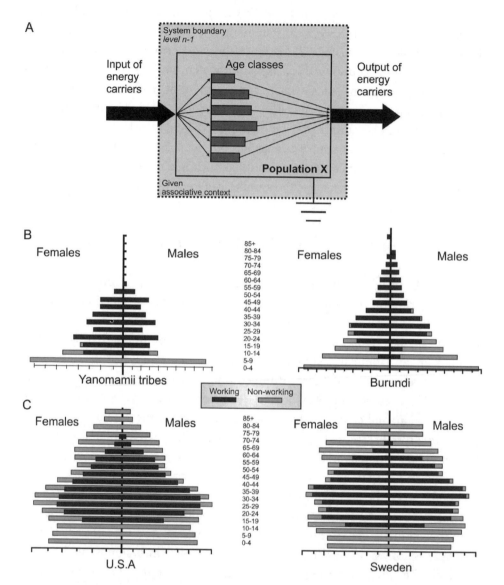

FIGURE 6.1. *A.* Representation of Lotka's insight as reported by Giampietro. Organisms are energy carriers. If there is not sufficient energy to carry, the vessels remain empty. *B.* Giampietro graphs the Yanomamii tribes in South America and in Africa (Burundi) to show the ages at which people in those societies work. Everyone of working age works, and nobody retires in either place. There is less societal investment in children learning for the Yanomamii tribes because almost everyone over the age of five works. *C.* The lower graphs show the United States and Sweden. Retirement starts in both industrial nations at age sixty, with more or less everyone retired by seventy-five. The difference in the industrial nations is that a greater proportion work during the working years in Sweden than in the United States (Giampietro 2003).

Different societies have different social rules as to who works and who does not. Figure 6.1B is from Yanomamö tribes who live in the Amazon forests on the border between Venezuela and Brazil. Giampietro reports that they have children working very early, more girls a bit earlier than boys. There is little specialization in learning. Learning is done on the job. By contrast, data from Burundi in Central Africa show work starting just as early for some children, but with others not working until later. Everyone is working by their early twenties. Nobody retires in Burundi (figure 6.1B). In the United States, more or less nobody works until their midteens. There are child labor laws that limit children working significantly. Unlike the Yanomamö tribes and Burundi, all ages have some not working. In the United States, most are working by their early twenties. The greatest proportion is working by age thirty. There is a decrease in the proportion working as of age fifty-five, with more or less full retirement by seventy years of age. Sweden is similar to the United States, except that almost everyone is working by age forty. Retirement is about the same as in America, but more Swedes live into old age. However, the oldest in a declining population live more into their nineties in the United States. The productivity of society can thus be well captured and understood if the various contributions to energy throughout are measured (figure 6.1C).

Often, population biologists will work with sectors of the population if there is metamorphosis. Eggs are the vessels that are an opportunity for population growth. Insect larvae are the means of resource capture. The chrysalis is the stage that consumes nothing, but effects an efficient and measured transfer of resources to the adult phase. The adult may capture some resource, but it is generally short-lived, is not a big consumer, and it is the package that turns into eggs.

Giampietro (2003) has devised a scheme to look at energy flow and investment in a population or a society. It deals with the different phases of energy transfer in the society or population. It cleverly translates energy into units that are required to transfer one facet of growth and work into others. It is a two-way crisscross chart that looks at energy entering the system, being used, and coming back out again in an effort to capture more resource. The left-right division is between inside and outside the system (figure 6.2). The top of the chart deals with entry into the system, while the bottom half deals with exit from the system.

The top-right quadrant translates effort in capture from the environment and translates it into food inputs. The top-left quadrant translates inputs and converts them into time offered for living and work by that sustenance. That quadrant divides the inputs into that directed at the future, education and rearing and sustaining young, as opposed to that used for doing other work. Work capacity from the top-left quadrant is received by the bottom-left quadrant. The lower-left quadrant divides that work into work on producing luxuries (perhaps play) as opposed to work directed at the environment to capture more resource. The output of the lower-left quadrant is work. The lower-right quadrant is next. It is now fixed because

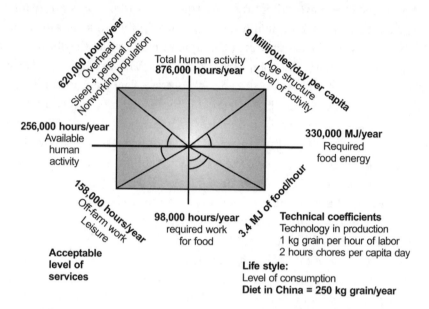

FIGURE 6.2. Giampietro raises an example that uses pedagogically convenient round numbers. His population is on an island, so he can ignore invasion and emigration. In a population of one hundred people, he uses acceptable values for conversion of energy input to give work inside the population. His scheme divides the four sectors each into two, by inserting a diagonal line from the center. The angles of those lines capture the patterns for input and output of a given quadrant. In the top-right quadrant, the upper triangle has a vertical edge that is the input to the society. The hypotenuse of the top-right triangle is an expression of the conversion rate of energy outside to give input to for work inside the population. The top-left quadrant also has a diagonal line that divides energy actually used for work (below the diagonal), as opposed to energy invested in the future (above the diagonal). The area above might be invested in education or feeding babies. In the lower-left quadrant, the angle of the diagonal line takes the energy invested in work and divides it into work on luxury, as opposed to work for capturing more inputs (e.g., farming, irrigation, or for drilling rigs in an industrial society). In the lower-right quadrant, the angle of the diagonal converts the work directed to the environment into production. The lower-right diagonal line is akin to a marginal return curve, where more effort gets less capture per unit of effort. The lengths of the diagonal lines indicate energy throughput. The angles show how investment is made. Ecuador and Spain were compared from 1976 to 1996 as to growth and strategy (Giampietro 2003:185). Both rectangles got bigger (growth), but Ecuador kept the same angles showing more of the same. Spain changed the angles to reflect increased industrialization (Giampietro 2003).

all degrees of freedom are spent unless the population changes size. The bottom-right quadrant is a marginal return curve coming directly from economics. Marginal return is in contrast to average return, which is how much you get for how much effort. In average return, you get less per unit effort for each increment of effort. Marginal return is how much extra you get for extra effort. The more extra effort you put into an enterprise, the less return you get on that individual increase in effort. In economics, there is a "law of diminishing marginal returns." Try harder to get more, and you will get less for the extra effort, although perhaps more in average return early in the process. Tainter's argument on the collapse of societies pivots on diminishing return of effort directed at problem solving. The lower-right quadrant in Giampietro's scheme feeds into the input quadrant that demands more land, for instance, if enough resource is to be captured to balance the budget.

The rectangular crisscross chart has axes of a St. George cross going to the edges of the rectangle showing how much energy is involved in transfer. Energy is transferred into, within, and out of the system. The fourth side is how much energy output turns into resource to be gathered and used. Giampietro's scheme also has a distorted diagonal St. Andrews cross that goes from the center to the corners of the rectangle. The distortion on the crosses comes from the fact that all four quadrants do not meet plumb center. That is, the upper and lower, and left and right divisions do not exactly bisect the space. The angles on the St. Andrews cross tell of the conversion factors that take input to a quadrant to its output. The angles tell the proportion of the population that engages in various acts of work or consumption. The scheme identifies what juveniles in the future may cost in terms of education and rearing. South Africa, in its fight for majority rule, could not afford to educate young blacks, and is now paying a price in an uneducated generation that was young when the society was particularly violent. A more egalitarian postapartheid society did not come free. The same happened as Britain paid the cost of industrialization, as children worked instead of attending school. In times of stress, hamsters eat their young. The evolution of specialized queens in insect colonies stops the workers from being totipotent. In an instance of diminished potency, workers do not do everything an ant of that species can do. This evolved because of ant police who would eat eggs or steal them from rank-and-file ants. Eggs need not be a source of vessels for the new generation; they can be simply a good source of protein. In times of stress, the angles of the St. Andrew's cross change.

The angle of the diagonal on the lower-left quadrant distinguishes luxury from directly useful work. Any population that has exactly the food input it needs, with no waste, is doomed. Waste is essential for stability. Let us draw attention to the parallel construction of two of the great pyramids in Giza.[14] A pharaoh has use for only one pyramid. He can only die and be buried once. The second pyramid he built was to hold things in reserve for when there was a biblical seven lean years. Take the workers off the pyramid and put them on irrigation. In our society, diverting corn to make ethanol in general is an energy loser. It generally takes more energy put into agriculture to raise the corn than appears in the ethanol, particularly if you do full cost accounting. So why would we be so crazy as to do such a thing, particularly since it has raised the price of corn globally with disastrous effects on third world food prices. Corn ethanol is a matter of food independence for industrial nations that produce it. It preserves farmers and a farming infrastructure for when we really need them. Well, that is the plan anyway. Waste is crucial so that there is a buffer. Populations of humans and animals do that exactly. Any that did not were wiped out by the first crisis.

Thus, the angles of the diagonal separators inform us as to whether or not a society is industrial with services or is entirely marginal, with little planning for the future. As the top two sectors lose depth, the society is hard-pressed, and it does less planning and playing. If the top two sectors are large, there is an excellent return on effort. Oil as a fuel is not what it used to be, but in the best days, it gave a 50:1 return on effort. A society like that must invest in the future and waste resources; otherwise, it will grow fast and overheat. The oil boom created waste and gave big investments in the young and growing society. Had the first and second worlds not wasted so much in an arms race, the growth would have been much bigger, and we would now be in much worse shape because of even greater overpopulation.

DIFFERENT WAYS TO STUDY POPULATIONS

There are several ways to study populations. MacArthur did field studies, observing different species of warbler, noting how they used the resource distinctively, each compared to the other species.[15] Robert May studies the equations for populations, doing a sort of natural history of mathematical forms.[16] Differential equations are one of his favorite devices. Park performed experiments on populations in the laboratory. Fruit flies and flour beetles lend themselves as model species in this sort of work. There are equations in work of this type, but it turns on experimentation and aggregate properties of populations such as density and numbers in the population. Finally, there are demographic studies, sometimes with animals in the studies of Deevey, but later and more abundantly of plants where individuals are measured and counted.[17] John Harper was the champion here.[18] The plants are commonly weeds or crop plants, and physical size and proportional reproductive effort is recorded as part of studies of resource deployment. Age and fecundity here are natural measurements to make on plants. There are equations in this work, but description of the progress of individuals is central. Statistical devices such as regression are common tools in plant demography, with graphs of dependent and independent variables. Competition is a favorite of most of these types of population biology. While environment is varied to show different population patterns, it is fair to say that most population studies have only one or two environments in which the populations perform. Population work is exacting, and it would be tiresome to repeat experiments using many environments. In addressing competition, Park looked at species interaction for two temperatures and humidities and their intermediates.[19]

This chapter considers the types of processes making organisms and demes attributes of populations, as opposed to any other type of ecological assemblage. Lotka (1956) did warn of the limits of counting organisms, but we cannot simply cut the Gordian knot of the mass of population work performed since his classic book. It could have been better, but we do need to discuss what it achieved.

TABLE 6.1   Main Styles of Investigating Populations

CHAMPION SCIENTIST	PRACTICES	PLACE
MacArthur	Comparing activities of different populations. Niches. Wildlife ecology maps onto landscapes here.	In the field.
May	Structural stability of equations to test consequences of assumptions (e.g., does population size protect survival?).	On the computer.
Park	Experiments where model animal populations are tested for stability. Coexistence, competition focused on winner. Competitive exclusion. Dispersal. Diversity.	In the laboratory.
Harper	Plant demography. Individuals are measured in experimental settings. Parametric statistics. Graphs. Focus on different life stages (e.g., germination, plant size, fecundity, death, proximate neighbors). Diversity.	Greenhouses, experimental gardens.

THE TAXONOMIC REQUIREMENT

With a few exceptions, the minimum requirement for members to be from one ecologically defined population is belonging to the same species. This is not a necessary restriction, for the population concept could be useful with members belonging to different biological species. The use of the term "population" in statistics captures this more general notion of population. Given the proviso of taxonomic homogeneity, ecological populations then revolve around two major organizing principles that give two types of populations. The first consideration is spatial contiguity: members are aggregated. The second consideration involves a shared history of some sort. Often, this echo from the past amounts to a level of genetic relatedness; members share ancestors. However, the historical connection could be some other bond that is not genetic, such as being in an infected population; the disease agent may be of one genetic strain, but the population of the infected need be neither closely related nor genetically homogeneous. Most populations amount to a mixture of these two types of prescriptions; there is a continuum between historically and spatially defined populations.

POPULATIONS, COMMUNITIES, AND THE RELATIONSHIP TO LANDSCAPES

Defining the population as a collection of individuals belonging to the same species says nothing about the time frame over which populations might exist, and nothing about the spatial coherence of the population on the landscape. While it is possible to consider individuals scattered through time as members of a population, for the most part, population refers to a temporal cross section, an instant in time. This does not preclude populations existing over time. Even when we define the population temporally, such as an epidemic, we still find it helpful to map literally the progress of the disease (figure 6.3). History books map where the Great Plague moved in waves across Europe during the fourteenth century.

Populations within species do involve interactions of time and space, but do not require an interference pattern because of the relative scaling of the respective parts of populations. Unlike community members, being all of the same species, members of a population occupy the landscape all at the same scale, or at least at a

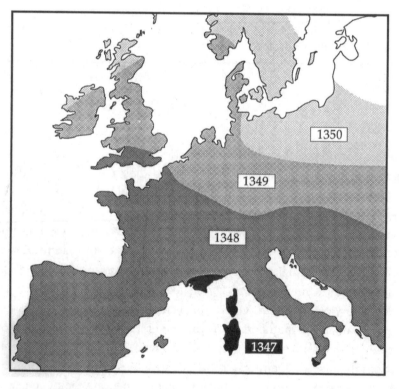

FIGURE 6.3.   Even though populations can be dynamic, it is commonly helpful to circumscribe them on the ground. The populations of infected individuals in the Great Plague of Europe serve as an example.

scale more similar than those belonging to community members. All members of a given population move at about the same rate. Therefore, space and time do not interfere with each other in populations.

THE PROCESSES BEHIND THE PATTERN

We have asserted two types of populations: spatially aggregated and historically explained. We have also noted a continuum of admixtures between the two types, namely populations that are variously both spatially aggregated and bound by history. Concentration of individuals in a local place often indicates reciprocation between individuals that makes them members of a meaningful population as opposed to just an arbitrarily defined collection. Often, these exchanges are processes that are associated with reproduction. In asexual reproduction of plants, the process may leave the members of the population literally connected to each other by roots, runners, stolons, or other such modified stems. Sexual reproduction involves exchange, often between neighbors, so that there are connections between members through ancestors that formerly occupied the same general vicinity. Thus, some spatially contagious populations can be characterized also by a certain genetic homogeneity. Some populations exhibit spatial contiguity with no real historical explanation, reproductive or otherwise. In these situations, the spatial aggregation is constantly updated. The aggregation reflects a continuing process. Cases in point would be herds, flocks, or a school of fish that are held together by behavioral cues between individuals.

Much in the way that we look at the landscape criterion at many scales, it is possible to see populations nested inside bigger populations. When we do see such nesting, there is no reason to expect the differently scaled populations to be of the same type. Consider the example of the pronghorn antelope. At the lowest level, it is a genetically defined collection of the doe and her fawns. The next larger aggregation is not genetically defined, and is barely even historically held together. This is the herd, an entity with floating membership serving to avoid predation. Conspecific cuing holds it together. The members are not necessarily closely related. Herds meet, intermingle, and then separate with different membership than at the meeting.[20]

At the next level is the antelope population, as defined by habitat. There is a degree of genetic homogeneity to this grouping. Because favorable habitat is discrete, there is little mixing between populations at this level. For example, antelope cannot survive in deep snow and only very occasionally will an individual cross the difficult, drifted terrain to join another isolated population. Most would not survive the passage and normal conditions would not lead an individual into the unfavorable habitat in the first place. Only individuals disoriented by a chase or an accident would even find themselves in a position to make the crossing. Note that although the herd is not defined on relatedness, the larger population can be so defined. Our newfound methods in molecular biology allow us to look directly for markers that can be calibrated for time of separation of genetically different groups. Since the upper-level population contains the herd, one might infer a contradiction here, but there is none. Relatedness is a matter of degree relative to that which occurs in a larger aggregation. Relative to the genetic heterogeneity across a collection of habitat-isolated populations, one habitat-isolated population is indeed more homogeneous than the collection of several. Therefore, the habitat-isolated population can be said to be defined by a certain level of genetic homogeneity, while herds are not.

Finally, there is the population defined by the range of the species. The species range need not be defined by the limits of favorable habitat. It is quite common for a species at the edge of its range to be well represented by a large, healthy, reproducing population. That local population will be habitat-defined, as are all collections concentrated in favorable habitats. Nevertheless, the fact that this population is peripheral is not a matter of that population occupying the last piece of favorable habitat in that direction. For organisms as diverse as antelope or tree species, it is well known that there is favorable habitat available at the edge of its range, separate but close at hand, but it happens not to be occupied. Prince found the individual southernmost population of the mopane tree in South Africa; it was vigorous, reproductive, and showed normal densities.[21] He was explicit about favorable habitat that was further south and close at hand, but not occupied by the tree. The range of a species is often a statistical matter. With no change in habitat at all, the range can move in at one point or out at another. The limit is set by the probability of local extinction in combination with the probability of accidental invasion of unoccupied, favorable habitat at the species periphery. The processes at this largest scale are different again from those that pertain to all the other more locally defined populations of antelope.

It is possible to erect climatic and topographic hierarchies that pass from local to global considerations. John Harper has analyzed climate and topography at scales so small that in his "microsites," seeds germinate or not depending on how they fit, at the scale of millimeters, into local soil, water, and air relationships.[22]

EQUATIONS FOR POPULATION DYNAMICS

Much of the work on populations is concerned with dynamics of population numbers. The critical processes modeled here are birth and death. We have already suggested that Lotka thought birth was less significant than the conventional models would indicate. Individual deaths and births are events and so, while fixed in time, they are not dynamic in themselves. A death is time dependent but rate independent. However, collections of these events run together to give a pair of dynamics, the birthrate and the death rate. In conventional terms, the balance between these processes gives the total population dynamics.

The equations in populations require additivity, and for organisms in a population, one can often assume that when you have seen one population member you have seen them all. Equations of the population type applied to either ecosystems or communities would generally be unwieldy and analytically intractable. Of course, there are models for ecosystems that consist of a large number of equations, but that is a different matter. Each equation there usually involves only one scale-defined relationship; that is what makes them manageable. In general, we quantify and model communities not with analytical solutions to equations, but rather with numerical, computer-driven solutions. In communities, we will sometimes model them with quantified Markov chains. There, each defined type or state of the community has certain probabilities of becoming vegetation in some other state. Marcel Rejmanek has used two-dimensional Markov chains to capture complicated changes in communities of the Mississippi Delta.[23] All of this is really a bookkeeping device, not so much an analytical treatment. The homogeneity inside populations allows an analytical treatment, where the form of the equation tells of processes in nature.

Some populations are in fact heterogeneous. They may all be sheep, but some of mature age can reproduce, while others are immature and do not contribute to the first age class, the newborn. Leslie built matrices for dealing with sheep of different ages.[24] Sheep within each year class were counted as the same, "you have seen them all," thus achieving the critical equivalence required for population work. For instance, old sheep tend to die, and more frequently the older they are. As with Markov chains, where the arithmetic is very similar, Leslie matrices are largely descriptive rather than analytic. There is some analysis, but not much, as the stable age structure can be calculated; that is, how many are needed in each class to stabilize age structure so that the proportion of those that are in each age class remains constant.

Tommi Lou Carosella moved beyond age classes to use stage structure in cacti on a Wisconsin sand prairie.[25] Age moves inexorably on. Instead of age, Carosella used stages, such that immature plants eventually moved on to reproduce, only to stop reproduction and fall back to a resting stage equivalent to the immature phase. Thus, there was postreproduction that would allow an eventual return to reproduction in these cacti. Clearly, Leslie's mathematics opens the door to making the approach of Giampietro (2003) and Giampietro et al. (2014) achievable. Giampietro went far beyond counting individuals of different ages. He parceled individuals into energetically defined roles allowing a quantified approach to human population resource use.

The value of analytic as opposed to descriptive models is that if you have an equation, each important facet is coded specifically. You can increase the birthrate and see what it does; or you can alter any process or relationship in an exploration of what happens. In descriptive models, you can get very good simulation, but you do not know exactly how you achieved it. Neural net analysis is a powerful descriptor, like the gradient analysis techniques, but you cannot say how it does what does it. The neural net is creating the equivalent of a very long polynomial, pages long, without spelling it out. They are enormously more powerful than equations and modeling phenomena, but you have no idea about specific mechanisms. Calibrate the net a second time and you may get the same predictive power, but it will be with a different if similar polynomial. By contrast, population biology is generally analytical, and its scientific culture tends to eschew actively more descriptive approaches, to the detriment of the whole discipline of ecology.

It is argued by mechanistic ecologists that analytic equations allow the general condition. The argument is that there is an infinite set of values for variables and initial conditions that can be treated by just one tractable equation. But with characteristic brilliance, Rosen points out that only a small minority of equations can be solved.[26] In fact, the general condition is not the infinite variants of soluble equations, but is that equations generally cannot be solved. As analytically tractable as the subset comprised of population equations may be, many equations yield only to numerical calculation on the computer. The method is to get a computed output for one version of the equation, and then do it again with the variable changed at random. That is what Gardner and Ashby did when they calculated the probability of various setups of small networks being stable.[27] Change the variables a thousand times, and you get an average estimate of stability, something not available to analysis and solution. That is just one device for getting answers without solving the equation analytically.

If birth and death respond to external influences at different rates, they demonstrate different relaxation times. A relaxation time is the time taken to recover in a population moved a unit distance from equilibrium. The interaction of disparate reaction rates can generate complex behavior, and population equations are no exception. Even simple equations can yield very complicated behavior if birth and death rates react over incompatible time frames. Remember the general rule that complexity arises from the interaction of differently scaled processes, an interaction between different levels of organization. Small differences simply yield nonlinear behavior, but the nonlinearity can be so great as to lead to folds in the surface. Then functions fail and leave only relations. We discuss functions in contrast to relations when we consider David Roberts's clique analysis in chapter 4.

The instincts of conventional population biology will be to recognize the special features of birth as opposed to death and fine-tune them. That fine-tuning will have made population dynamics more realistic. Death is not the simple opposite of birth, but longer equations could describe the difference. In this way, instincts would lead conventional population biology to be tempted to make the equations longer. After Lotka and Giampietro have shown that birthrate is not as important as has been thought, further elaboration is only likely to obscure the mistakes rather than lay open population processes transparently.

Let us start with simple growth equations. With the appropriate scaling of variables, the exponential equation matches the geometric equation. The exponential expression of growth turns on an instantaneous growth rate term, r. The geometric expression turns on doubling time, how long it takes the population to double in size. The geometric expression can be applied more readily to situations where growth is discrete. Such would be the case of sheep, whose numbers only increase in the spring. As a result, the growth is calculated only once a year. Bacteria are generally modeled with continuous growth equations where r prevails. The continuous growth equations generally invoke e, the base on natural logarithms (figure 6.4A). When compound interest rates are calculated, there is the issue of when you get to apply interest on the interest. If this happens once a year, you go a year before you get interest on the interest earned by January 2. The shorter the time to recalculation, the more interest you get for a given percentage interest. If the interest is compounded instantaneously, you get the most return possible on a given percent interest rate. The base of natural logarithms allows the calculation instantaneous growth. That is how e applies in the exponential expression of growth.

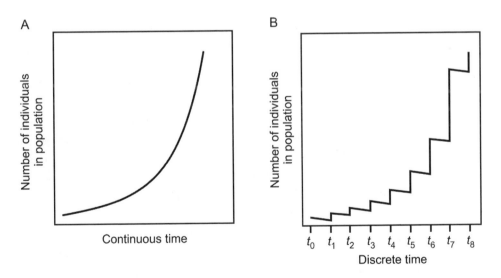

FIGURE 6.4.   A. When a population is modeled with a differential equation, the change in population numbers is continuous and a graph of the population over time gives a smooth line, in this case, of exponential growth. B. When a population is modeled with a difference equation, the population numbers are updated intermittently (discretely) and therefore a graph of numbers of individuals in a population over time follows a jagged line.

Where $N(t)$ is the number in the population at time t, and $N(0)$ is the number in the population at the beginning of calculations, and λ is the growth term, the two equations are:

$$\text{Exponential } N(t) = N(0)e^{rt},$$

$$\text{Geometric } N(t) = N(0)\lambda^t.$$

The two equations can be linked and shown to be the same thing by:

$$\lambda = e^r \text{ and } \log_e \lambda = r.$$

The reason we use higher levels of abstraction is that at the correct higher level of abstraction, the situation becomes simple and easy to grasp. So let us move to differential equations because they will make things simpler (figure 6.4*B*). Differential equations do not give the number in the population, but rather the change of population numbers coming from the given N. The *r* term gives unfettered growth, but of course populations do not grow exponentially for long. For low population numbers, growth is essentially what naked *r* will allow. But as N comes close to the limit K, the growth slows way below what *r* would allow unfettered.

The limiting term is called the carrying capacity, and is conventionally assigned to K. The growth-limiting equation often takes the form of:

$$dN/dt = Nr\,(1 - N/K).$$

In his general treatment of a single population's growth, Robert May (1981a) used this expression as his starting point, and gives a clear explanation as to how to build up to chaos theory.

As growth starts, K does not limit *r*. The difference in population numbers is positive and as large as r will allow. But as N gets larger, the term N/K increases in size. The N/K relationship is inverted by subtracting it from one to give (1 – N/K). The naked Nr term is multiplied by a number less than one, and this slows the growth rate. When N = K, (1 – N/K) becomes zero, and growth is zero. The population sits stably at K, no matter how big is *r* .

If a lag is introduced, we do not use the present N, but some N that occurred earlier. That is a routine protocol. We might need to adopt it because, for instance, the number of phytoplankton is not determined by the zooplankton today, but rather by the number eating phytoplankton some time earlier. The number of zooplankton some time ago left what phytoplankton you see now. In continuous growth equations, you have to put the lag in explicitly. This is done by changing (1 – N/K) to N(t – T)/K, where T is the amount that the lag is set back in time. But in discrete growth equations, say for populations of sheep that only give birth in spring, the lag is built into the period that you wait before reapplying the equation next spring. The wait embodies a lag.

If the lag is very short, the equation asymptotes to K. If the lag is a bit longer, the equation over- and undershoots K, but the oscillation damps down. By using an N in N/K from earlier, a population now at K might still grow (overshoot) because the equation has not picked up that N is already at its limit of K. It will only find that out after the lag has passed. Therefore, it can grow larger than K because of using the earlier N. In a couple of iterations later, N/K is greater than one because of the overshot numbers coming from earlier. Thus, (1 – N/K) becomes negative and the growth becomes negative.

This is a delayed negative feedback. We can use a dimensionless expression of lag and growth rate to predict what will happen. The essential difficulty is that the lag is in units of time, but the growth rate, r, is not. We have already mentioned relaxation time, the time for recovery of a disturbed population. In fact $1/r$ is a good approximation to the relaxation time. With $1/r$, the trick has been to convert r into units of time, the time taken to relax. Lag time divided by relaxation time gives a dimensionless number that captures the essence of the effect of the lag.

Both increasing the lag and increasing r has the same effect on the behavior of the equation. A bigger r allows more to happen in the time of the lag. A longer lag gives more time for r to do its work. So let us get back to predicting the behavior of the equations. If the relaxation time is less than the lag, that is, the dimensionless expression is less than one, the equation will asymptote to K. If the lag is greater than the relaxation time but not twice the lag, the dimensionless expression will be between one and two. In that case, there will be some oscillation around K, but it damps down. In the end, the equation will rest at K. But if the lag is more than twice the relaxation time, that is, the dimensionless expression is greater than two, something remarkable happens. The equation cannot find K. It does stabilize, but if it is below K, the lag is long enough to let the equation overshoot above K as much as it was below it. The equation goes into a two-phase cycle of perpetual over- and undershoot.

But the complications do not stop there. A further increase in r makes the lag more important. Whereas in the two-phase cycle the stability point at K bifurcates to above and below K, at higher values of r, the equation cannot find the stable over- and undershoot values. They too bifurcate. The equation goes into a four-phase cycle: up a lot, down a lot, up a little, down a little. At slightly higher values for r, the equation moves to an eight-point cycle, and then a sixteen-point cycle. Notice that the bifurcations increase the number of stable points by two (figure 6.5A).

With only a slightly bigger r, the split is into three points. When that happens, the system is chaotic. Chaos has some interesting properties. All the previous bifurcations were to points that are regularly visited. By contrast, a chaotic equation

A

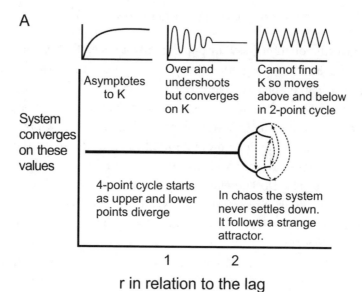

System
converges
on these
values

Asymptotes
to K

Over and
undershoots
but converges
on K

Cannot find
K so moves
above and below
in 2-point cycle

4-point cycle starts
as upper and lower
points diverge

In chaos the system
never settles down.
It follows a strange
attractor.

1   2

r in relation to the lag

B

Models, states, and derivatives

- The horizontal axis gives
  the state.

- The vertical axis gives
  change—1st derivative

- Pendulum moves in a
  phase-space attractor

- In chaotic systems we
  use the acceleration as
  we plot the phase space
  in 3D...strange attractor.

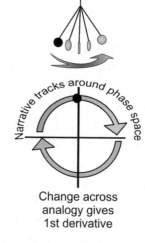

Narrative tracks around phase space

Change across
analogy gives
1st derivative

State

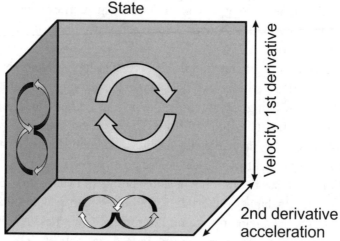

Velocity 1st derivative

2nd derivative
acceleration

never visits any point more than once. That is, it never settles down to any discrete pattern. Another manifestation of that phenomenon is that chaotic equations are infinitely sensitive to initial conditions. You cannot ever repeat a complex situation with sufficient care so that it will not turn out differently, and soon too. Run the Roman Empire again, and it will turn out differently. Grow a tree from a seed, and it will never repeat the exact form of any previous tree of its species. There is an infinity of points that the chaotic equation can visit, but it is a bounded infinity.

With this bounding of the infinity of possibilities, chaos does not mean anything can happen. If we plot a graph of the value on N this time as opposed to next time, a pattern emerges. If the equation settles down to K, this time and next time in the end will always be the same, K. You can predict that without running the equation because equations of that sort always settle down. A similar graph for the two-point cycle will also settle down, but to two points, not one. The four-point cycle settles down to keep visiting four points. And all this you can predict from the two-and four-point equations, without having to run the equation on. But in chaos, the plot never settles down. Even so, this time to next time is not a random change. The pattern settles down to clusters, not points. You can predict the next point, but only with a certain probability inside the cluster. It will move to the next cloud cluster, but you cannot predict where in principle. All you can do is run the equation out to see what happens. There is no fixed point. In the nonchaotic oscillators with one,

FIGURE 6.5.   *A.* The ordinate plots the stable value(s) of population numbers visited by the equation. The abscissa plots the dimensionless number that captures the relationship between r and the lag time. The inset graphs above show the respective behavior of the equation over time at the pertinent dimensionless value. Below 1 on the abscissa, there is insufficient effect of the lag to achieve any overshoot. Between 1 and 2 on the abscissa, the equation demonstrates a damped oscillation, where each over- and undershoot is smaller until the value rests at K. Once the dimensionless expression is above 2, the equation bifurcates. It stably over- and undershoots K, never resting at K, but following the two-point arrow. At slightly higher values, the stable over- and undershoot points themselves bifurcate, and the cycle has not just two but four repeating values. This cycle is diagrammed with the four arrows, as shown on the right-hand end of the main graph. Further increases in the dimensionless expression lead to chaotic movement, where the equation not only ever rests, it never repeats any value. While chaos offers an infinity of values, it is a bounded infinity. Values change within certain limits and patterns (May 1981a but significantly altered). *B.* Equation output can be plotted as N over time. Another way to express equations is in what is called "phase space." The two-dimensional expression of a frictionless pendulum plots the position of the pendulum against its change in position, its first derivative. In that 2-D space, the pendulum follows a circle. We can create a 3-D expression. To do this, insert the second derivative; the acceleration of the pendulum and the circle becomes a figure of eight on the other facets. A pendulum fits on an attractor. A pendulum with friction spirals in to the point of hanging straight down; it would spiral in to the center. A chaotic equation moves in phase space following a general pattern, but the track never repeats. The phase space tracks behavior in a three-dimensional space. In 2-D, chaotic tracks appear to cross, but the track is always different because of the third dimension of acceleration. Such attractors that never repeat are called "strange attractors" (May 1981a).

two, four, eight, or sixteen stable points, you can predict next time from this time in principle exactly, and without having to run the equation out. Chaotic equations do settle down, but to movement along a track, not a set of points. Even when the system has settled down to its particular chaotic tracking chaos there is always a stochastic aspect to "next time."

Now let us reintroduce phase diagrams (figure 6.5 B) for simple systems. Remember the plots of system state, against system first derivative, its velocity. One can do the same for the second derivative, its acceleration. The second derivative matters because it implies an environment, a surrounding force field. A plot of phase diagrams for systems that asymptote (figure 6.5 A) move simply to a point. Systems that oscillate may spin around before reaching the equilibrium point. A pendulum with friction spins around in phase space until it settles hanging straight down (figure 6.5 A, middle panel). It eventually finds its equilibrium point. A pendulum without friction, if there could be such a thing, would simply keep spinning around in a circle in phase space (figure 6.5 B, right-hand panel). It is a continuous version of a two-point cycle, with two equilibrium points. Four-point cycles settle down to two cycles that alternate. Notice in all of these systems—one-, two-, and four-point cycles—that the settling down makes them variously some sort of equilibrium system. They are various versions of an attractor. With chaos, the system follows what is called a strange attractor. In chaos, the cycling never settles down to a point, but it still does not go anywhere or everywhere. Start somewhere in the phase space, and soon enough the system settles down not to a point but to a chaotic track. The attractor is now called a strange attractor because it has a general path but it never repeats. In a two-point cycle, the system cannot find K. In chaotic systems, the equation cannot find any stable distinctive point or set of points, but it can find the strange attractor. After the system is at equilibrium, you lose all information about how it got there. Similarly, once a chaotic system gets onto its strange attractor, you cannot see how it got there. So elaborate and infinitely rich as chaos is, it is still an equilibrium theory. It is just that equilibrium keeps moving.

Chaos taught us a lot about stochastic and deterministic models. Chaotic equations are deterministic; they spell out all subsequent states once you fix the starting point. In contrast to deterministic models, stochastic models always have a random component somewhere. So the quandary is that chaotic equations are deterministic, but deliver results in probabilistic terms. The translation of determinism into stochastic terms goes both ways. Alan Johnson created a cellular automaton that had rules for how new pieces would appear on the board.[28] It was, however, a stochastic automaton, with pieces appearing within the rules, but only with a certain probability. If he got his specification just right, he could get it to always create a checkerboard pattern. In other words, he could get a stochastic system to give results that are apparently deterministic. Probabilistic patterns of stochasticity can be simply alternative expressions of the same thing, counterintuitive as that may be.

Devotees of chaos theory declare that chaos is about complexity. Our rebuff is that it is only trivially so, and not really interestingly so. There is so much more to complexity theory than chaos. The important contribution of chaos theory to complexity is the way it shows the minimal condition of complex behavior. Chaos applies to fractals, and as we said in chapter 2 on the landscape, fractals take zero and infinity seriously. Fractal geometry shows patterns that are consistent, even when they invoke those two undefined numbers. The general condition of complexity arises when the situation has not been defined, and indeed is undefinable, but we can still deal with it. The nub of it is that complexity invokes the undefined and the undefinable. If you have not defined the situation with a paradigm, it will be complex. That is why scientists must use paradigms for everyday business, where complexity has been defined away. There are lots of reasons for something being undefinable, but chaos shows us the minimal condition for that. Complex systems would not work out the same if you ran them twice. But they are still regular. We are not very good at predicting in the face of complexity, but sometimes we can to a degree, exactly because of that regularity despite undefinability. Almost certainly a given measured complex situation is not strictly chaotic; it does not have an equation, although it may in some conditions yield to chaotic description. But chaos shows why we should not be surprised at infinite sensitivity to initial conditions in measured systems. Chaos shows the minimal conditions for an inability to run the system twice and get the same result. If short equations of chaos can be regular in a funny way, then that is the minimal condition for more impressive and general complexity.

We have noted that the analytical methodology of population equations allows us to pinpoint causes of phenomena in equations. For instance, r, the growth term, is more important for phenomena associated with pioneer species quickly occupying open space. Weedy species therefore have been called r-selected. Meanwhile, species that stay close to the carrying capacity are said to be K-selected. With a lower r, the K-selected species tend not to overshoot the carrying capacity, which would despoil their environment. In general, these ideas have been used crudely to say things like whales are K-selected, while bacteria are r-selected.[29] At one level of analysis this is true enough, but it is trivial. Ecologists like the big conventional hierarchy in biology to write something large across the sky about levels of organization. We have argued that the convention is not so much wrong as it is not general. The crude use of r- and K-strategies invokes a similar mind-set of stuffing everything into one conventional bag. The problem with comparing whales to bacteria is that they have almost nothing in common. We have critiqued diversity in the same terms; most places do not have the same causes for their diversity. All statements in science have at least an implicit partner that asks, "As opposed to what?" Sometimes the counterpoint says "ceteris paribus," that is, "with all else equal." There is very little equal between whales and bacteria, so there is very little context to comparisons of them. Much more interesting is that minke whales are small and their populations grow relatively fast, as opposed to the slow growth rate of blue whales. To say that minke whales are r-selected actually says something. Similarly, some bacteria from rock strata deep underground appear to stay dormant for astonishingly long periods. They are K-selected bacteria. That also notes a significant phenomenon.

We have already introduced notions of high and low gain in chapter 2 and earlier in this chapter. They map onto r- and K-selection to a significant degree. The particular power of high and low gain is the introduction from economics of the notion of profit and marginal return. Average and marginal arose as we discussed Giampietro's model for the economics of ecological systems. Over time, a convex curve for return applies in both average and marginal return. When the average return is still increasing at forty-five degrees, the marginal return is flat. That is normally the point where biological and human resource extraction quits because the increasing effort is not seen as worth it. The literature of high and low gain is clear on the point that the two patterns of resource use are interchangeable, depending on how you look at the system in question.[30] Whereas r- and K-selection are commonly taken as absolute statements, high and low gain are distinctly relative.

We have a nice comparison available for two species of booby seabirds, where one is high gain and the other is low gain. There is a lot of all-else-equal in the comparison. Boobies are coastal birds that fish inshore and offshore. The masked booby fishes offshore and accordingly must put more effort into resource capture than the blue-footed booby that fishes inshore. The offshore fisher must use what resources it can capture frugally. Therefore, its first hatched chick commits siblicide, tossing out eggs and other chicks. "Obligatory siblicidal masked boobies virtually always lose their second of two hatchlings to siblicide within days of hatching" (Lougheed and Anderson 1999:12). Thus, the limited food supply is used efficiently so that only one chick is raised, but with a good chance of survival. The inshore fisher is high gain because less effort is needed to capture food. Its chicks do not commit siblicide so that there is more potential for population growth in a good year, with several chicks surviving. High-gain inshore fishing cashes in on a high rate of procurement, and so is a rate-dependent description. Siblicide is rate independent, it just happens. Thus, the efficiency of raising just one chick is a rate-independent, low-gain phenomenon. Rate-dependent thermodynamics has no values and is just what happens. Efficiency works with some preferred state, and that is a rate-independent value. High and low gain capitalize on that distinction.

Most interesting in the boobies is that context is crucial. If eggs are switched between species, whether or not there is siblicide depends on the nest, not the species of the egg. Cues coming from the food regime and its rhythms determine chick behavior. So with that example of low-gain siblicide, we turn now to cuckoos and cowbirds and their nest parasitism. The parasite lays an egg in the nest of the host. It hatches quickly and throws other chicks or eggs out, so that the chick of the parasitic species gets all the food and attention. But note that nest parasitism is a more high-gain activity, but at a higher level of analysis. In nest parasitism, the high gain is getting other birds to do the rearing; it increases reproduction rates for the parasite. Recent work on cuckoo nest parasites of carrion crows shows greater survival of host carrion crow chicks with parasitism. When disturbed, the cuckoo puts out a noxious "repellant secretion." This appears to offer parasitized nests protection from predators such as cats.[31] We have here another level of analysis. The high- and low-gain strategy distinction is much richer than its r- versus K-selected counterpart. It looks at the economics of the situation, not just the r and K in the equation. High and low gain are about choosing the level of analysis; it shows how to choose the equation. It is not just a simple disaggregation of the equation into r and K.

EQUATIONS FOR TWO SPECIES

There are many population-based relationships involving two species. While we could give some of the equations from the literature as to how these relationships have been modeled in particular cases, we prefer to dwell on how there is a general architecture to two-species population equations. Given that almost all these equations are counting the wrong thing, organisms instead of vessels filled with available energy or other resources, there are several specific ways to model particular relationships in the conventional use of equations, so just choose one. For us to trot out examples of particular forms seems gratuitous, more a history of inappropriate uses of birth as a term in equations.

The culture of population biology may be offended by our position, their champion Lotka notwithstanding, because the discipline is populated by scientists who are often enough limited to understanding only through equations. The Wikipedia accounts of Fourier transforms and neural nets are of little general help because they consist of pages of equations. We have presented much more intuitive explanations of both those devices, with no equations at all. We have done the same for neural net analysis. The best of the best think just as effectively in geometric terms and minimize algebraic approaches. For instance, R. A. Fisher thought about analysis of variance in geometric terms that create deeper insight than do its equations. Herr refers to Fisher's early geometric approach and goes on to laud "William Kruskal's elegant 1975 paper on the geometry of generalized inverses."[32] Herr goes on to say, "the relative unpopularity of the geometric approach is not due to an inherent inferiority but rather to a combination of inertia, poor exposition, and a resistance to abstraction." A colleague complained of our last edition that it did not have many equations. That misses the point that the nub of the issue is conveying understanding, not a slavish devotion to a particular device for exposition. With regard to mathematics for ecology, our advice is learn enough about handling equations so that you can learn new mathematics as you need it. Even mathematicians specialize in the mathematics they do, and that gives ecologists permission to do the same.

But, with regard to equations, there are general principles that do apply across particulars. Those principles are worth presenting here, and we do so because they can easily be used as a general guide as to how to model the right things, those that Lotka and Giampietro recommend. When you have seen a couple of two-species relationships, you have pretty much seen them all. Parasitism, predation, and competition can all be cast in terms of the simple, constrained single-species population equation.

After the equal sign, the front of all these equations is something like Nr, the growth expression for the population being modeled. The rest of the equation is some version of $(1 - N/K)$. The trick with multiple-species interactions is that the N in N/K comes from the other species—the prey, the predator, the parasite, or the competitor. The bigger the predator population, the more the K for the prey is forced down. The bigger the prey population, the more food there is for the predator, whose K moves up. Predator-prey relationships are negative feedbacks with a delay. The delay comes from the prey meeting its end instantaneously, while the predator equation to which the prey is linked behaves more slowly. It takes time for a predator to die of starvation. Living longer because of a meal by definition takes longer than the demise of the prey item. It takes time for more food to translate into more reproduction. The predator is in a cycle that is about half the phase behind the prey cycle (figure 6.6).

Competition is basically a positive feedback. As one competitor succeeds, it puts the brakes on the other. It does this by once again imposing its N on the N/K equivalent in the competing species. The more constrained species can apply less pressure on the winner. The positive feedback may continue until one species is extinct, or until a position of mutual constraint standoff (i.e., diminishing returns on energy invested) is achieved in an emerging negative feedback on effort, in a sort of diminishing economic return.

Predator-prey equations are negative feedbacks that involve constraint at the level of direct interactions between small numbers of populations. Predator-prey equations are generally stable in themselves, and need not involve invoking constraints from the community. It is possible to look at complex food webs as community structures, but that is a different matter.

Mutualism, however, is a positive-feedback relationship where different species benefit each other. There has been little work involving formal equations on mutualistic systems. One reason is that the mathematics of mutualism is not as

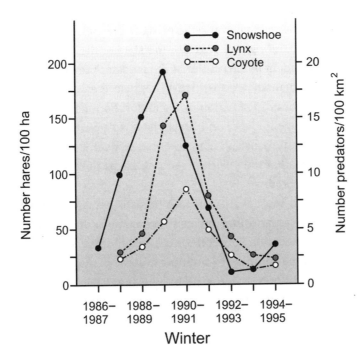

FIGURE 6.6.   Lynx, coyote, and snowshoe hare population cycles, and their estimated densities in northwest Yukon (O'Donoghue et al. 1998).

transparent as the simple difference and differential equations of prey and predator. The other reason is that being a positive feedback, mutualism is inherently unstable. Positive-feedback systems are hard to study because the very relationship one seeks to study would be transient if it were left to its own devices.

Inside every negative feedback is a positive feedback trying to escape. Mutualism is sometimes that positive feedback. The negative feedback is the constraint that holds the mutualistic system in check, the context of the mutualists. The limits of resource for filling the vessels come to the fore. By the time we find a mutualistic system, its dynamics are not principally determined by the mutualism. Rather, the positive feedback of mutual benefit is held in a context that limits the system state. If the constraint is unchanging, then so will be the state of the mutualistic association. This constancy would prevail even if the parties to the mutualism were changing their relationship by intensifying it or weakening it. Sometimes the mutualist flirts with an ambiguous parasitism or commensalism. For instance, the mutualism embodied in the root mycorrhizae is a case of downright predation on the root by the fungus in the spring. So mutualism is not straightforward and may readily change. Any change in state of the parties to the mutualism will be a reflection of a change in constraint that gives more resources to, or takes more resources from, the mutualistic system as a whole. The direct effect of change in mutualism is not to produce change but to keep the system constantly pressing up against resource limitations. The same is not true for a predator-prey system because the predator-prey relationship itself provides its own constraint. Accordingly, changes in the state of the prey and predator can sometimes be independent of the context of the predator-prey system.

Equations for mutualism are distinctive because they are positive feedbacks without an obvious bound. Extinction or a standoff bounds the positive feedback of competition. DeAngelis, Post, and Travis have written the definitive work on the quantification of mutualism.[33] The essential difficulty in the mathematics of mutualism is the way that the positive feedback keeps going until it destroys the situation to which it pertains.

All this means that mutualism, and not predator-prey systems, are the more suitable devices for relating communities to population considerations. The community is the accommodation between community members and it represents the set of constraints on the community members. Because they have their own built-in constraints, predator-prey systems are to an extent isolated from constraints at the community level. Certainly they represent a more complicated situation compared to mutualistic systems.

PREDATOR-PREY PATTERNS

We have spoken of predator-prey equations in general. The Lotka Volterra equations are linear and exhibit neutral stability.[34] In other words, the oscillation that is set at the outset persists (figure 6.7). C. S. Holling and his colleagues have used nonlinear equations that capture more of the reality, as he puts it. Holling makes a general statement about simulation with population equations.[35] There are four aspects of the modeling and you have to choose. The model can be realistic. It can be simple. It can be accurate. It can be precise. However, Holling insists that you cannot have it all. In his simulation of predation (the mechanics of insects grabbing and eating), he abandoned simplicity:

> Recent studies of predation, however, have shown that it is possible to achieve great analytical depth and simulate whole systems in the form of realistic and precise mathematical models. This is accomplished by ignoring the degree of simplicity traditionally required of population models and by emphasizing the need for reality . . . the form of the explanation and the resulting equations is hence dictated by the process itself and not by the need for mathematical neatness. The considerable complexity of the predation model arose from features common to many biological processes i.e. the prevalence of limits and thresholds, the presence of important discontinuities and the historical character of biological events. (Holling 1964:335).

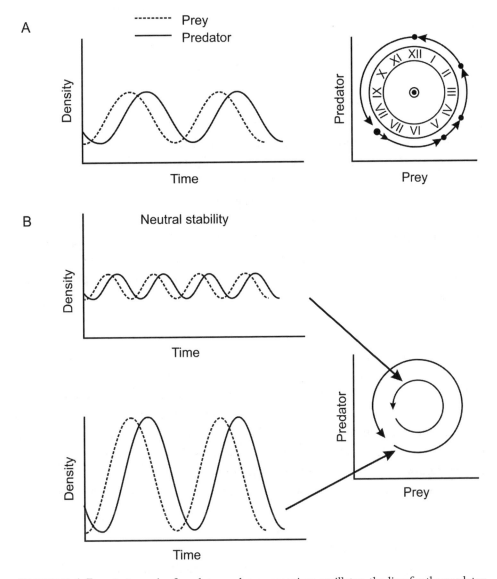

FIGURE 6.7.   *A.* A graph of predator and prey over time oscillates, the line for the predator lagging behind that of the prey. Plotting prey numbers against predator numbers, the trajectory of predator-prey relationships follows a roughly circular pattern. Once around the circle brings the predator-prey densities back to their original numbers. We are interested in general patterns here and therefore have normalized the numbers of prey and predator to their minimum and maximum values. That way, we can get rid of differences caused by loss of energy in transfer to predators, making their aggregate biomass smaller. With the normalization, we also can get rid of differences in numbers due to prey and predator organism size. *B.* In predator-prey equations, there was neutral stability. In this situation, the initial deviation from equilibrium, predator and prey numbers persist through time. The pair of predator-prey values never reaches equilibrium together. They continuously oscillate the same amount above and below the unachievable equilibrium condition.

Holling investigated the effects of manipulating system stability by changing the efficiency of the predation. He tinkered with the lag in the system by altering its fastest reaction rate.[36] His particular findings are related to the form of the equation that he used to study host-parasite systems, but there also appears to be a significant generality to his results into prey and predator.

At the lowest level of abstraction, one graphs the numbers of the predator and prey over time. However, a more insightful image takes the same system states and puts them onto something close to, but not exactly, a phase plane: predator and prey numbers plotted against each other. The trajectory of the system on the predator-prey plane is roughly circular. We put a clock face on figure 6.7A and have also labeled the respective time that corresponds on the left panel. In this way, the two sides of the figure can be easily related. Let us start at seven o'clock, where both prey and predator populations are low. With few predators, the prey increase in numbers, and the hand moves to five o'clock. With more prey, the predators move up in numbers, to four o'clock. With yet more prey, the predators increase further. The prey stop growing and then reverse in a diminution of the prey population. This is the move to two o'clock. The declining prey start to limit the predators, which themselves stop growing at twelve o'clock. A period follows when both prey and predator decline, leading us back to seven o'clock where we started. At that time, with low predation pressure, the cycle starts again.

In the most common linearizations of predator-prey equations, there is a neutral stability (figure 6.7B). In Holling's more sophisticated equations, there is the potential for damping of the oscillations until the system comes to rest at coincident equilibrium populations for both predator and prey (figure 6.8A), like a pendulum at rest (figure 6.5B, but spiraled inward). However, in Holling's equations, if initial conditions involve very low or very high populations of either prey or predator, then the oscillations amplify until one or the other population, or both, is driven to extinction or increase without bound (figure 6.8B). In terms of the predator-prey plane, there is a safe region in the middle of the plane, and any initial condition inside that region leads to damped oscillations and equilibrium (figure 6.8C). There is, however, a peripheral region where either or both populations are extremely large or small, leading to amplified oscillation. Initial or perturbed conditions then take the system to these peripheral regions, leading to amplifying oscillations and extinction. In figure 6.8C, the waving lines are not prey and predator, but a prey population stable in a safe regime (staying in the gray zone) and again prey populations that are spinning away from the safe zone. There will be respective predator populations in both cases, but they are excluded for clarity.

Holling introduced a yet higher abstraction as he expressed the predator-prey plane of the response surface as a cup and ball. Sectioning the cup shows three equilibria, one stable and two unstable. The bottom of the cup represents a stable equilibrium to which the system tends to return. The lips of the cup indicate

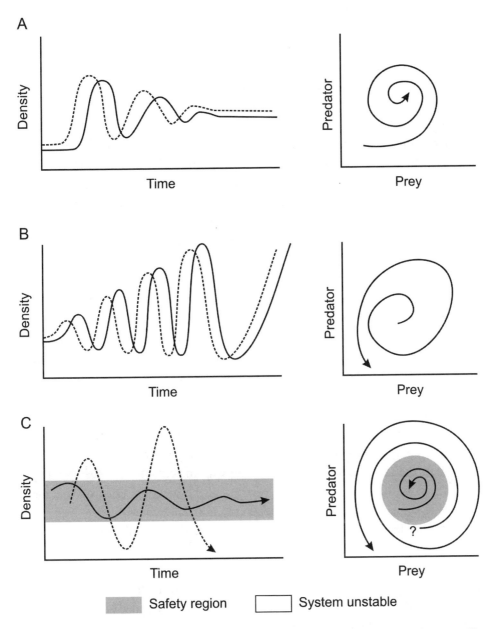

FIGURE 6.8. *A.* In Holling's more sophisticated equations, the predator and prey oscillations may diminish until the population values damp down to an equilibrium value. *B.* Again using Holling's equations, very large or very small predator or prey populations can produce amplifying oscillations until either the predator or prey goes extinct, or both. *C.* Inside a range of very high and very low values for predator and prey, there is a stable region within which the population will spin down to an equilibrium condition. Only the prey are plotted on the two runs of the system. The question mark indicates the point on the trajectory where there has been a critical move outside the region of stability.

unstable equilibria, in that the ball can balance on the lip, but will tend to move away on one side or the other if it is perturbed. The gray safe zone of figure 6.8C is captured in the cup in figure 6.9A. Notice the question marks on these two figures. They indicate the place where there is a transition from the safe zone to the unsafe zone whereupon the system spins out. Holling worked with host-parasite equations that are very similar to predator-prey. Holling altered the efficiency (k in figure 6.9B) of his parasite. With k, he manipulated the implied lag in the system, and that changed the shape and size of the safe region. The equilibria in the system changed in interesting ways (a stable equilibrium for the bottom of the cup, and two unstable equilibria at the lips of the cup). We can pair Holling's movements on the host-parasite plane with an implied shape of the cup. The movement across the response surfaces is on the left side of figure 6.9B, while the respective implied cup and its stability properties are paired on the right. The changes in the shape of the cup sum up the changes on the host-parasite plane as Holling changed the efficiency of the parasite. The steepness of the sides of the cup refers to the speed with which the ball moves to equilibrium. Once in a flat-bottomed cup, the ball may roll around for a long time before settling to equilibrium (the top panel of figure 6.9B). With increased efficiency of the parasite, the safe zone narrows and the walls of the cup come together. The cup loses resilience. In a steep-sided cup, the ball plops into equilibrium in a single cycle of host and parasite. The narrowing cup refers to the speed with which equilibrium is found, not the area on the phase plane; thus, the phase plane widens while the cup narrows. One expression shows propensity, while the other indicates response to disturbance. The cup representation is of system dynamics. The bottom panel of figure 6.9B shows how the safe region disappears along with the right side of the cup. In the bottom pair in the figure, notice the reversal of the diagonal track on the plane, and of the small arrow in the cup representation. The reversal is the loss of the unstable equilibrium on the right lip of the cup and therefore the loss of the stable equilibrium in the bottom of the cup.

A confused population literature uses words like "population resilience" (capacity to bounce back from a disturbance), "resistance" (the ability of the system to hold its state despite outside influence), and "stability" (the speed with which stable equilibrium is achieved). Inconsistent usage prevails; definitions we suggest here are switched between writers. The dilemma is in identifying resistance to being displaced, as opposed to what the system does when it is displaced. The point of tension is displayed in figure 6.9B, where beginning to lose the right unstable equilibrium (edge) of the cup corresponds to a steepening of that side of the cup. Meanwhile, the response surface expresses further movement away from the bottom of the cup. The steepness of the side illustrates the speed of recovery to equilibrium, while the corresponding widening of the safe zone on the response surface indicates fragility of the situation in that direction.

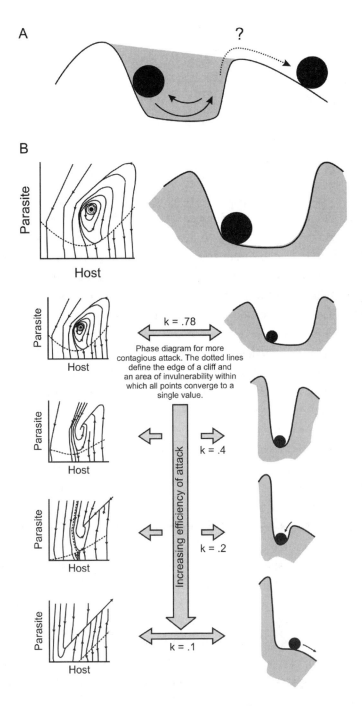

FIGURE 6.9.   *A.* It is helpful to picture the "safe region" in figure 6.8 as a response surface plane and a paired ball and cup representation. The "safe region" is the cup. The question mark indicates the place where the ball moves to a new region, from stable to unstable. The move to instability occurs with a move over the edge of the cup. *B.* Holling and Ewing modeled the host-parasite relationships changing the efficiency of assaults (k) on the host. This produced a series of response surfaces ranging from a local stable attractor at the top to the loss of all stable eqilibria at the bottom of the series. The top image is larger only for detail and clarity in the first point of reference. The lower parts of the image show successive increases in efficiency of the parasite. At first, efficiency allows the system to reach the stable point faster. But later increases in efficiency erode one wall of the safe zone, until it disappears (Holling and Ewing 1971).

We can apply these host-parasite ideas to predator-prey, and human use of resources. If the predator is inefficient, then the cup was wide, flat-bottomed, and is clearly bounded on all sides (figure 6.9B), a resilient system in Holling's terms. Holling suggests that natural ecological systems exhibit this form: a wide, well-defined range of states, across which it often passes, and from which it is unlikely to be thrown by less than catastrophic disturbance. The steepness of the side of the cup indicates more control and efficiency, but the corresponding expansion of the safe area on the response surface expression indicates vulnerability in that direction. Thus efficiency and speed of return from disturbance are counterintuitively related to vulnerability to disturbance. Conversely, slack in the response corresponds to safe stability, despite intrinsic oscillations. Holling notes that traditional tropical slash-and-burn agriculture of the Tsembaga in New Guinea follows a track involving slack and oscillation.[37] They have a pig taboo that causes the pig population to rise unchecked, storing rich protein resources, until the tribe goes on a pig binge in times of strife. The advantage of inefficiency is that it leaves lots of slack in the system. The last thing the Tsembaga need is a well-meaning infusion of Western-style, efficient cropping, for that would reduce their slack. Tsembaga inefficiency leads to their long-term stability because of resilience.

If the equation of host-parasite relation is adjusted to make the parasite more efficient, then the cup narrows and its steep sides become close (figure 6.9B). Equilibrium is achieved more quickly, but at the cost of the area of the safe region on the response surface. As long as nothing is seriously wrong, such a system tracks the equilibrium closely, but at the cost of certain fragility. This, Holling asserts, is like the Western industrial agribusiness, where efficiency leads from wood lot to cornfield in one season, a system where government programs can reclaim or set aside vast tracts of marginal land with the stroke of a pen on an incentive program.

As Holling's parasite becomes more efficient again, the sides of the cup become even steeper, as the system reaches the stable equilibrium faster. The price is that one side of the safe region begins to erode away. Eventually, further increases in efficiency remove one side of the cup altogether. Note the reversal of directional arrows at the bottom of figure 6.9B. At that stage, the system has only an unstable equilibrium, and no stable equilibrium; extinction or uncontrolled growth is the rule. Holling is anxious that the industrial global village is approaching that condition.

We would wish to modify Holling's assertion that natural systems have plenty of slack and suggest that natural systems can exhibit unstable architecture with no stable equilibrium. We would say that the system is stable as long as there is at least one scale at which the system exhibits slack and inefficiency. We can generalize Holling to our scheme that invokes level of analysis. Consider Paul Colinvaux's observation that lions are inefficient predators that rarely strain the capacity of their prey to provide a food resource. In contrast, Colinvaux notes that insect predators are deadly efficient. A single predator like a ladybug finds a prey population of aphids and proceeds to kill its prey as if they were sitting ducks.[38] Furthermore, the predator reproduces quickly to generate a large, ravenous population of beetles that finally eat the aphid prey to local extinction.

Clearly, the lion exists with lots of slack, even slack enough to ride out extended periods of drought on the Serengeti. Lions live in a relationship that has a broad flat cup in figure 6.9B, where the population is neither large nor quickly responsive to abundant prey. Meanwhile, the ladybug is ruthlessly efficient and given to explosive population growth in the face of abundant food. With the arrival of one predator, the local prey population is doomed to extinction in a short time. There is no stable equilibrium because the predator is too efficient. If, however, we aggregate to a larger system in a move to a higher level of organization, then the problem for the insect predator is in finding not items of prey, but new prey populations. There we find the requisite inefficiency. Ladybug slack is in the inefficiency of finding new prey populations. When it comes to killing a single item of prey, the pride of lions is reliable and efficient. There is no means available to lions to be prudent and only half-kill a single prey item, for it is all or nothing; big cats gorge themselves if they eat at all. Thus, the insect prey population corresponds to a single lion's kill, and the population of ungulates comprising the lion's food corresponds to a metapopulation of insect populations scattered across the landscape. Stability at all levels is not a requirement for persistence, for only one stable inefficient level of organization is necessary.

CONCLUSION

In this chapter, we show that populations are not merely the attributes of communities. They are ecological entities in their own right that deserve their own approach. Only in some situations and some conceptions is the population the level below the community. Conversely, there are things about communities that cannot be readily laid bare by population approaches. Populations have homogeneity of scale in their attributes, for all members are usually from the same species or are at least equivalently scaled for the ad hoc reason of electing the population in question. This makes populations tangible and manageable in a way that communities are not. As a result, there is a depth of insight into the working of populations through formal representation with equations that is not available to the community ecologist. However, that very homogeneity of scale in populations is exactly why multispecies population work is not community ecology. Population ecologists are well advised to play to the strengths that their conceptualization allows. By all means, intellectual hybrids somewhere between community and population conceptions will be worthwhile. However, an insistence that the precision and tangibility of populations makes them a somehow intellectually superior conception is ill-advised. There is robustness in a diversity of approaches to complexity in ecology. Only because populations are simpler on the inside can we make the important strides of Giampietro that were ushered in by Lotka. There seems to be more profit in trying to relate ecosystem energy dynamics to populations, as opposed to the population community exchange. The internal consistency of populations greatly facilitates their expression in energetic terms. To do thermodynamic calculations, the system has to be close to nested. The easy nesting of populations leads to a natural thermodynamic approach to population calculations. We hope for a new particularly relevant population biology as one of the main tools in dealing with the complexity of ecology and society.

7

THE BIOME AND BIOSPHERE CRITERIA

As we suggest in the introduction to our organizing scheme at the beginning of this book, biomes are conventionally considered to be spatially large, at a high level of organization. Folk wisdom in ecology says biomes are the major parts of which the biosphere is composed. However, advantage is to be gained from dissecting questions of scale away from other aspects of the biome, in the same manner that we did for other criteria. The biome criterion can be insightfully applied to systems that are in fact small in area and ephemeral. Biomes are not so much systems above a certain size as they are a way of looking at ecological material. All the other criteria invite a certain method, and so do biomes.

COMPARING BIOMES TO OTHER CRITERIA

To show how the biome model is not the same as our other criteria but is distinctive in its own right, we need to make explicit contrasts. Biomes in the conventional view belong in certain regions on the surface of the planet. Nevertheless, they are not simple landscape entities. There is a distinct causality for biomes that only secondarily circumscribes biomes in space.

Biomes are distinctly climate-mediated. Despite the central involvement of the physical environment of the biota in defining a particular biome, the biome is not just a big ecosystem. An ecosystem, by our definition, includes the soil and the local atmosphere as being explicitly inside the system. The biome is primarily defined by its biota. In our framework, large ecosystems would offer a better account of the soil and biota as an interaction over vast tracts of land. Let us emphasize again that we do not erect our framework as being singularly correct, but in this case, it does allow certain distinctions that might otherwise be lost. Note how ecosystems by our definition are studied specifically using mass balance, and their components are pathways.

The emphasis on a multispecies biota does not make biomes into large communities. While a biome is defined by its biota, life forms and not species are the biological units. In biomes, there is primarily an accommodation of the biota to the physical environment. Of course, a community must accommodate to its physical environment, but there is more slack in community responses to climate. In biomes, the relationship is closer to one-to-one. The response of communities to microclimatic variation is manifested as interesting species differences, not as changes in life form. The same vegetation can be seen as either an exemplar of a community or a biome. The causal entailment gives primacy.

It is no coincidence that it was Clements who coined the term "biome,"[1] and that he also anchored his community concept to climax.[2] Climax is the only time when environment does determine community. All the major community ecologists before Roberts tried to make environmental determinism work, and failed. Clements came the closest to making it work in that he locked climax, a community consideration, to biome. All biomes are environmentally determined, while successional communities are not.

A biome would be incomplete without its animals, whereas most community considerations are of either plants or animals, but not usually both at the same time. While the other kingdom is present, it is not involved in the tight species-to-species association that makes the plant community or the animal community. It is possible to consider particular plant-animal accommodations between species, but that is a separate issue discussed in the following section, "Plant-Animal Accommodations."

Animals play a special role in holding biome components together and produce the remarkable homogeneity that we find across the vast tracts inside a single biome. Biomes are animal groomed. C. S. Holling suggests that the movement of animals over large distances relates otherwise widely disjunct stands of vegetation.[3] He notes that the spruce budworm moves several hundred miles, and in doing so produces the physiognomy of the boreal forest. Certainly, grazers play a large role in giving the grazed biomes their distinctive vistas. Thus, individual grasses, as constituents of biomes, are less importantly embodiments of a genome and are more fuel for the grazers who beat back encroaching forest shade. Fire is the only other system attribute that can fashion vegetational physiognomy as animals do. It is no accident that pestilence, grazers, and fire all move at about the same speed over the same sort of area. They glue biomes together.

For a biome, the very essence of the situation is the manner in which the physical environment, mostly climate, determines what the biome shall be. The life form is all that the climate will allow. If there were more water, there would be more shrubs or trees. It is the quantity of water available or the critical limiting temperature that determines the biome. There is a certain cost and benefit in being a tree. The costs are susceptibility to fire damage and grazing at the sapling stage, as well as respiratory load and critical water demand that must be met; the benefit is overtopping herbs and shrubs. The costs that a tree incurs are minimized in grasses, and become the benefits of being able to live in a dry climate under ungulate grazing. The benefits of being a tree are exactly the cost of being a grass; herbs suffer shading if the climate permits trees. The physiognomies that characterize grasslands, shrublands, deserts, and deciduous and evergreen forests are all direct reflections of what the climate will allow.

Now consider changes in climate and how they affect organisms. Individual plant organisms adapt to physical conditions with devices like closing stomates to conserve water in drought. However, there are limits to adaptive responses, particularly when the whole life cycle is considered. Trees may be able to withstand drought when seedlings cannot. The frequency of rain determines if the wet season is long enough to allow establishment. Thus, in Ronald Neilson's assessment, vegetation physiognomy is a stable wave interference pattern between climatic periodicity and tolerances of critical life stages of dominant life forms.[4]

Very small shifts in the character of the component waves produce radically different patterns as a period of no rain exceeds seedling tolerance. A continuous change in climate might gradually lengthen the longest period between rains; nothing happens to the vegetation until an increment crosses a critical threshold. Changes in periodicity can have great effects quickly. Clarence Cottam took Ronald Neilson to see the northernmost hillside covered in Gambel oak in Utah.[5] The tree reproduces vegetatively, and as such was happy at the site. The next hillside had no trees because it would take seedlings to make the crossing. Some two hundred miles to the south, a squeeze in the temporal placement of drought and frost forbids seedlings of the tree. Local newspapers have accounts of old-timers in southern Wisconsin puzzled as to why forests had grown up so fast since Euro-American settlement a few decades before. Human-contrived change in fire frequency, releasing burr, white and black oak grubs, was the cause of the shift in vegetation physiognomy that was unrecognized by locals.

Also related to sudden changes in vegetation is the relationship between landform and climate. Even as the global climate changes, the major topographic structures on the earth remain in place, casting a climatic shadow that is the context for major biomes. Neilson has data that suggest the boundaries of entire biomes can be remarkably robust. He reports that the entire Great Basin can switch between shrubland and desert grassland with alternating millennia of wet and dry periods. The snap can apparently be across the whole biome region at once. We have little reason to suppose that gradual encroachment of biomes into one another's territory is the rule. The explanation is changes in wave interference, discussed earlier. In the drought of the 1930s, the prairies started to break up wholesale. This surprise shook the faith of the orthodox Clementsians, who had always thought their climax vegetation was more stable.

Given the massive areal extent of some biomes, they are remarkably homogeneous. The previous discussion indicates why. A given climate sets the basic rules. In general, trees cannot persist in areas where the soil moisture is completely depleted for a month every year. Soil can hold water between rains, but only to a point. Human grooming of the landscape, as in Israel, can collect water to give trees the extra water between longer periods without rain, including droughts (figure 7.1).

Ecological human management is often of biomes. In drought cycles, trees will completely deplete the water supply and the weak will succumb. Without

FIGURE 7.1.   Negev Desert biome transition zone with the Mediterranean region where the Jewish National Fund installs terraces to capture runoff water during rainy season (October–March) to supplement soil moisture for use by trees during the dry season (April–September). The area receives 250 millimeters of rain and has 1,000 millimeters of evaporation. Any rain that occurs during droughts is also captured as runoff. Trees shown are fifteen years old and about four meters tall. Thirty-year-old trees continue to grow in similar conditions. Terrace height and upslope runoff area are calculated to receive nearly 100 percent of the potential runoff. Six thousand hectares of the northern Negev have been planted with this management method. (Photo and explanation courtesy of Itzhak Moshe.)

widespread catastrophe, the trees could persist for their natural life span of several centuries (as do Gambel oak in northern Utah). However, catastrophe through pestilence or fire causes wholesale change in physiognomy (probably the cause of the treeless hill to the north that Clarence Cottam showed Neilson). At that point, limits on establishment and the animals of the new biome maintain the new state of affairs. Physiography sets the critical limits that follow from the climate, and the forces for internal cohesion keep the response homogeneous across wide climatic regimes (figure 7.2).

We are now in a position to press the point of small biomes. In northern Wisconsin, there are open areas in the middle of forests. They occur in slight depressions into which cold air drains. In general in Wisconsin, even in the north, the temperature in winter does not go to forty degrees below zero, the temperature where Fahrenheit and Celsius meet. But in frost pockets, cold air from surrounding areas pools in the pocket because cold air has greater density. While in the unforested pocket, the air then loses even more heat to the clear sky. As a result, frost pockets go below minus forty degrees where water will freeze no matter what you do to it. This was the problem of the British Antarctic expedition to the South Pole led by Scott. Temperatures were so low that the snow would not melt to float

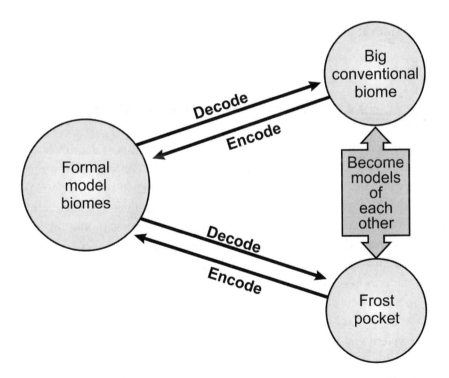

FIGURE 7.2.   A restatement of Rosen's modeling relation in figure 5, but this time used as a device to explain how small frost pockets can be validly seen as biomes, their size notwithstanding (Rosen 1991).

the sleds. Scott and his colleagues were pulling sleds over what might as well have been sand. So tree seedlings in frost pockets experience these temperatures and die. Field trips from the University of Wisconsin would go north in the spring, and frost pockets were a convenient place to camp because there were no trees. Smoke from the morning campfire would go straight up and then spread out in a thin layer. That was the clue that the local atmosphere was inverted, with the cold air pooling in the campsite. In most places, the air is warmer next to the ground. But not in frost pockets on a still night.

Biomes are (1) climatically determined, (2) physiognomically recognizable, (3) disturbance created, and (4) animal groomed. Frost pockets experience all of those things. We do not care about the species present, but we do care about how it looks. The vegetation is a physiognomy, not a set of species. The frost pockets are herbaceous and look very different from the surrounding forest. The pockets are animal groomed, and this also keeps trees out. Furthermore, the surrounding forest is perfectly stable. The pockets only exist because they have been disturbance created. Apparently, the four causes of biomes make a stable set that can apply at many scales. The biome is a stable syndrome and as such deserves its own specific treatment, just like flows and fluxes deserve the appellation of ecosystem. Thus liberated from the size constraint, biome cleanly emerges clear from other methodologies for ecological investigation. We need studies of other examples of size-independent biomes so that the full utility of the biome point of view can be released.

PLANT-ANIMAL ACCOMMODATIONS

A critical part of biomes is the relationship between plants and animals. We said that plants and animals do not form intimate relationships in community with each other because they are differently scaled. In biomes, plants and animals have generic models of each other. Some readers at this point will be thinking of many exceptions to the general statement that plants and animals do not relate as species, and they are right. So we turn briefly to some of those special relationships, noting that they are very different from the plant-animal relationships in biomes, and are not reason enough for plant-with-animal communities.

We do find specific animal species tied to specific plant species, but it is almost always between plant and animal species one-on-one, where there is coevolution. It is no accident that plant-animal interactions as a subdiscipline attract population biologists with their proclivity for genetically and evolutionarily based definitions of groups of organisms. We confess that this type of plant-animal entity receives little attention in this book, partly because it lies unrecognized as a distinctive criterion in more conventional views. It has no name, although it probably deserves one. If we were to extend this book further, we could write a whole chapter on the "evolon," or some such term for entities with highly coevolved parts.

The evolon has its own distinctive relationship to time and space. Clearly, a long time line is important here because the tight accommodation between the species is the very essence of the structure. The accommodation in communities is a fine adjustment to the species that happen to be there in this interglacial period. In the evolon—such as a highly specialized pollination system, parasitism, or commensalism—the species involved owe their identity to the relationship: no interglacial happenstance collection here. The evolon must have a longer history than the community, as evidenced by related species where the plant and animal have continued to evolve while keeping the relationship; whole genera of plants and animals exist in paired, highly evolved relationships.

Another distinctive characteristic is that the evolon is usually held in a very narrow scale range. For example, in an evolon of pollinators of cacti, the animal ignores the plants as if they were rocks except when they have flowers. The critical relationship is held in the vise of a very particular time and space framework. This is probably required for the stability of an evolon. For evolution to get a handle on something as canalized as a moth's tongue and a petal spur so that the two are the same length, the scaling of the parties to the interaction would have to be precise in time and space. If scales were different, various signals would pass only with difficulty and with much noise. Slack in scale of the relationship of the parts would almost certainly break the chain of coevolution. Coevolution has to be a continuous process to produce the highly specialized structures that we commonly see in evolons.

AGROBIOMES

In the comparison of ecosystems to biomes, we drew clear distinctions. So this raises the issue of agriculture, commonly defined as agroecosystems, a perfectly good word, but it focuses on processes in agriculture, of which there are many. Almost all terrestrial ecological systems are nitrogen deprived. They work hard to keep their mineral nutrients. Deprivation is particularly true of agricultural systems. So humans step in with nitrogen subsidies. Water is added or removed to increase primary production, so agroecosystem is an apt term.

But there are many things going on in agriculture that are not particularly process focused. First, we can recognize agriculture from its physiognomy, even if as laypeople many cannot tell the crop. And fields are disturbance created; they are certainly animal groomed, by humans, that is. And agriculture is set in particular favorable climates unless they are contrived by technology, as in irrigation. The Corn Belt occurs in a certain climate. All this opens up the notion of an agrobiome.

In the context of climate change, agrobiome as a concept is much more adept than agroecosystem. It is the agrobiome that humans will adapt under climate change, not so much the agroecosystem.

As to climate change, we might prefer to see the megafauna extinction after the last ice age as a change of biome.[6] The changing climate compromised the biome, so the big animals had no place. In Holling's terms, the lump of their size was disallowed. In the face of melting permafrost, big animals could not take full advantage of their stride. So mammoths could not reach the next bonanza of food required to meet their huge demands. The paradigm of geologists dislikes big, sudden events that change things. Their favored view is that there is some under-lying process. As a result, there has been much chafing over the suggestion that a meteor on the North American ice sheet drove the megafauna. It would certainly have changed the biome, so we do not object.

BIOMES IN THE BIOSPHERE

In their book, *A Hierarchical Concept of Ecosystems*, Robert O'Neill and his col-leagues suggested a pattern where physical constraints alternate with biological constraints in an ecological hierarchy. The argument is that when a biologically based advantage gains ascendancy, it is pressed by reproduction to a new limit that has a physical basis; this is the same line of logic invoked by Darwin for the resource limitation underlying natural selection. The physical limit on some raw material is then broken by some economy of resource use. The old limit becomes irrelevant as some other biological advantage is evolved. The process is then repeated, as bio-logical and physical constraints are interleaved while the systems evolve to higher levels that incorporate increasingly general limits. This alternation of biological and physical constraints is the organizing principle we use to relate the biome criterion to the biosphere. Grasses escaped grazing by developing silica spicules. The limit becomes the difficulty of silicon being largely insoluble; a physical limit. But then horses evolved to eat grass spicules anyway: the limit becomes biological.

The four main paradigms of contemporary biology are: organism, species, evolution, and mechanism. Together these paradigms fix the conventional view. Orthodoxy says that the environment interacts with the genome to produce a pheno-type of the organism. That phenotype performs variously well in the environment, and through an environmentally mediated reproductive success, the organism con-tributes to population evolution. And it is a useful metaparadigm. In all this, the point to notice is that the physical environment is always the context that acts on the biology. This is not a necessary point of view; it is arbitrary, and it exists because many biological phenomena fit the paradigm well. But it is only a paradigm, a point of view, which is taken by adherents to be more necessary than it is. This is an easy mistake for anyone devoted to excessive realism. The error is understandable because paradigm devotees wrap themselves tightly in the security that the paradigm offers. We are not against evolution, but not everything needs to be cast in its terms.

This basic model has been the orthodox view of the functioning of the entire biosphere. However, Lovelock proposed the Gaia hypothesis.[7] Enthusiastic followers have overextended Lovelock's hypothesis to levels of great mysticism that are entirely unnecessary. Simply, the Gaia hypothesis invokes a different causality from the conventional wisdom. Instead of life responding to and being held in the context of the physical environment, Lovelock sees life as the context of the physical environment in the vicinity of the earth's surface. It stands against the useful conventional view that supports evolution, and some of the resistance to Gaia will come from that conflict. With the switching physical and biotic constraints, which alternate with the move upscale, there is no reason to reject Gaia's position. As an alternative point of view, it can answer important questions that conventional wisdom studiously avoids.

Consider the very unlikely scenario of the earth's atmosphere. Despite the fact that the sun has increased its energy output by 30 percent since the beginning of life, the temperature of the atmosphere at the surface has remained within a narrow range. For the last three and a half billion years, somewhere on the earth's surface it has remained warmer than freezing and well below the boiling point of water. This narrow range is far too unlikely were physical forces driving the system. It would seem that life has been controlling the atmosphere at the scale of the biosphere, and not vice versa. Life is the top holon, and brings with it a certain purposiveness. Purposiveness comes from the observer, not the material system. The atmosphere is no more aiming than is evolution by natural selection.

Given the model of O'Neill et al., if the entire biosphere is under biological control, we might expect large ecological subsystems immediately below the biosphere to be under physical constraints: remember that biological and physical constraints appear to alternate. The biological control of the entire biosphere does indeed appear to be in stark contrast to the control of the biomes. Biomes are held in the vise of physical constraints. If O'Neill and his colleagues' assertion is general, then all very large but sub-biospheric entities should first be seen as held in the context of a physical environment. That is the way to address biomes.

BIOSPHERIC SCALE–APPLIED ECOLOGY

Very large-scale ecology has only just become an object of study and has been left for meteorologists and paleobiologists to impose their own particular points of view. In general, ecologists are not comfortable at a global scale, but plant physiologists are. The reason is that in the biosphere, most movement is some version of three-dimensional diffusion. In plant physiology, movement is also largely a matter of diffusion, so plant physiologists are at home with global systems. By contrast, there are barriers and special communication channels in most ecological disciplines. Ecologists are not used to diffusion being an ecological explanation.

The destruction of the ozone layer has serious consequences. It does appear to be something new, for we have measurements of ozone in the stratosphere that go back far enough to indicate that there was no hole sixty years ago. The most important destroyer of the ozone layer is Freon. It is a harmless chemical at our layer of the atmosphere, the troposphere, such that chlorofluorocarbons were used as the propellant in spray cans and in refrigerators.[8] Freon is a class of compounds that resemble the smaller hydrocarbons, like methane, but they have fluorine and chlorine in place of some of the hydrogen atoms. The class of compounds is functionally inert in most circumstances, although it is hazardous to smoke cigarettes around Freon.

Ozone is the unstable molecule consisting of three oxygen atoms. Even isolated from other reactive molecules up in the stratosphere, ozone is so highly reactive that it has always degenerated to ordinary oxygen molecules with only two atoms. Ozone is replenished by the sun's ionizing radiation that creates ozone from oxygen, reversing natural degeneration. In the stratosphere, Freon is broken down by the sun's rays to release its chlorine. Chlorine from Freon is a highly reactive arrival to the stratosphere, as it reacts with the ozone to make oxygen, reverting back to chlorine gas again after it has done its destruction. Chlorine therefore acts as a catalyst in the destruction of ozone. It would appear that the damage will continue for at least a century or so, until the chlorine is lost into space.

If ozone is so reactive, one might ask why other chlorine-containing gaseous materials do not wreak the same havoc in the ozone layer. The answer is that chlorine reacts and is immobilized in some large molecule before it can reach the stratosphere. That chlorine falls back down to earth. The estimates are under some revision, but it takes between two and ten years for tropospheric Freon to reach the stratosphere.[9] For a molecule of its size, Freon is remarkably inert, so it is a vehicle like no other for transporting chlorine up into the stratosphere.

Underlying the above chronicle of events leading to ozone destruction are some general principles about relative scaling. Together, they form a predictive framework that indicates far-reaching strategies for human design in an ecological setting:

1. Predictions are not that something will change, but are that nothing will happen—we can only predict that whatever orders the system now will remain in place.

2. The effect of nothing happening is likely to be the same across a wide range of situations; for example, if something is inert, its behavior can be predicted to be nothing for a large number of conditions, whereas if

something is reactive, one reaction is a poor indicator of the effect of other reactions.

3. Doing nothing includes not going away, so we can expect accumulation. Inert material accumulates. This explains a lot about the way we find the world, the composition of air, for example. Nitrogen is the major gas in the air because it is mostly what is left over after a time. Run the global system for a few billion years and something as inert as nitrogen gas must accumulate, for everything else has reacted and been removed in the process.

4. "Inert" is a scale-relative term. In that light, being inert now means that only the passage of time, albeit a long time, is needed for there to be a reaction. Wait long enough and inert will react, and if it has accumulated, there will be a big reaction.

Let us expand upon and weave together these statements. One very effective way for the human social system to choose elements that will not harm us is to select inert materials. If a substance is inert, it is scaled so that it is not capable of involvement in our own chemistry or the chemistry of anything else near us. Therefore, Freon appears completely safe.

For a student of relative scales, "inert" is a relative term. By temporally scaling Freon away from the scale of increments in seconds for our biochemistry, we have scaled it up to the decades and centuries on which total biospheric time proceeds. The general principle for human design in ecological systems is as follows: things that are inert at a small scale can often be expected to be critically reactive at a large scale. We can expect large-scale, long-term effects (figure 7.3). Being inert has a generic effect that leaves unspecified the universe to be perturbed. With such a large target, the principle is predictive. As predictions go, it is a robust forecast to say that something that has absolutely no effect today will have an effect some time later. At first it appears an uninteresting prediction, but with examples like Freon destroying the ozone layer, it appears to pertain to unexpected and important situations.

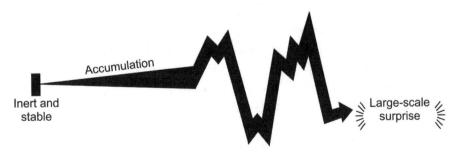

FIGURE 7.3.   The production of stable material over time necessarily leads to an accumulation. Eventually, this will lead to a change in an undefined upper level producing a surprise.

SCALING STRATEGY IN NATURAL SYSTEMS

In human industrial systems, there is an active decision-making process. Premeditated action is a standard operating principle in industry. In natural systems that is not so, but evolution does produce systems that can be described as goal directed, even though evolution itself is not a system with its own cognition and innate purposefulness. Purpose emerges when long-term information is transferred to small, local realizations such as organisms. We might expect living systems to find inert materials helpful for the same reasons that industrial systems actively choose them. For organisms, using inert materials that scale their immediate context away from their own level of behavior has very general advantages. First, the material will be decoupled from the general level of functioning of the organism. Second, if it is inert for one critical process, it will probably be inert with respect to many critical biochemical processes. Accordingly, many organisms use inert materials in their homes. That is the reason diatoms make the frustules in which they live out of silicon; sand is inert stuff. Another inert material used by microscopic organisms is calcium carbonate. Calcium bicarbonate is soluble and reactive, and it is the vehicle that delivers calcium to microscopic homemakers. Respiration and photosynthesis involve carbon dioxide, which changes the local pH up and down depending on the light conditions. This causes the precipitation of calcium carbonate, a relatively insoluble form of calcium. It has been a favorite building material for microbes ever since the blue-green bacteria started forming calcium deposits about three billion years ago. Many aquatic organisms now use calcium carbonate as a protective covering—mollusk seashells and microscopic rotifers, for example. It is predictably safe with respect to biochemistry, and keeps doing the same good job throughout the organism's entire life.

Particularly interesting in the present discussion is the way that fossil accumulation of calcium does indeed have an effect on the functioning of the present biosphere. This would be a prediction that we might take from the way Freon has biospheric-level properties. Fossil calcium deposits are even larger players in the modern biosphere. Recent work on watersheds suggests that calcium plays a role in increased transpiration of forests.[10] Given new ideas that transpiration is a surrogate for work, it would appear that calcium lets forests work harder. It takes millions of years for calcium to become an active player: changes in sea level and a rising of the seabed to become land. Calcium gives lime-rich soils that determine the flora of entire regions. Calcium influences the solubility of most other plant nutrients and so the inert homes of ancient microbes set the ground rules for all contemporary terrestrial plant communities by being abundant or otherwise in a given locale. Given its effect on other nutrients, calcium is directly and indirectly responsible for the concentrations of productivity in modern oceans close at hand. Sedimentary rocks react to water erosion differently from igneous rocks, and these differences create physiographic structures of ecological importance, for example, Niagara Falls. Landform in general can have the placement of calcium sediments as one of its causal agents. Formerly inert calcium is a big player in the modern biosphere.

In the same light, consider the contemporary crisis in rising levels of carbon dioxide and the projected global warming. One of the main drivers of change in this scenario is deforestation. It fits our model perfectly: trees use cellulose, a relatively inert material for building; carbon trapped in forests represents an important part of the global carbon budget; the present levels of carbon dioxide are homeostatic with immobilized gigatons of wood carbon as part of the equation;[11] humans harvest wood because it is relatively inert and therefore good for building; if we use wood only for building, all will be well, but synchronous destruction of large numbers of trees and the burning or rotting of their wood products seems likely to precipitate global climate change.

There is a quandary about global warming that is instructive. Wixon addressed the uncertainty of the effects of the thawing of soil carbon in mires. Will the bacteria in those soils start to release vast quantities of carbon dioxide, or will some other constraint moderate the process? In terms of high-gain use of this carbon, bacteria would pile into the resource and release carbon in a positive feedback: more carbon dioxide, so more warming to release more permafrost. The low-gain scenario involves nitrogen, which would be used quickly and limit the release of carbon. The smart money is on the constrained low-gain scenario, but there is still uncertainty on this hugely important potential positive feedback. Wixon showed that the mass of experimental results is generally measured in sufficiently different units that the large integrated model is not feasible.[12] The limiting organizing factor in the scientists' view is generally what they respectively study. One paper argues that earthworms must be understood for the problem to be solved, and the argument in the paper is plausible, too. But there are hundreds of papers that argue coherently for many other crucial factors. This is a classic medium-number specification of the problem. There are many reasonable constraints and we cannot tell which of them would apply and when.

Wixon did the only thing open; she did a metaecology analysis of the bulk of the literature. The specific problem was how to weight the likely dominant processes that would apply. She teased apart changes in the community as a response to intransigent carbon. She addressed temperature change as a component of it all. She looked at the issue of the identity of immobile carbon, when there are so many types recognized and defined. The scientists prefer their own particular definition, even if it puts their work at odds with other valid calibrations of some other sort of immobile carbon. Wixon noted the focus on bacteria breaking down the slow carbon, and then raised the issues not really addressed so far. How do bacteria themselves contribute to slow intransigent carbon when they die? The critical point is that her problem was not a data problem. Most ecologists think that more data will solve most problems. Seasoned researchers are generally aware that the next study right around the corner will not come because the problem is at a higher level. Science needs to be specific to get the right answers in the chosen framework. The pressures for vulgar careerism to publish incremental work, and the short time taken to achieve a PhD make it almost impossible to frame the big, important questions. Wixon did not settle the issue on climate change and thawing carbon, but it is one of the few studies that had courage enough to go for the big answer.

We should emphasize that Wixon herself does not criticize the experimentalists whose ways of measuring carbon efflux are incompatible or even contradictory. They are not fools and most are well aware of the problem of incommensurate measurements. If you want big solutions, you should ask big questions, but the politics for success in science discourage that. Accordingly, most ecological research is locally valid, but globally trivial.

SCALING PRINCIPLES IN APPLICATION AND DESIGN

We might have predicted the general reactivity of ancient inert material in the modern biosphere. Now we can generalize to other human influences. The accumulation of plastic is a serious civil engineering problem. The very reason we use plastics is the reason why they are such a nuisance. They are inert, that is what makes them useful, and that is what makes them a waste disposal problem. This we could have predicted. Even great expertise is not sufficient to make particular predictions. In the decade before we knew we had a hole in the ozone layer, James Lovelock himself, the hero of Gaia and a fully respected atmospheric scientist, wanted to see if manufactured material would disperse though the atmosphere. He chose to measure Freon well west of the coast of Ireland because there is no source of chemicals in the atmosphere there, since the air always comes from America. He found Freon, so his hypothesis was confirmed.[13] But even before the ozone crisis, there were questions as to whether there was need for concern. Lovelock said there was none because Freon is inert. If Lovelock can get it so wrong, we clearly cannot expect experts to get global questions right with enough reliability for us to depend on them. We can use general principles about reaction rates and surprise might alert our best investigators to look in places they may not have thought to do so.

Inactivity is predictive at the scale of real-time biology, but activity is not. Unlike inactivity, which is generalizable, reactivity can take many forms, some of which will give unpleasant surprises. A beneficial reaction only portends other reactions

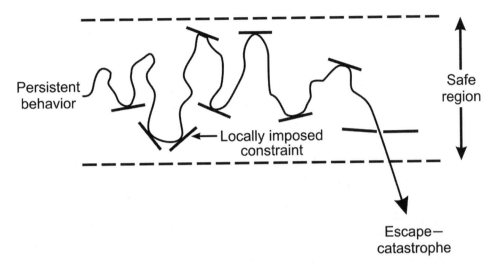

FIGURE 7.4. Production of material that is permanently reactive requires constant monitoring and preemptive action to avoid undesirable effects. It is to be expected, however, that eventually some preemptive action fails and an undesirable consequence follows.

in the human-made material; it does not indicate the favorability of the unknown reactions (figure 7.4).

Consider what happens when one tries to make plastics exhibit rapid behavior after they have been used and are ready to be discarded. Note that under combustion, plastics immediately lose their desirable properties of existing close to human beings without doing damage. We can drink coffee from polystyrene on a camping trip, but try to burn the cup in the campfire afterward and the local environs become unhealthy. The inert, stain-resistant materials used for airplane interiors become sources of deadly gases in the fire following a crash.

Biodegradable plastics amount to a rescaling so that the materials still have the low level of reactivity to make them useful, while making them sufficiently reactive to avoid the present waste disposal problems. However, that low level of activity makes such plastics less generally useful because they can influence the flavor of foods stored in them. We cannot expect biodegradable plastics to be a panacea. As they degrade in landfills, we can only hope the end products that percolate to the water table are harmless. It is not that we think biodegradation of plastics is pointless. Rather, we raise the issue that we have been surprised before by the long-term effects of apparently nontoxic or action-specific chemicals like DDT. Therefore, caution should replace optimism.

Plastics and Freon raise a class of problems that emerge with a vengeance in pesticides and other similar organochlorine chemicals. From the point of view of the pesticide users, they want it stable enough to complete its work without the expense of reapplication. The price paid for the pesticide being even moderately stable is that accumulation is bound to happen. Being a pesticide involves some biological activity, so they are persistent but not exactly inert. Accordingly, the deleterious upper-level effects come more quickly than the effects of plastic or Freon. We have now come to expect these midterm effects of pesticides and are apt to forget that we were very surprised by bioaccumulation when it first manifested its effects high in the food chain. Of one thing we can be certain, new classes of compounds are going to surprise us again.

We also can expect the problem to worsen in cases where the product is not one material. Like Freon, PCBs are not just one type of molecule, but a class. In general, the amount of PCBs in our environment is going down because they are getting buried in sediments that take them out of circulation. They were prevalent in their day since electrical insulators were made of them. One horror story is that they were used on kitchen aluminum foil to keep the shiny side shiny. The bad news now is that the variant forms of PCB that linger are becoming disproportionately those that are more toxic. That is because the more toxic versions are the biologically more active forms, and that keeps them in circulation.

The ecological problems of nuclear power appear as related models. In conventional nuclear power in the United States, the material at the end of the process is active in obviously deleterious ways. Reprocessing those materials can alleviate the problem, but there are implications for nuclear weapon proliferation if we go that route. Jimmy Carter was a nuclear engineer, and so was aware that reprocessing made weapons-grade fissionable material. That is the reason the United States does not reprocess its spent fuel. The French do reprocessing anyway, and do not seem to have had a security problem so far, but it is worrying. We have the technology to reprocess and use the spent fuel in storage to generate a lot of electricity, while creating only about four cubic feet of waste per year.[14] That waste is very nasty material, full of radioactive cadmium, but it seems likely that modern industry is sufficient to handle that small volume.

The problem of nuclear waste is related to the DDT issue. When DDT was first used in World War II, it saved many lives by stopping insect vectors of typhoid and other killers. The argument was that there was no need for concern because there would be only a few tons of it dispersed, which would be diluted in the large environment since it would go everywhere. It did go everywhere, but it is fat soluble. It therefore ends up stored in fatty tissues. Predators scavenge for fat, and this was the unrecognized factor. Fat solubility caused organochlorines to bioaccumulate. Bioaccumulation is a problem in the waste products of fission. Animals scavenge for rare useful material such as iodine and calcium. Strontium is below calcium in the periodic table, and so is absorbed collaterally in the acquisition of calcium. Radioactive isotopes of iodine concentrate in the thyroid gland, where they cause cancer. Strontium-90 gets deposited in bones, the slowest part of the body to turn over, so it too persists in the body, giving chronic radioactive exposure. Products of fission therefore have the same bioaccumulation problem as DDT.

Fusion, if we can get it to work, is likely to present less of a problem. The escaping radioactivity is likely to be deuterium or tritium, double- and triple-heavy hydrogen. Remember the original argument for there to be no concern about pesticides; they will go everywhere and will be so diluted that they will be no problem. In fact, that argument does hold for all isotopes of hydrogen. They will get into water and that really will go everywhere.

The essential problem is that nuclear waste is persistent in its potential to do harm. The permanent storage of nuclear waste is probably tractable. The strategy is to immobilize the waste, thereby making it predictable for the hundreds rather than thousands of years it takes to decay to safe levels. This is worrisome, but probably manageable as long as we are cognizant of the scales that matter.

Consider the large number of materials made by the chemical industry that are persistently dangerous. The general principles raised here place a heavy burden of proof on industrial engineers. The only thing needed for disaster is time; and that is not a speculation, for we can predict it from the basic principles laid out earlier. The planners have to answer all possible scenarios, while the problem is free to take any unlikely course. We have enough faith in engineers to say that the disaster, when it happens, will take a very unlikely course; but it will happen. We cannot escape the general condition with an economically expedient list of special controls; implementing the list will bankrupt us.

All this highlights the genius of recycling. It is no accident that the metabolism of healthy organisms and the functioning of natural ecosystems both use recycling. It is a strategy both for escaping the constraints of scarcity and for regulation. Organisms control the whole oxidative process by limiting the amount of ADP, the low-energy form of fuel in the ADP/ATP cycle. ADP accepts energy as it changes to ATP. ATP then acts as an energy supplier and regenerates ADP. Biological systems use recycling to avoid both the problems of supply and of waste disposal.

The cycle itself gives the properties of predictability through persistence that are required for human artifacts to do their job reliably and safely (figure 7.5). Since there is reactivity in a cycle, we might be concerned about the generally unpredictable nature of reactivity. However, the positive feedback that would generate the unpredictable behavior is held constrained by the upper-level negative feedbacks in the cycle. The cycle is always in place to outcompete any undesirable, unpredictable side reactions. Thus, the reactivity is predictably favorable. There are, of course, bad cycles, and they present the difficulty of persistence that comes with cycling. We just gave the example of the PCB cycle, but phosphorus in eutrophic lakes is another.

The advantage of expressing the problem in terms of only relative turnover times is that we are not locked into the details of particular scenarios (figure 7.6).

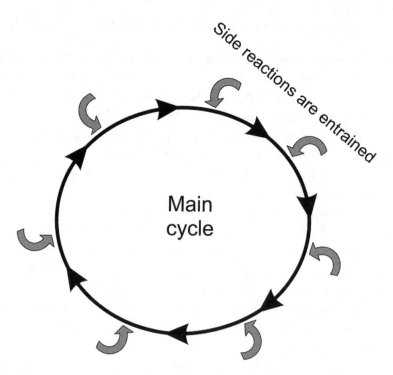

FIGURE 7.5. The cycle is a special case of constrained dynamics. It never runs out of inputs, it outcompetes side reactions, and it does not accumulate waste. It will continue to be predictive. One can therefore run it until one has had enough and then just stop it.

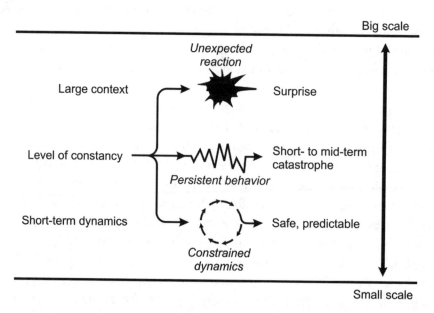

FIGURE 7.6. If the middle level is defined by some constancy or structure, then the general characteristics of system predictability depend on the level at which the important dynamics occur. Production of stable material leads to accumulation and the expectation of upper-level surprising dynamics. Constant activity at the middle level can be expected to produce undesirable effects in the midterm. The particulars of these effects will be difficult to predict. If the constancy is a cycle, then the reactivity occurs at a lower level and the system is predictable in its behavior.

Therefore, experience with apparently unrelated materials could be used to steepen the learning curve. The Babylonians knew about particular right triangles for laying out fields with square corners. The contribution of Pythagoras was to generalize the problem so that it applied to all right triangles in principle. All that we have said earlier about Freon, plastics, and pesticides was already known. What is new here is the general statement of principles with broad application in unfamiliar circumstances.

CONCLUSION

In this chapter, it has been profitable to see the biome independent of the large size that is usually ascribed to it. From this treatment, it has emerged that the biome is an autonomous intellectual device that is more than a large-scale loose compromise between other criteria. A good deal of this chapter is concerned with basic science issues.

Despite the advantages of considering the biome independent of the large scale usually associated with it, we have spent effort in discussing large-scale questions. Some of the most pressing issues for ecology are extensive. We have therefore used the biome concept to address large-scale, global questions of an applied sort.

8

NARRATIVES FOR COMPLEXITY

W E HAVE now worked our way through the chapters assigned to the ecological criteria that we set out at the beginning of the book. But at this point there are a lot of loose ends. Therefore, we need a chapter that shows how to achieve unity, or at least order in ecology. We have been at pains to put the posture of the observer in front of what we see. Our observers are human. Humans tell stories. Their special intelligence is an ability to deal with things that are at odds, and they do that with narratives. Scientific narratives are not to be dismissed as just myths or mere cartoon representations. Narratives are the bottom line in science. Yes, there are hypotheses, predictions, theories, and models, but all of these devices are in the service of achieving compelling narratives.

Let us ease into the context wherein different types of investigator tell their stories in their many ways. The reference for narrative in science focuses on observer activity as a starting place. Observation is much more abstract than is generally understood. Rosen said that the most abstract thing a scientist can do is make a measurement because one has to remove everything else in the universe to get it.[1] This is anathema to modernists, who believe better science is closer to reality and observation is the concrete way to get there. When models fail, narrative is the way out. The end product of science is a story improved by models and made convincing by predictions.

This chapter is probably the most abstract, but also the most rewarding for the unification of ecology. The authors here do not have the talent of practicing empiricists. We cannot take our insights into empirical work as empiricists can, so we are keen to explain well how what we do is different but valuable. In chapter 6, we quote Herr, who complained about a "resistance to abstraction."[2] In narrative, we move beyond the relatively concrete world of algebraic equations in models to a higher level of abstraction. We became theorists because we find empirical work too hard. So there is no pecking order here, and almost all of us have blinders somewhere. Only a few in ecology live in all three worlds of field experimentation, of algebraic formulation, and the intuitive geometric narrative.[3] So most of us need our excuses, with the exceptions having shown that all of it can be done by one scientist.

So we invite empiricists to be patient with our abstraction beyond equations. There is no need to trot out the myths of the scientific method that nobody actually observes. Modernism distorts its subject matter so as to make the situation appear real. Picasso's modernist cubism distorts so that the reality of a full face and profile are both in the picture; television before the postmodern 1960s series *That Was the Week That Was* universally distorted the process so the technicalities of television production were obscured.[4] Scientific papers present the logic of the scientific method so that it appears tidy, but they distort the actual process of research being messy, intuitive, and ad hoc. With the touch of postmodernism we offer, we expect that experimentalists will return later to their day-to-day testing with a greater understanding of what they are about. The change we recommend is not in methods of experimental practice, but is in encouraging a different agenda that acknowledges when there is distortion to make things appear neat and logical. Then we can choose to twist things around if we still wish to do so.

Hierarchy theory is not a theory in the conventional sense.[5] It makes no hard predictions, but rather it clears the minds of those who do make predictions. Hierarchy theory is a metatheory. It helps the theoretical modelers develop better models because it unpacks what is bundled up and taken for granted in the modeling process. As theory, this chapter may be unfamiliar to most who would call themselves ecological theorists. There is an underlying mathematics to what we present, but it is not the differential equations or statistics that anchors what is generally taken as ecological theory. Behind the veil, we are using category theory. Rosen is central to our thinking and presentation on this.[6]

HARD AND SOFT SYSTEMS

An ungenerous view would say that anything that puts "science" in its title is not science at all: social science, political science, library science, domestic science, and so on. All these "sciences" deal with messy systems where controlled experimentation is often impossible. One alternative is triangulation, as in Holling's work on lumps,[7] or the macroecology of Jim Brown.[8] One perspective often cannot offer confident understanding, but many perspectives that somehow coincide can increase confidence. All this was captured in the notion that science is interested in that which is robust to transformation, something that is somehow still there when observed from different perspectives. To do that we need stories.

In the soft sciences, the critical problem is to find an effective way of looking at a system. The same applies to hard science, but that is not commonly understood because hard things offer a deceptive self-evidence. The subject matter of soft sciences does not offer self-evident entities. Even something intuitively obvious, like "family" in social science, is not without tension: extended family; a single mother and daughter; membership in more than one family after divorce; villages; a Victorian household unified above and below stairs; and so on. Finding workable structures and explanatory principles is the main task. Accordingly, measurement has to wait for a definition of what is to be quantified. Data collectors and modelers are always in danger of taking the qualitative side of the problem as self-evident. The process of entitation needs to be taken seriously.[9] A mistake at the outset makes for much fruitless quantification.

Narratives are semantic. Giampietro speaks of hard and soft semantics with great precision.[10] We rely on him, so the discourse here is not soft. In hard semantics, the process of definition does not influence the definition. He uses the example of a dog as something that yields to hard semantics; four legs and a tail are concrete observables, the definitions of which are not changed in the process of measurement. Having defined something in general terms, a set of measurements follows, which firms up what meets the definition and what does not. There is the definition, but then there must be a check to see if this thing we called dog does in fact meet the criteria for equivalence in the class dog we have erected. In contrast to the hard semantics of a dog, soft semantics would apply to something like the value of biodiversity. That depends on the context and on criteria that may not be shared by many independent observers.

Semiotics covers the issue of description, meaning, and symbolic representation. Hard and soft semiotics are distinguished on what Giampietro calls predicative and impredicative definitions. In the definition of tall people, there are available predicative and impredicative definitions. A hard semantic predicative definition might say that Hoekstra is tall because he is six foot one, while Allen is not because he is only five foot nine. Everyone as tall as Hoekstra is tall.[11] By contrast, a soft and impredicative definition of "tall" might say that the tallest 80 percent of people in the group are tall. You can see how the soft definition depends on applying the definition. For instance, Hoekstra would not be tall in a professional basketball team. Some guards might be shorter than Hoekstra, but he would not make it into the top 50 percent of the team for height. Hoekstra would be a short basketball player, his tall stature on a predicative definition notwithstanding.

Giampietro takes these arguments into the notion of loops. He refers to chicken/egg questions as invoking impredicative loops. Which comes first—the chicken or the egg—depends arbitrarily on where each independent observer chooses to start. That is not to say that the chicken/egg loop is without value in solving the problem of chicken farming, but it is still arbitrary. We assert in chapter 9 that management falls somewhere between the landscape and the ecosystem criterion, and it integrates them. The landscape tends to avoid impredicativity, as it often uses hard semantics. By contrast, the loops of the ecosystem have chicken/egg written all over them. Management of ecological systems is therefore down-to-earth and concrete on the one hand, while being equally a matter of arbitrary design and measurement scheme on the other.

The concrete achievements of the hard sciences and their applied partners are so great that they distort our judgment. Hard science achievements can make the insecure in the firm but not hard disciplines reject narrative. Like it or not, the social standing of scientists roughly corresponds to a ranking from hard to soft. Accordingly, the temptation is to move to the hard science end of any given discipline; sometimes this strategy can be productive, but it can degenerate to mere calibration for its own sake. In ecology this problem is exacerbated by modern gadgetry. Often, we measure only because we can. And the review process of granting agencies accepts data collection over thoughtfulness about what to measure, a pathology in funding that needs fixing. Certainly, making easy measurements is cheap and sometimes data collected for its own sake reveals surprising insight, surprising because there was not really a plan. The downside is that we are buried in data that do not matter.

Premature hard definition can slow progress mightily. Grime (1985) cites himself as he quotes:

In plant ecology as in golf there is a time for precision and a time for progression. Only in fog or acute myopia can the hazards of driving justify putting from the tee.

And he is right. We insist that a mistaken quest for the hard has slowed ecology to a snail's pace, when the large problems at hand demand swift action and bold moves. With narrative we can pick up the pace.

Ecologists are not trained to measure social factors, and they are simply at a loss as to what to do. Soft science requires rigorous training. Social scientists learn taboos, of which biologists and ecologists are unaware. But ecologists wade in anyway, particularly in management of human ecological systems. The big biological taboo is that "evolution does not plan ahead." The important social taboo is that outcomes in social situations are neither good nor bad; they are simply there. Joseph Tainter has noticed that this insight is rarely achieved without explicit training, which social scientists get but biologists do not.[12] Looking on the bright side, Tainter identifies that societal collapse may not be a bad thing.[13] Lower taxes and greater freedom for the rank and file is a good thing for that group, but that would not be the opinion of the elites. Inside any society is a set of values taught and agreed upon, as to what is good and what is bad. For instance, most of our readers would feel that the collapse of the first world would be a bad thing. The counterpoint is, "Well, that depends." Tropical nations, whose markets are dictated in ungenerous terms by first world cartels, would see a lot of advantage in industrial collapse. Bananas at 38 cents per pound in U.S. gas station convenience stores must be costing third world peasants significant income in return for long, hard work. Imprecision that inserts goodness into soft science settings is as bad as slovenly experimentation in hard science. Ecologists doing good should beware.

We were impressed when we engaged with our anthropological colleague, Joseph Tainter. As ecologists, Allen and Hoekstra with James Kay set about identifying what biogeophysical factors would give indications of sustainability as to whether or not a national forest was working well with its content and surroundings. We found workable but unremarkable factors and suggested some measurements. The biophysical factors were subsequently incorporated into a North American test on monitoring sustainability of managed forests and grasslands. Tainter made a separate contribution. In the space of a few hours, he suggested a long list of social measurements that could enlighten us in our quest. The list will open ecologist's eyes as to what constitutes social science rigor and clever measurements. Tainter's list was divided into structure and process. It focused in a way that resonates with the landscape/ecosystem tension we have already mentioned. For the sake of maintaining the flow in this section, we present only two examples of Tainter's suggestions here, and put the rest in appendix 8.1.[14] Any readers who are biophysical ecologists will be struck with how different is the social scientist's point of view. In the meantime, we can capture much of that difference with just one structural indicator and another that is a process indicator. Even the structural indicators include change of structure. As is the example we choose:

Is the rank-size distribution of communities within the ecological system management unit changing in a direction unfavorable to smaller sized communities?

The process indicators may not necessarily indicate change, but they are often about human processes, not their structures.

Are people's expectations of their communities future being met?

We note that the structural measurements that Tainter suggested were on the side of being predicative. Meanwhile, the process indicators have more than a touch of the impredicative in Giampietro's terms.

Joel Cohen said that biology has physics envy (see comparison between Cohen and Rosen in note 15, with the tension between predictability and impredicativity with real players).[15] The sad thing that makes Cohen's rebuke strike deeper is that mainstream ecologists envy Newtonian tidiness, not the penetrating logic of quantum mechanics. Confident in their material systems, physicists are free to press hard with soft devices like narrative. Hamilton worked in the 1820s. John Brinkley, a contemporary astronomer, said of Hamilton: "This young man, I do not say *will be*, but *is*, the first mathematician of his age" (emphasis in original). Hamilton tried to reduce optics of lenses to Newtonian particle behavior. He failed. So he had to write a dictionary with analogies to translate, rather like the translation of water pressure to voltage in the pedagogical analogy used in teaching electricity. When he had finished, Hamilton then performed a reduction on optics and found a waveform. This kept physics busy until Einstein. Rosen points out that if Hamilton had only gone to the Newtonian side of his analogy and performed essentially the same reduction, in the manner of Schrödinger a hundred years later, he would have found quantum mechanics.[16] So physicists regularly use metaphor, analogy, and narrative, when ecologists would not like to appear so soft.

Ecologists often mistakenly stick to models when what they need is a good story. Other ecologists have developed whole outstanding careers on powerful narratives; for instance, Paine's narrative of the intertidal animal community, or Likens's narrative of Hubbard Brook clear-cuts. There are few models that can map ecosystem processes onto a landscape, but narratives can handle the changes of perspective that are required. Likens kept changing his story as the forest kept revising the tale it was telling. It recently told him a rattling good tale about calcium and transpiration.[17]

In physics and chemistry, the observations are hard won, but once the complicated detection devices are working, hard semiotics appears to reflect very general situations. For most purposes, one does not need to make allowances for the differences between electrons. By contrast, at the soft end of investigation, human individuality dogs the investigator every step of the way. In more positive terms, it is exactly those human quirks that make the soft sciences such rich indicators, as was Malthus to Darwin.

The methods of the soft science disciplines may hold the key to dealing with the more awkward aspects of ecological systems, like developing methods for their management. There is a way to be rigorous and address larger issues, but it will be with a different sort of rigor. The best have been cleverly analyzing big data in ecology already. Tom Swetnam was faced with masses of tree ring data.[18] There are many reasons for trees to show a narrow ring in a given year; it might be drought or it could be pestilence. Some trees show epidemics in pestilence, such as conifers hit by spruce budworm. By separating out tree species that are susceptible to such focused causes of poor growth, the years of stunting with climatic causes can be identified. Clever use of data can give deeper insight than more or harder data.

We have already referred in chapter 1 to Wald's work on aircraft damage in World War II.[19] He suggested reinforcing places that were not often hit because planes that got hit there did not come back. Data are always set in a context that can be used as a lever. Ecology is learning about big data, but it is still, for the most part, only at the stage of collecting data without much forethought as to how it might be used. That oversight blunts how the collection might be refined. The National Ecological Observatory Network (NEON) large-scale data collection needs to avoid simply aggregating normal practitioners who have benefited from narrow reductionist agendas. More of not much is still not much.

Clearly, ecological management has a large soft system component to it, particularly if joint production of, for example, wildlife, timber, and recreation is envisaged. A tree almost always imposes itself on our senses, for it takes little effort to see it standing out from its background. Easily won data come directly through unmodified human senses, and each one of us has a style of personal observation full of values that cannot be suppressed completely. In his research on data collection, Grant Cottam showed that some of his students and colleagues were includers and came up with higher plant densities, while others were excluders.[20] It is well known in plant systematics that some taxonomists are splitters and others are lumpers. That is why the rules of data collection in plant ecology are so specific as to be ritual. The sheer number of reasonable approaches and questions that ecological material can generate can be overwhelming. This means that investigator preference is a major contributor to the results.

Lynn White described science as being intellectual, speculative, and aristocratic.[21] By contrast, he said that technology is action oriented, empirical, and plebian. He acknowledges the melding of the two these days, and suggests that unifying technology and science has been part of the cause of our ecological crisis. Science that is focused on action has large effects. But the general distinction still holds; so engineers still think differently from scientists (figure 8.6). They tell their stories in a particular way. A problem for managers is the political setting in which management is set. In general, political traps stymie effective long-term management based on applied quantitative research. Government reports are rarely read and are not often the basis of effective remedial action. But the volume of government reports would suggest a different motivation. As long as someone named is producing a gray literature paper, the supervisor responsible has cover for inaction: "I have someone working on that." And the scientists have an interest in inaction, too. They will always accept more funds to continue what they were doing. It takes agencies like the U.S. Department of Agriculture (USDA) Forest Service about five years to build a management plan that is based on detailed negotiations between the public interest groups and the U.S. Forest Service (USFS) managers. But that is about the time it can take for a change in which major party controls the purse strings, and the policy is sent back to the drawing board. The Forest and Rangeland Renewable Resources Planning Act of 1974 has never been properly implemented for this reason, despite competent and effective applied scientists continuously "working on that." Legal suits representing differing political views coming from both sides complicate the issue.

Treaty-mandated salmon runs on the West Coast of the United States now have legal costs that exceed the value of the fishery.[22]

An exception was the policy put in place for Lake Erie in the 1970s. There is never certainty as to the cause and remedy for a problem. Those who would prefer not to pay the price of action can always use that as a block on action. The only thing to do in that situation is to lay out some simple, unequivocal deadline. It was suspected that phosphorus was the critical pollutant coming into Lake Erie, and that phosphorus was coming from household detergents. The major detergent manufactures had a short-term incentive to do nothing, but there was a mandate that started in the International Joint Commission giving a specific date to remove phosphorus from their products. When special cleaning is needed, as in washing walls as preparation for painting, then trisodium phosphate (TSP) is still allowed. But phosphorus was removed from mainstream washing detergents, the source that was large enough to be making the difference. There was no certainty that it was indeed phosphorus, but it seemed a reasonable course. No more funding to prove it, just a deadline for action. And it worked.[23] Lake Erie was in enough trouble that the humming fish said in Dr. Seuss's children's book, *The Lorax*, "I hear things are just as bad up in Lake Erie." But within a decade of the ban, Secchi disk measurement for water clarity greatly improved. The line about Lake Erie was removed fourteen years after publication when researchers wrote to Dr. Seuss about the improvement. There is an irony that it became even clearer when the zebra mussel got into the lake. That is another example of how social happenings are never unequivocally good or bad. Clear lake from humans introducing zebra mussel: good. Pipes clogged with the invasive species: bad! That means the decisions for action must always weigh competing values, as social scientists would expect. Stories can deal with conflicts like that.

NARRATIVES IN RESEARCH AND MANAGEMENT

The various parts of the engine that we recommend here have been presented in the earlier chapters of this book, but the whole—as it occurs in Zellmer et al., and figures 8.1 and 8.2—has not been put together in this work until now. Figure 8.1A identifies the levels of analysis from the observed object at level n up to level n + 3, the level of the metaobserver whose narrative fixes the level of discourse. Level n + 1 provides the model building and level n + 2 gives its working context. Figure 8.1B provides the vocabulary for what occurs at each level. The top level provides the paradigm or grand narrative. There, the metaobserver anchors or defines the whole scheme based on its unique perspective (level N + 3, in figure 8.1B). If the system dies, if the metaobserver dies, or especially if the grant money is gone, everything stops. Plenty-coups, Chief of the Crows, says at the end of his biography, "When

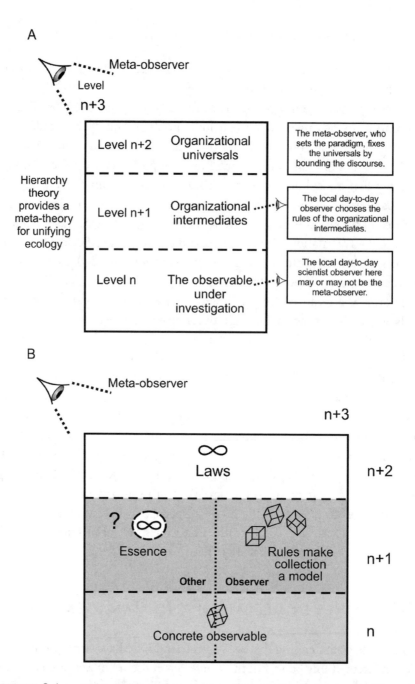

FIGURE 8.1. *A.* The modeling scheme expressed as a four-level arrangement. The metaobserver at level n + 3 bounds the general discourse, and in doing so, fixes the universals of the enterprise at level n + 2. At level n + 1, a model is created from generalizing the observed structure at level n. The observers at level n and n + 1 perform the second compression down to a model. *B.* Filling in *A*, *B* introduces the terminology. A concrete observable occurs at level n. It rests inside the set of universal laws at level n + 2 that follow from the purview of the metaobserver. At first, the concrete observable will be simply an experience with no name or type. But through past experience, the observer will recognize a class at level n + 1 to which the experience belongs. The model is that equivalence class created under local rules. Note the right-hand side of the figure is in the realm of observer decision. The left-hand side is the realm of the other, that which is in the observations but not at the choice of the observer. "Other" is a technical term in postmodern philosophy that is purposely noncommittal because it is the realm that might be observed, but is not strictly understandable or definable beyond being simply experienced. Behavior that is a free outcome is on the side of the other.

the buffalo went away, the hearts of my people fell to the ground, and they could not lift them up again. After this nothing happened." Cronon (1992) comments: "All that happened afterwards was part of some other narrative, of which there was neither point nor joy in telling it."[24] So it is in research.

At level n + 2, "law" is the term used as by Pattee (1979a, 1979b).[25] Laws come from the perspective of a metaobserver at level n + 3. The metaobserver chooses the arena of discourse, which comes with its set of laws. Laws are universal, but only in the universe chosen by the metaobserver. For instance, if you are doing social work, people are a requirement, but this law does not apply in physics, where people rarely pertain.

In Pattee (1978), laws limit possibilities; they are: ubiquitous, inexorable, structure independent, and rate dependent. Laws are universal context to the discussion. Rules come from what the observer brings to an observation within the context of the laws. They delimit what the day-to-day observer allows. Therefore, by contrast to laws, rules are: local, arbitrary, structure dependent, linguistic, and rate independent. Models are made with rules. The rules are manifested in equivalence classes that amount to models. Equivalence defines relationships that make the model. The rules occur at level n + 1 on the side where the observer makes decisions about how to observe the named categories and their relationships. Rules provide decisions about the structure and process components of the model (figure 8.1B). They define what is allowed.

Models are defined. The essence is the upper-level undefinable at level n + 1 above the concrete observable at level n. The essence is responsible for the equivalence in the defined equivalence classes of the model. The observed concrete entity at level n is a realization of the essence. The essence is the generalization of the realized observable. Figure 8.2A has four numbered arrows. Arrow 3 connects the essence to its realization. Notice that the concrete observable sits astride the region of the observer and the other. Arrow 3 is on the side of the other. The essence keeps changing, perhaps under evolution, and is therefore undefinable. The essence is the general condition of the concrete observable. The trick in the whole scheme is to iterate between the model and the observable to see how the essence explains what happens in the model. It explains the equivalence in the equivalence class of the model. Arrow 4 on figure 8.2A notes the attempt to link the model to its respective undefinable essence. The structures in the model are named and defined by the equivalence in the equivalence class. The essence does not exist without a model to explain, so we are not being idealist.

Figure 8.2B links the parts through a set of four actions or happenings that created two connecting loops. Figure 8.2C labels the loops in their complete track and introduces the final top loop that represents the iteration between models and observations that links the model to a supposed set of material causes (which can only be known through inference).

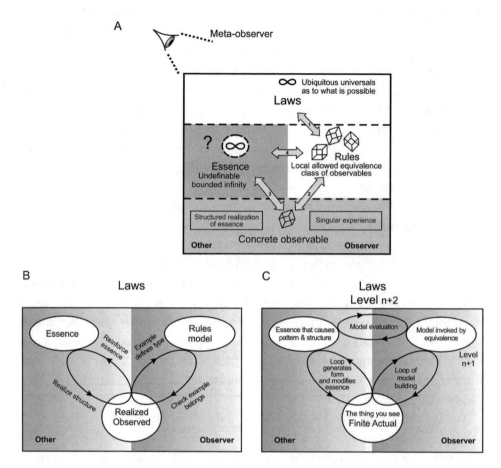

FIGURE 8.2. *A.* The path of connections in figure 8.1*B* is identified here. Arrow 1 shows that the universal, rate-dependent laws are the general condition, the super set above the rate-independent rules that are chosen by the observer in making the model. Arrow 2 shows how the rules based on a definition chosen by the observer offer the general class or type to which the concrete observable at level n belongs by definition. Arrow 2 is the iterative naming and check-ing relationship between the modeling rules and the concrete observable. The concrete observ-able is the tangible foundation for both the realm of the observer and the other. On the side of the other, Arrow 3 shows how the constantly changing, undefinable essence is realized in the concrete observable. Arrow 4 is the most important. Through working back and forth, the aim is to understand how that which is outside the observer's choice explains the equivalence in the equivalence class of the model. The laws are infinite and universal in the chosen discourse. The essence, which does not exist in the absence of the model, is bounded, like a role, but is undefin-able. Thus, the essence is a bounded infinity, like a strange attractor in chaos theory. The model created by the chosen rules is a finite equivalence class, a collection. The concrete observable is named on the side of the observer decisions by assigning it to the equivalence class. *B.* A recast-ing of figure 8.1*A*, but this time showing the loops that relate the parts of the whole conception. The labels on the loops indicate the action responsible for that part of each respective loop. *C.* This scheme labels the loops themselves as wholes. The loop of model building achieves discrete updates in the modeling activity of the observer. The loop on the side of the other updates con-tinuously, and shows the processes of realization as well as the influence of the realization back on its essence.

The metaobserver moves into the arena of discourse and establishes laws at level n + 2. A paradigm is imposed. The laws are infinite and inexorable in that they are universal to wherever the discourse might go. At level n + 1, there are two considerations: (1) essence or role independent of the observer; and (2) the rules or equivalence class, which embodies the local observer's model. The observed entity at the bottom straddles both the side of observer choice and the side of the other. That is what allows the inference between the observer's model and the essence. Movement that is not decided by the observer is on the side of the other. The rules are on the side of the observer's modeling decisions.

The essence might be a role in social science. It is on the side of the other. The essence is the generalized condition of the realization. Not only is it undefined, it is undefinable because it keeps changing. In following text, we use the U.S. presidency as an essence in the social realm, to show the equivalence of roles and essences. While presidents can be defined, the presidency keeps changing, depending on what the U.S. populace thinks it is, and as different categories of person assume the role (figure 8.3A). It only pertains once the rules are set, also at level n + 1. At the bottom of figure 8.2, diagrams A and B, at level n, is the concrete observable, which links the side of the observer and the other. The other is not the metaphysical external real world; it is still under observation of the realization, and is therefore an epistemological characterization, not a matter of metaphysics. The rules are what help us model the chosen observed thing that was realized in the other. The essence, particularly in social settings, could be the role of the observed thing at level n. For instance, the presidency is a role occupied by a specific incumbent at level n (figure 8.3A). The rules might, in that case, generate a list of all the presidents ever to serve. The essence is what all those members on the list share in common. The whole point is to see the extent to which the model embodied in the list of all presidents is explained by the essence. The essence is responsible for the equivalence across the list of past incumbents.

The lower concrete observable is linked, on the side of the observer on the right, with a loop that asserts the observable belongs in the equivalence class coming from the rules. "What is the thing at level n? Ah, it looks like it is one of those on this list." The rules are a generalization connected by the observer to the thing being observed. The returning side of that loop on the right checks to see if the thing observed is a worthy member of the equivalence class. Each trip around the loop on the right improves the model with thought experiments and some observation.

There is a second loop on figure 8.2B between the concrete observable and the essence. We have already discussed this side with regard to realization, and the transfer of information down to the realized thing being observed (see the introduction, figure 10 and surrounding text). The foal is physically and behaviorally a realization of its essence. That essence might be all the foals and mares that came

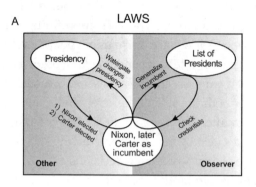

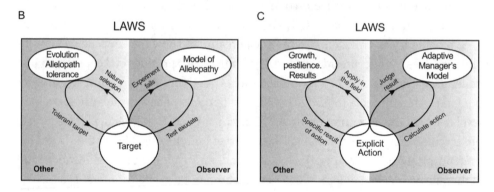

FIGURE 8.3.   *A.* An application of the scheme to a social setting. Sometimes an essence can be a role for the realized person, the incumbent. *B.* An application of the scheme to allelopathy. This shows that in biology, any time zero has a history that matters, the time for a chosen time zero is not zero. Allelopathy forces evolution on the target species such that its poisonous effect becomes blunted. Experiments fail because we see only the ghost of competition past. You can only show allelopathy in the evolved response of targets. You see it clearly only after it has been effective and then stops working. *C.* An expression of adaptive management in terms of the scheme. Unlike so much population biology, which just keeps spinning around the modeling cycle, adaptive management is careful to follow the modeling cycle with the cycle of observed material changes.

before, carrying information accumulated by natural selection. One side of the left-hand loop coming down is realization through the development of the embryo, guided by DNA. In other situations, the realization might be a process of election to realize an incumbent president. The loop might be closed as the realization influences the essence, perhaps through natural selection. Richard Nixon having his Watergate crisis and leaving the presidency might close the loop back to the presidency essence (figure 8.3*A*). The point of all this is to map the model, the equivalence set, to the essence, the cause of the equivalence in the equivalence set. The cause does not sit there externally without human observation and modeling. It is a putative cause for what the observer has created. Observer values are all over the cause embodied in the essence. How that cause exists in an external world independent of human observation remains mysterious, and is best treated agnostically in science. Going further than that is a problem for philosophers, not scientists. The usefulness of this whole conception can be shown with allelopathy, that is, chemical warfare in plants (figure 8.3B). It is a well-known phenomenon, and in prairies it is clear that sunflowers are beating on the surrounding plants. We should be able to take water washed off the allelopath's roots and leaves and see it damage the target species. But those experiments do not work out.[26] We can concentrate the eluded poison and then it does work, the target is debilitated. But in that case, the concentrate also damages the allelopath. The experiment that needs to be done is to find two target populations of the same species, one exposed to the allelopath, the other naïve.

Then use a set of different concentrations of the poisonous solution on both populations. If the naïve population ever shows damage while the other does not show damage, or shows less damage, then you have shown allelopathy and can suggest its mechanism.[27] The experienced population has evolved resistance, such that the allelopath can only increase the dose at its own peril. The message here is that the experimentalist cannot show allelopathy unequivocally until it does not work as well as it did previously.

In biology, there never is a fully satisfactory time zero, before which history does not matter. Much of physics can rely on some workable time zero, but not in biology. And that is what is happening in allelopathy. Geoff Sagar suggested the experiment that shows you can only prove allelopathy when it is part of history. The concept coined by J. H. Connell of "the ghost of competition past" captures the same idea.[28] It is hard to show competition because most of the evidence is character displacement in species that have already minimized competitive interaction. The ecologist does not often see evidence of competition until so-called competitors have evolved to avoid it.

The scheme we lay out is a device that cleaves material consequences away from conceptual devices. In biology there are many pitfalls that make it appear that a normative value judgment is in fact a material cause. Many aspects of biology that are viewed as material are instead normative value judgments. The mechanics of the performance of a dominant gene does have a material basis rooted in protein synthesis. But the decision to call it dominant or recessive has already been made. Notice we never call sickle cell dominant for malarial resistance, although it clearly is that exactly. That is because sickle cell is a lethal gene and can never become the normal condition because of selection against it in the homozygote. As mutants first occur, they are recessive. If they become advantageous, then they are selected and become the normal condition. At that point, the context has changed and the previously dominant gene becomes recessive. But it is not because of any material change; it is because of a human decision to change what is considered normal. The old normal that was dominant makes what it always made, as does the formerly recessive gene. The difference is that the former dominant is no longer considered normal or advantageous. All genes are dominant for the protein they make, and all genes are recessive for what all other genes make. Thus, a dominant gene performing the mechanics of dominance is material, but the decision to call it dominant and the concept of dominance is not mechanical.

There are many situations in biology that fool us into thinking something is material when it is not. Our scheme deals with the time zero problems. It teases apart what is normative or semantic from what is demonstrably material. By separating out the modeling process, our method shows with its two sides what is human decision and what is a material happening beyond scientists' decisions. We are explicit about having taken a point of view. Others also have taken that point of view, but often not as a conscious act. We offer self-conscious thinking.

A common relative of these difficulties is working out what a data problem is versus what is merely conceptual. Different logical types speak of roughly the same thing, but they employ different levels of analysis: carnivore and herbivore are the same logical type (animal defined by what it eats); cat and dog are not the same logical type as carnivore. If the problem is expressed as an inappropriate logical type, then more data are not going to help, except as it clears the mind by offering frustration that the new data do not help. Most ecological research and management problems are not data problems. Particularly in large modern ecological challenges, the next critical experiment that will sort it all out is never just around the corner. In chapter 7, we refer to Wixon's findings about the research in soil and global warming, where she found that there are plenty of data on the issue, but still no resolution. It is not so much precision in measurement that limits us, as it is lack of precision of thought. Our scheme serves that need.

It is from this argument that we can see the power of what is called adaptive management, a superior practice in applied ecology. The ecologist sets up a reasonable model generated from observables and experience—once around the modeling loop, and then over to the other side to see what happened in practice (figure 8.3C). In light of the observed productivity or regeneration or whatever, adaptive managers use the last cycle of management as an experiment to adjust the model. It is like a human management version of the learning that goes on in a neural net with the learning data set. Adaptive managers keep cycling around from one loop to the other, and in this way improve the model.[29]

Population biology is particularly susceptible to spinning around the modeling loop with insufficient attention to what happens in nature in the other loop that allows for evolution in changing environments. The equations get longer and more subtle in the symbolic world, separate from material checks on concrete observables. Notice that Cohen raises Lotka above most population biologists by emphasizing his reliance on observation and action (see quote in note 15).

Mistaking the model for what it models is a well-known error, but it is not so well known how easily that can happen, and in how many ways.[30] Our scheme puts up flags to guide the way.

The cycling in Figures 8.2A to 8.3C is, in fact, the way science is informed by models. There is an underlying strategy, which is an exchange back and forth between the specificity of the model and the slacker approach of narrative. The strategy is like a series of filters. (Later, we cast Checkland's problem-solving engine as a set of filters, figure 8.12.) It is a strategy of loosening the discourse in narrative, followed by the tightening of the focus in a model. The following sections demonstrate that to be the case. We show how the one-to-one specificity of a mathematical function interplays with the slack of a many-to-many relationship of a mathematical relation. Using characterizations found in the humanities, we show how the expansion embodied in the form of a metaphor alternates with the compression of an analogy. We present the mathematical tools of a narrative. Much of the rest of the chapter is devoted to dissecting the way science comes to know what it does know. We compare and contrast a set of devices that science uses in that quest. By laying out the process, we can then see how science helps managers, and how management is a spur for science in complex situations.

RESEARCH LATTICES

The process of research starts with a primitive narrative that tallies first impressions. We have found that by looking at a series of two-by-two lattices, it is possible to link to the aforementioned scheme in a way that informs us how models are built and used to improve the primitive narrative. In the end, a higher-quality narrative is generated through modeling. So in the next few pages, we investigate and link together a series of two-by-two tables that address different contingencies. The first two-by-two table addresses categories of ecological investigation (figure 8.4). In that table, the research that underlies ecological system management issues can be conducted using applied or basic questions. We have already seen how our scheme maps onto adaptive management (figure 8.3C).

In figure 8.4, basic research in ecology involves a style of scaling the question very differently from the style used in strictly applied research. First, we deal with basic ecological research in the bottom-left quadrant. There would be a muddle if we discussed more than one quadrant at a time. For the moment, let us keep applied ecology out of it. The original questions in basic ecological science do not usually involve a consciously held scale of operation. The scale only becomes fixed as the questioning (problem definition) becomes operationalized. The scale becomes fixed through closing options in basic science until a constrained model is built. Different scales are pruned away as the science is moved forward. Human

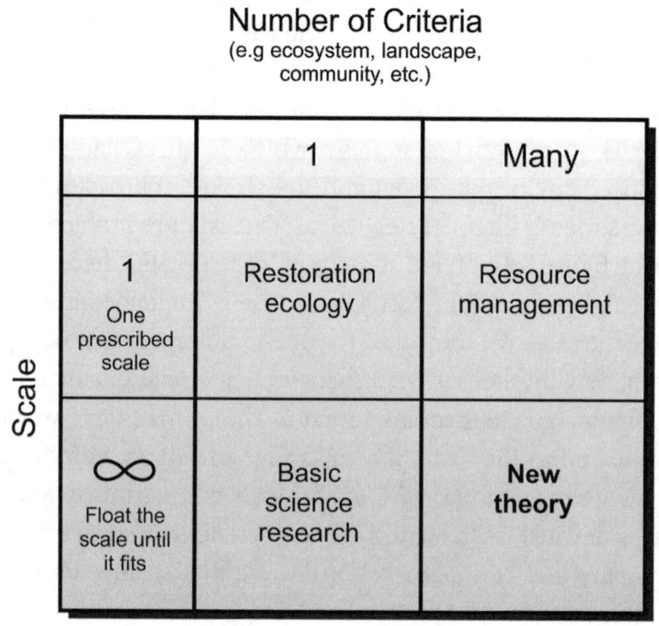

Number of Criteria
(e.g ecosystem, landscape, community, etc.)

	1	Many
1 One prescribed scale	Restoration ecology	Resource management
∞ Float the scale until it fits	Basic science research	**New theory**

Scale

FIGURES 8.4.   A two-by-two lattice that contrasts floating versus fixing the scale against one versus many criteria for observation.

children get smart by pruning their neurological options. That is why a child of age twelve or so moving to a new country with a new language picks up the language and becomes fluent, but speaks the new language lifelong with an accent. Slightly younger children have not yet pruned their networks, and so are still open to the sounds of the new language, coming to speak the new language without an accent. The process of scientific investigation prunes away alternative experimental and conceptual pathways. In the end, in ecology, the model is explicitly scaled.

Basic science ecologists almost always have one subdiscipline that is their default. Ask them what sort of ecologists they are, and they will give you just one type of ecologist. For instance, Allen feels he is a plant community ecologist, although he clearly does much more. Given a plant community research issue, he researches in that way, only later coming to other ways of looking at the problem. Basic ecology is almost always of only one type per investigator.

The original research question is not conceived as being scaled, but the answer to the question often has a convenient scale at which it can be resolved with reasonable effort. In ecosystem science, the scientist chooses boundaries that coincide with the boundaries of an ecosystem, such as a watershed, not a community or an organism. In organismal ecology, the scientist usually scales the research so that it looks at one organism, not, say, two and a half organisms. As the system is observed, the scale is floated so the boundaries of unit structures are changed.

The rescaling is so that the scales used remain convenient for the protocol of just that one type of ecology. That is why scaling is important for finding that which is robust to transformation. It also indicates why a unified ecology is so difficult.

Rescaling of the research protocol is not dishonest or bad in any way, for the original conception of the problem is only implicitly scaled, and the readjustment of protocol just brings the mechanics of the field or laboratory work in line with the original ideas. Basic science ecologists rescale as it is convenient.

THE COST OF FLOATING SCALE/ONE-CRITERION BASIC RESEARCH PROTOCOLS

In floating the scale to fit the empirical protocol, and choosing one ecological criterion, the benefits are clear. The sharp focus of the question that comes with such a research strategy is important. Less obvious are the costs of this conventional approach to basic research. Some examples of basic research protocols using a floating scale and one-criterion might illuminate this point. In all these examples, there is a rescaling such that the measurement system loses power. Of course, one can always rescale by taking a different point of view, and scale to that.

1. Calibration of models that look to soil runoff is quite often calibrated to fit the local results.

> Model 'calibration' amounts to converting a 'parameter' to another 'variable' without adding another equation, so one has essentially an underdetermined system with more variables than equations. Anyone who has solved simultaneous mathematical equations knows that an underdetermined problem does not have a unique solution.[31]

Usually the recalibrated model does not predict soil runoff in other fields because the model now applies very well, but only to the place from which it was calibrated. The calibration data are not representative of the general case of soil runoff, but only of the local data collection happenstance. We need data, not to show the local time and place of the measurements, but to represent the phenomenon. Over-trained neural networks have this problem because the net has remembered the details of the measurement on that one day and compromises the general patterns of the phenomenon in play.

2. By normalizing data collection to damage done by herbivores over a given area during a defined period of time, one produces a neatly scale-defined study. However, differences in leaf longevity and replacement are in differently scaled units across species. Grasses replace loss by leaves growing from their leaf bases.

Meanwhile, forbs must regenerate from new buds. Forbs not only lose capital but must also replace the means of regeneration. Linear loss in grazing is the one-to-one mapping of material consumed by the animal and the biomass loss of the grazed plant. But plants can deploy below-ground material as a replacement, which, if one-to-one, is still a constant as to loss of the whole.[32] But this introduces the complication of the new leaves often being more productive than the leaves eaten. This changes the exponent on growth and so amounts to a non-linear change. Faster recovery can change competitive advantage for space and light, a second nonlinear term. The scaling of the measurement (how long after grazing leaves are measured) changes the scale of the phenomenon measured. Thus, different scaling quickly becomes a qualitative difference; the ecologist is no longer talking about the same thing, but may not recognize the change that has taken place.

3. The cornerstone of the conventional view of Darwinian evolution is competition applied to variations expressed in form and function. One of the problems with Darwinian evolution is that it is capable of explaining almost anything that we observe by invoking some mutation or contrived competitive contest. Anything that explains everything explains nothing. There has emerged a fixation among adaptationist ecologists to put everything in terms of competition. It took repeated articles by Dan Simberloff to disabuse the competition experimentalists that nobody had actually shown competition.[33] Sometimes other processes come to the fore, but the situation is rescaled until what is happening can be expressed in competitive terms. The simplest case might be mutualism; it is possible to rescale the relationship such that it is expressed as giving a competitive advantage to all parties to the mutualism over anything that is not in the association. In this case, the rescaling seems innocent enough, and it makes intuitive sense. However, an insistence on rescaling so that the context is always a competitor is in danger of organizing the world experience around the concept, instead of the concept around the experience. That would be an example of spinning around on the modeling side of our cycle diagram (figure 8.2C) without going to the other side to make a test, as would occur in adaptive management (figure 8.3C). The fact that one can make it fit the data is beside the point, for there is a hidden cost.

4. One last pair of mixed scaling occurs in some gradient analysis or ordination techniques. Despite Peter Minchin publishing several papers that prove unequivocally that one would not want to use detrended correspondence analysis (DCA), the technique refuses to die.[34] Its appeal is that it orders species and stands of vegetation at the same time, so one ordination may be directly translated to the other in a transposition of data matrix rows for columns. This occurs because the technique normalizes by columns and rows iteratively until the data matrix stabilizes. In principal components correlation analysis (PCA), there are two averages around which the points to be ordinated can be rotated, the column mean and the row mean. A correlation between stands is stated relative to average species, the

species that occurs in average quantities over all stands. That is what the $x - x^{bar}$ in the correlation coefficient is saying. There is a normalization around a mean. All depends on which average (which x^{bar}) is used, the average stand or average species. Conversely, the correlation between species is stated relative to the average stand, the one with an average value for all species. PCA ordinations of stands are achieved by rotating around the average stand. Species ordinations rotate around the average species. That is why species and stand ordinations using PCA cannot be directly transformed from one to the other; different means are used as the reference for rotation. The x^{bar} used is from the other edge of the matrix, column means versus row means.

The scaling issue is that the average stand has a different ecological scaling from that of the average species. They do not relate to each other in any straightforward way. In fact, how they relate would be an interesting study that has not ever been done, as far as we know. It could be done by subtracting a species ordination away from its corresponding stand ordination. DCA iteration rotates around the average stand and species. It is bad enough that PCA stand ordinations rotate around the average stand, which is too even and too diverse to occur in nature. But at least the average stand means only one thing at its own scale of operation. Who knows what an average stand species means? The use of DCA is inexcusable. It was a good try, but it is time to move on. As an after note, the detrending is unstable, but that is another problem.

FIXING THE SCALE

Restoration ecology, like basic research, uses only one criterion at a time. Town parks or golf courses are landscapes restored to some mythical form where Greek gods might have walked. The great manor gardens of England are a contrived natural landscape that is not natural at all. Preservation of rare species is the restoration goal under the population criterion. Prairie restoration is of communities. Creating functional soil on abandoned urban land is the restoration of ecosystem function.

Restoration is like basic research, but it differs from basic science in that it is in a prescribed area of a certain size (figure 8.4). The scale is prescribed, not chosen. Restoration ecology offers the acid test of basic science. It challenges that if you have sound ecosystem theory, make one! But make it not at the convenient size of a single watershed. Most of the Curtis Prairie in the University of Wisconsin arboretum is a tall grass prairie restoration. But it is not big enough to include bison. Small suburban front yard prairie restorations cannot even include fire. Neither of these restorations is inadequate; they are only to a certain scale that includes only processes of a certain size. If they were inadequate, then by the same logic, the Alaskan wilderness would be inadequate by virtue of not embracing tropical forest, too. Restoration is always at a certain scale.

The real heroes in all of this are the managers. Not only do they have a certain size imposed, like a U.S. Forest Service management unit of one hundred thousand acres, but they must also embrace multiple criteria. This is not just a quirk; the Forest and Rangeland Renewable Resources Planning Act of 1974 imposes multiple land use criteria. Land must be managed for wildlife, recreation, and ecosystem function, also with an eye on production of a certain number of board feet of lumber and the withdrawal of as many cubic feet of water as possible (figure 8.4). An intellectual flexibility is required.

The last cell in the two-by-two lattice is to float the scales so they fit phenomena, but to do it simultaneously for many criteria. This is what we are doing in this book. It is theory to unify ecology. In management there are many models, at least one for each expert in the arena of figure 8.5A. The context of the different models will be different. For instance, the context of a toxicologist starts just before the arrival of industrial production on the landscape. Toxicologists can be apologists for the chemical industry, who validly point out that many of the chemicals in the environment blamed on chemical companies were already present in nature. The fisheries biologists' context might start with the arrival of Euro-Americans, or with a context of a failed fishery and the creation of dams. The Native American's context might go back to the creation of the world in their creation myths, and reach seven generations into the future. We do not use "myth" here in a condescending or pejorative way, but are rather guided by Barry Powell from the University of Wisconsin's Classics Department. He said to Allen, "The big bang, now that's the stuff myths are made of!" The ecologists' context might go all the way back to Native Americans first crossing the land bridge from Asia. For the manager, the context starts with the time frame of the current problem that needs fixing. For the Native American, the issue is that the Great Spirit is offended by careless use. For the toxicologist, the issue is that the blame for pollution is being meted out fairly. For the fisheries biologist, the issue is bringing back the fishery. For the ecologist, the issue is the repair of the functioning of the whole ecosystem. All these players have their own model that is bounded and internally consistent. They each have a different narrative that they are respectively trying to improve with their models. There is a lot to be said for creating a narrative through negotiation and getting to respect each other.

Overarching all these models is a set of stakeholders who must be given account (figure 8.5A). The various players mentioned previously would be part of that. Our measured view is that science gets us all to agree on something. Science creates commensurate experience. A balanced scientific discourse addressing some management issues needs narratives.

In science, it is the commensurate experience that counts, not some mythical verity. Better to rejoice in the power of commensurate experience and narrative than ascribe dubious relationships to reality. The process of finding commensurate experience is diagrammed in figure 8.5B. The same players are present that

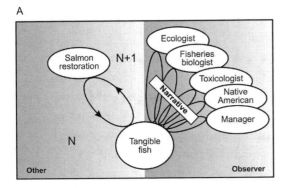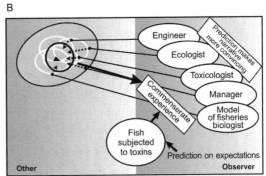

FIGURE 8.5.   *A.* Multiple observers can come as different sorts of people, each with their own perspective and predilections. Each will have a model, one for each type of observer, derived from their personal set of rules for the model. But all can address the same material entity in a way that pertains to different but related essences. Including many stakeholders is a way to generalize what is said to be experienced. The inconsistencies of the various models are handled well by a common, negotiated narrative that arises from coalescing the different essences. That generalized narrative is the consensus of and rapprochement between all the stakeholders, as when green politics can support constrained business interests. *B.* Showing the way prediction makes the narrative compelling such that a common understanding occurs. Commonality can lead to a strong commitment to action. Predictions work on expectations to make the narrative convincing so the models converge to reinforce agreement.

are in figure 8.5*A* where each player has their own model. Each player's model maps to the same general region of essence as all the others. But since the models are different, the respective essences will be in a slightly different part of the neighborhood of all the players' essences.

Most of the player's essences will have some overlap, at least to the point of including fish, water, and an expression of something being wrong. If there is no significant overlap, the parties will simply put each other's paradigm down and get nowhere. But if there is any overlap in essence, it can often be increased by adjusting models. The solid arrows across to the region of the essence show the relative position of the essence of each respective player. Each essence is illustrated by a white oval. The small black rowboat-shaped region shows the point of intersection of all the essences, the overlap from which larger agreement can proceed. The dotted arrows show how the change in perspective of each player causes a shift toward a common essence among all the players, illustrated by the black line oval. That congruence of essences allows all the players to agree on what they are saying. As a consequence, when each of them sees the general situation, they all experience significantly the same thing. This common agreement is illustrated by the broad arrow coming back to the side of the observer, so as to give commensurate experience. Independent observers come to agree on a unified narrative as to what is the situation. Prediction does not so much say that the science is right as it makes the story more convincing. Science is about achieving commensurate experience, and prediction helps that happen. Appreciating accurate prediction will be of selective advantage to our species. It keeps us out of trouble.

Managers attempt to develop narratives when they bring the scientific recommendations to public meetings to establish the consensus needed to implement a management action. Unlike the models of the stakeholders separately, the narrative of their commensurate experience does not need to be internally consistent. Old-school modernist scientists reference proximity to external reality, of which there is only one. This encourages them to be intolerant and inflexible. One is reminded of the line in the movie *Ghostbusters*: "Back off man, I'm a scientist," a counterproductive posture. Scientists do not become working stakeholders until they achieve respectfulness. Wise scientists know that the lay public usually has more extensive data, as when they know about an aunt who moved away and got cancer later. The aunt would not be in the scientists' data. Listen to the lay public.

Humans solve problems by changing the reference to accommodate rich environments. Engagement to help such situations needs a narrative that slips and slides, as it must in order to be able to create acceptable compromise. After the American War of Independence, there is now the "special relationship" between the United Kingdom and the United States. Similarly, Queen Elizabeth recently met Gerry Adams with civility, although he is the leader of Sinn Fein in the Belfast parliament. He denies ever being in the terrorist Irish Republican Army, but many accounts question his position on that point. The compromise that brought Protestants and Catholics together in Northern Ireland was forced by an impending tax from Westminster, which could only be stopped by a rapprochement that would reopen Stormont, the Northern Irish parliament, so it could take over taxing authority. The story that must be told in such agreements is of necessity contradictory, but it is a narrative, and they allow that.

Engineers are funded in a much more reasonable fashion than scientists. The openness and indecision at the outset in science makes writing and reviewing grants an object lesson in futility. As a side note, nobody does or even intends to use the methods they propose, but that is where grant proposals are pecked to death. Therefore, we suggest that methods sections in National Science Foundation (NSF) proposals be limited to something like two hundred words as a statement of feasibility. And given the passionate and jealous devotion to respective schools of thought, it would be a good thing to deny reviewers access to all citations as well, until they have made their decision. When they cannot withhold funding because the decision had been made already, we say that only then do they get to see the references and make helpful suggestions. Given the rejection rate of grants, the measures of quality are completely saturated and therefore give no signal with which to discriminate between who actually gets the funds and who does not. A system that takes the best scientists and wastes half their time writing grants for a random assignment of funds might as well be devised by some malevolent force that wished to bring science in general, and ecology in particular, to its knees. The funding agencies are staffed and assisted by honest, good people, but they appear trapped in a context of full dysfunction.

In figure 8.4, we have already looked at the process whereby science and management have different options as to the number of types and scales that are involved on a given project. But we have not run the contrast between engineering and science. We see that science and engineering start with opposite postures. The options for science are wide open. By contrast, engineers have a client before they are involved in a project. The engineer starts with some problem in particular that must be solved. Meanwhile, science is almost idly speculating about what it might think of addressing. But then, once science and engineering actually start, things change and even reverse.

When the scientist arrives at a solution, it is usually very tightly circumscribed. Solving the problem becomes increasingly constrained as the process moves forward in science. Very specific outcomes derive from the tight restrictions of a particular experimental design. Of course, there is often generalization after the fact, but the practice of discovery is a process of increasing constraint. By contrast, the engineer will have a specific goal, but the path to it is wide open and full of invention. The creativity in engineering comes from choosing which line to take as the problem is solved in a series of compromises between cost and quality. The scientist tries to create something closed in an open system. Meanwhile, the engineer tries to create something open in a closed system.

The one-to-many contrasts can be employed in the contrast between science and engineering. Figure 8.6B presents a two-by-two chart of one to many and a contrast of model and narrative. That captures the relationship between science and engineering protocols discussed earlier.

Figure 8.6C goes further to look at one-to-many relationships in both directions as it addresses standard devices for description in the humanities and in mathematics. We may be at odds with some definitions of metaphor in the humanities, but our usage here captures something useful in folding word usage from the humanities into science. Definitions are all arbitrary, and we choose to define metaphor so that it only goes one way. Katherine N. Hayles, the post-humist, has pointed to the working metaphor "the morning is an open door" while the inverse "the open door is a morning," does not work. In an email communication on August 28, 2014 she encouraged us "I think you are on the right track to insist on the importance of metaphor, including its one-way directionality." Another example of one-way directionality is , "Time is money" works. "Money is time" does not, or at least does not converge on the same concept. Metaphor is a representation. When Aristotle used the metaphor that makes the prow and keel of a ship equivalent to a plow, we would define that as an analogy because the comparison goes both ways. As an

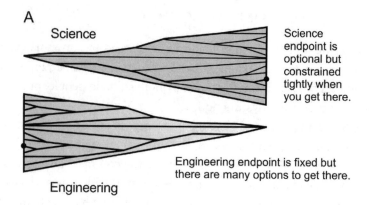

A

Science

Science endpoint is optional but constrained tightly when you get there.

Engineering endpoint is fixed but there are many options to get there.

Engineering

B

	Model	Narrative
1	Fix goal early on Engineering closed	Narrative is fixed in the end Science closed
Many	Goals are open early on Science open	Open narrative to get there Engineering open

Engineering has one model at the outset. Science has one narrative in the end.

Science has many models at the outset while engineering has many narratives available to it to achieve its one model.

C

To	Many	1
1	Metaphor pertains to other (expansion) Scale independent	Function focus on specifics
Many	Relation look to generalities	Analogy pertains to model/observer (compression) Scale comparative

FIGURE 8.6. *A.* Schematic for the difference between the creative processes of science as opposed to engineering. Science starts wide open to all possibilities, while engineering starts with a client who needs a particular problem solved. The scientist goes on to impose restrictions on the model as it is created, until all wiggle room is removed. Meanwhile, we can be impressed with how many ways there are to solve the engineer's specific goal. *B.* A two-by-two lattice that lays out the message in figure 8.6*A*. *C.* A lattice that looks at the devices for handling concepts. Two of them are from mathematics—a function and a relation. The other two are devices used in the humanities to make comparisons—metaphors and analogies. The metaphor maps one situation onto two or more. Thus, it is in the one-to-many corner of the lattice. The analogy is a sort of the reverse of a metaphor. Analogy is a compression down from two or more things with regard to the subset that they have in common. Analogy is therefore in the many-to-one corner, opposite to the one-to-many corner of metaphor. A function maps one variable onto some other one variable. It lives in the one-to-one intersection. A relation in mathematics is a many-to-many mapping. One-to-one mapping fails in a relation, but it is still bounded. A scatter plot showing correlation is a many-to-many mapping.

analogy, it is a compression down to an object that cuts through a soft material, the particular properties that both would have.

Formal models in Rosen's modeling relation are representations and may apply their relative scaling to various observables. For instance, the model embodied in the equations for aerodynamics applies to any flying object (see the introduction, figure 5). We use the Rosen modeling relation again in chapter 7 to show how biome is a scale-independent notion that can apply to small biomes embodied in frost pockets (figure 7.1). Here, in figure 8.7, the emphasis is on metaphor as a mode of representation in the formal model, and how the analog models are compressions down to commonalities.

While metaphor is a one-to-many mapping, an analogy is the reverse in that it is a many-to-one mapping. If more than one observable can be encoded into and decoded from a formal model, then the two material observables become analogs of each other. That is what happens in any experiment. The two analogs are locked together in a compression down to only what they have in common. Distinctions in the humanities, that is metaphors and analogies, are cast in terms of one to many and many to one (figure 8.6C). The contrast between metaphor and analogy is captured in the top-left to bottom-right diagonal of the figure.

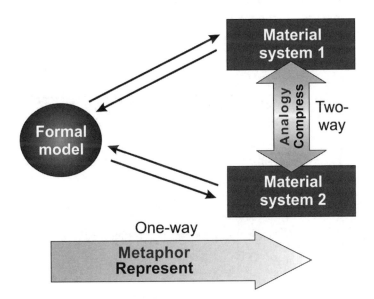

FIGURE 8.7.   Rosen's modeling relation in austere terms. A formal model is a set of scaling equations. It is a representation, and works like a metaphor. Metaphors, as we use them, map one device onto more than one situation. Formal models can map onto more than one material situation. It says that all is the same except scale (that is not true, of course). When two or more material systems can be coded and decoded into and out of the formal model, the two material systems become analog models of each other. Analogies comprise the subset they have in common. That is needed for all experimentation.

The bottom-left to top-right diagonal in figure 8.6*C* is a contrast of mathematical forms. A one-to-one mapping is a mathematical function. A many-to-many mapping is a mathematical relation. In a function, one point on one variable maps to one point on the other. A many-to-many mapping is a relation. A scatter plot of two variables together gives a many-to-many relation. We have addressed the contrast between functions and relations when discussing Roberts's clique analysis in chapter 4 on the community (figure 4.26). Functions are particularly focal. Relations are constrained, but have slack in them. Relations are therefore a more open expression likely to relate to essences.

We have used a series of one and many, and narrative and model matrices. In attempts to operate on the lattice in figure 8.6 (comparison of cells), we found a lattice that takes model to new model or narrative, or narrative to new narrative or model (figure 8.8). With all of the one and many and model narratives, the new matrix seems an obvious thing to try. To our surprise, this is what is going on in the loop diagrams of figure 8.2, diagrams *B* and *C*. When an experiment is performed (a practical test of a model), it comes from a general narrative that is brought to some local point of focus in a model. So an experiment lies in the narrative-to-model sector. The focus of the experiment is akin to a mathematical function. The result of the experiment is a specific result, but it is immediately used to inform and update the former narrative. Therefore, experimental result lies in the model-to-narrative quadrant.

The relationships in the realm of the other lies on the other diagonal of figure 8.8. The model-to-model quadrant sees the realization as the model of the

	Narrative	Model
Narrative	Role, Essence, Bounded infinity	Experiment
Model	Experimental result	Refinement

FIGURE 8.8.   When we create a lattice of narrative and model leading to both narrative and model, the scheme of the loop diagrams appears (figures 8.2 and 8.3). Narrative to model is an experiment. The open narrative is fixed in the model of an experiment. Model to narrative is embodied in an experimental result. Having focused the narrative in an experiment, the result, specific as it might be, is thrown back into the narrative. Model to model is a realization (a model of the essence) that influences the essence to produce a new realization, a new model. Narrative to narrative is the narrative of the essence being updated via the realization to make a new evolved story for the essence to tell.

essence. The realization updates the essence that then creates a new realization. Thus, the model leads to a new model. All this involves focus, as in the mathematical function. The other end of that loop is the narrative-to-narrative quadrant. The essence has the slack of a mathematical relation. An essence offers a realization, but that realization, perhaps through natural selection, changes the essence. Thus, the open narrative of the essence is updated in a new narrative.

An experiment takes an analogy and casts it in terms of the singularity of a model. The best experiments often take out all but one point of tension. It can tell us in a singular way how at least part of the analogy works. The experimental result is focused, but it is immediately hurled back for interpretation into the openness of the scientific narrative.

The realization is the model of the essence. The essence has a narrative (perhaps of the species) that is then concretized in the realization (a member of a species). The concrete foal carries with it a model as to what to do about its mother, that model coming from natural selection in the essence. The model, perhaps an organism, contributes to the essence of its species by natural selection, or some sort of like process. Thus changed by natural selection, the essence generates a new and different realization. In this way, the model of the essence, the realization, is refined by evolution. The narrowness of the function indicates a particular realization, a model from the essence. The model-to-model sector of figure 8.8 pertains to the updating of a realization.

Table 8.1 collates all the relabeling of the arrows across all versions of figure 8.2A. The rows of the table give the path of the analysis on both sides, the observer and the other. The columns of the table give the alternative interpretations of each activity in modeling and experimentation, as well as the happenings in the other.

Figure 8.9B shows how the side of the observer is updated in iteration between experiment and experimental result. Each part of that cycle is autonomous. Conduct the experiment and get the result. Separately, the result is used to conceive of a new experiment. However, on the side of the other, updates take going around the whole cycle. The update is either of the realization in refinement or of the essence through evolution. One update takes going around the whole cycle from either end of the loop or the other. It takes a full circuit to achieve the update.

So now we have new labels for the sides of the processes that drive the cycles of figure 8.2B. Figure 8.8 shows how narratives and models can be mapped onto the arrows of figure 8.2B. It explicitly notes the alternation back and forth between model and narrative in the process of modeling and in updating the essence and realization.

The discourse immediately above involved moves between different views of the modeling situation. Table 8.1 suggests different ways to label the arrows on the cycle diagrams. The top row shows the original labeling, as the cycles were first introduced in figure 8.2. The right-hand cycle was the model-building exercise; the

TABLE 8.1

LABELING SCHEME	UP ARROW ON SIDE OF OBSERVER	DOWN ARROW ON SIDE OF OBSERVER	UP ARROW ON SIDE OF THE OTHER	DOWN ARROW ON SIDE OF THE OTHER
Original labeling in figure 8.2B here	Example defines the type	Criterion is verified in the example	Reinforce the type or essence	Realize the structure
Labeling on figure 8.9A showing experimentation	Conceive of test	Do experiment	Reinforce essence	Realize the structure
Labeling on figure 8.8, after the model/narrative lattice on figure 8.9B	Make a model from a narrative	Apply a model to improve a narrative	Nature's model changes nature's narrative	Change the model using nature's narrative
Labeling from figure 8.9C where metaphors and analogies map to functions and relations	Give meaning to a function, set example in context	Associate a function with an analogy	Give an analogy meaning	Map nature's metaphor to a function
Moving between specifics and the general, figure 8.9D	Make specifics from the general	Check to see specifics fit	Relate realization to its general situation	Realize the general condition into a structure
Focal and tacit attention, figure 8.9E (Needham 1988)	Tacit to focal	Focal to tacit	Focal to tacit	Tacit to focal

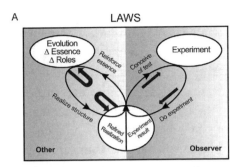

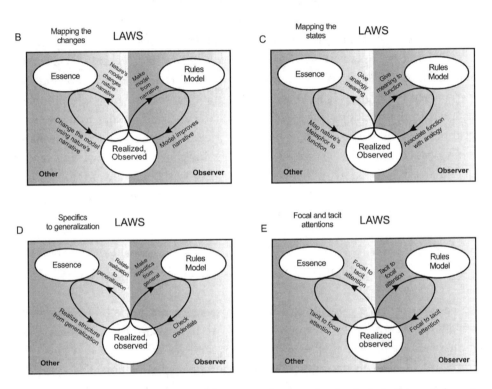

FIGURE 8.9.   *A.* Reintroduction of figure 8.2*B*, but translated to include the equivalence to Figure 8.8. The model-building loop of the earlier figures has as its ends the experiment and the experimental result. The realization loop ends are the essence and its realization. The straight shaded arrows in the right-hand loop of the observer indicate a straight connection; model to narrative; narrative to model. The recursive arrows, in the loop of the realization, indicate that the mapping arises from going right round the left-side loop. In model to model, realization maps to a new later realization. In narrative to narrative, essence maps to essence. The side of the other pertains to dynamics, not so much structure. In the left-side loop of the other, we are looking not at experiment moving to result but at a dynamic loop where realization is regenerated as a later realization. *B.* Relabeling of the loops in figure 8.2*B* to capture the way models improve narratives. *C.* Relabeling of the loops in figure 8.2*A* to the model/analogy relation in Figure 8.8. *D.* Relabeling of figure 8.2*A* in order to capture how each step is either general to specifics or specifics to general. *E.* Relabeling of figure 8.2*A* in order to capture Polanyi's arguments about tacit and focal attention. He suggested that hierarchical levels change as the observer moves from focal to tacit attention. An attempt to make the tacit/focal changes and vice versa.

left-hand cycle in the other creates realizations and drives evolution. The second row of Table 8.1 shows the relabeling to put experimentation explicitly into the scheme, as in figure 8.7. The third row of the Table 8.1 is the labeling of figure 8.8, where models and narratives alternate with each other. The alternation iterates to focus in models, and expands context into narrative. Narratives invite models, which then improve the narrative. The fourth row of Table 8.1 derives from figure 8.7, which contains one-to-one, many-to-one, one-to-many, and many-to-many relationships. The linguistic devices of metaphor and analogy interweave with the mathematical devices of function and relation. The fifth row in Table 8.1 shows how the labels on the cycling arrows can refer to specifics as opposed to generalization. Functions give specifics, while relations give generality.

The sixth row in Table 8.1 is related to the fifth row, but refers to focal versus tacit attention. Michael Polanyi pointed to focal and tacit attention, showing that only one level of analysis is focal, while surrounding levels are only tacit. Complexity arises in that tension, since a move to make something under tacit attention focal only makes what was focal now tacit. Needham, pointing to the limits of literary criticism, used Polanyi's device in showing how a piece of writing echoes up and down a hierarchy of literary devices.[35] He starts with a word, moves to a phrase, and continues on to major passages. And all of that reverberates inside the level of the whole. This is discussed at the end of this chapter.

PUTTING NARRATIVES TOGETHER

We have relied heavily on Rosen's modeling relation (in figure 8.7) and now do it again one last time. In previous chapters, we have also introduced the notion of phase diagrams (see chapter 6, figure 6.5), and we will fold in that idea. The formal model gives us description in the form of a representation. That gives the state of the generalization that is the formal model. As we have said, the encoding and decoding of the observable allows the recognition of an analogy between two observable systems that fit the formal model. There is no coding in the analogy. It is only the shimmering back and forth between the analogs that makes the analogy work. It makes sense to view the formal model as the state. It is a point in the parameter space of the model. There is one state at a time and one formal model. By contrast, it makes sense to view the analogy as the first derivative, there being two observables that are compared.

So what is the relationship between the narrative and the analog model and the formal coded model? Narratives are compressions down to what is in the story. Nothing happens after "And they lived happily ever after," or before "Once upon a time." And in between the beginning and the end is not everything that could be chronicled; it is only the events deemed significant. From this logic, a narrative is a

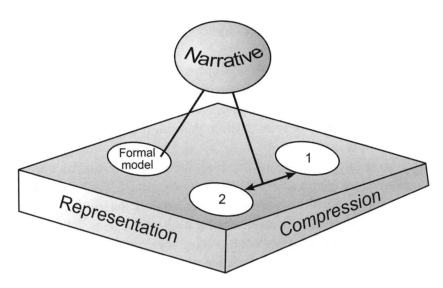

FIGURE 8.10. The narrative is a representation of a compression. The narrative is the whole Rosen modeling relation. Narrative is presented here as above the modeling relation because the narrative is a representation of a compression.

description of compression. If Rosen's modeling relation is on the geometric plane at the bottom of figure 8.10, then the narrative sits in the space above, holding the whole modeling relation together. Narrative is a three-legged stool with one leg to the formal model, the representation leg, and two more legs going down to the two sides of the analogy, the compression legs.

So if the formal model gives the state and the analogy gives the first derivative, then the narrative can be captured in the phase diagram in figure 6.5 *B* (see chapter 6, on population). Plot the state of a pendulum against its first derivative and you get a circular track. That is the narrative of the pendulum without friction. With friction, its narrative will spin down, spiraling inward, eventually to hang straight and motionless, a different story. If the model is chaotic, we get a strange attractor of chaos that never rests or repeats. That means that many stories never end. The big ones do not have the finality of an ending. Cinderella must return to rags between telling the story, so she can ascend to riches again in a new telling. The reversal back to rags is simple, as it is in a chaotic strange attractor; just backtrack. And chaotic strange attractors are infinitely sensitive to initial conditions. So although there can be a technical reversal of the story, running it forward again will follow a slightly different track. This is because the new beginning cannot be specified exactly. It cannot be exactly respecified in the same way that a repeated experiment has difference enough from last time for a chaotic system to follow a new track, very similar in the beginning but diverging quickly enough. That means you can never tell the important stories the same way twice. Rich stories cannot be repeated exactly, even if the attempts at repetition are all bound by the bounded infinity of the strange attractor that is the story's essence. The timeless stories are repeated in a way, but never exactly.

Great stories evolve, which is what happens in the central scientific canon. It is probably better to think of scientific progress as an evolution, as occurs in folk songs. Folk songs change with the times. The "Cutty Wren" is an ancient song sung from the Basque Country in Spain to Finland. It is about a ritual killing of the wren so nature can be reborn in the spring. In Ireland until recently, a few would still go to hunt the wren the day after Christmas (a festival on Boxing Day in Britain, but other names across Europe in many places). The song turns up in degenerate form in the nursery rhyme, "Who Killed Cock Robin?" It appears again in an American industrial song about striking workers not going to the woods but to the factory. And they are going not to kill the wren but to break up the factory machinery. In science, this accommodation of narrative over time is sometimes called progress, but the word "progress" has baggage we probably do not need. The big narratives (like the wren) never go away; we do not build bridges with quantum mechanics, we tell Newton's story again. This is why adaptive management is such a powerful device. It keeps on tweaking the narrative with models so the story improves, given human needs and desires. It is a mistake to think that adaptive management (figure 8.3C) ever gets it exactly correct. It simply keeps using models to improve the story. There is no "perfect."

For some time, David Snowden has been working on narratives for problem solving. He has patented methods of generating stories that work for those responsible for organizing institutions. These methods can be found on his website for "Cognitive Edge." Anyone who imagines that ecology is other than an institution is simply mistaken. Snowden notes that great storytellers have self-conscious methods, which resonate with the characterizations in this chapter:

> The non-repeatability of the story is not accidental; a good storyteller will weave variation into each retelling of the story so that they maintain power over the story and its telling, and thereby maintain control over the delivery of the message. (Snowden 2003: 4)

The local phase spaces are continuous and reversible, like the strange attractor in chaos. You can run a chaotic equation backward so as to find the initial conditions. One last wrinkle is that there are many strange attractors in most stories (figure 8.11). There is the part that starts "Once upon a time." But then it is immediately followed by, "And then the wicked witch . . . " "Once upon a time" is local and is continuous. It is also reversible, as you just play the equation of the strange attractor backward. But not all is continuous and reversible. "The wicked witch" would make no sense without "once upon a time." The children have to

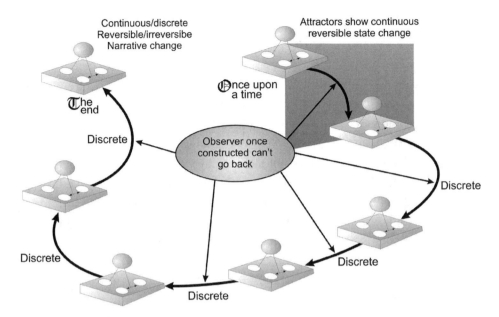

Continuous/discrete
Reversible/irreversibe
Narrative change

Attractors show continuous
reversible state change

Once upon
a time

The
end

Discrete

Observer once
constructed can't
go back

Discrete

Discrete

Discrete

Discrete

Discrete

FIGURE 8.11.   In the grand narrative, there are multiple versions of the phase space depicted in figure 6.5.

get into the woods so the wicked witch can work her mischief. In the move from "once upon a time" to "and then the wicked witch . . . ," there is a move between attractors. And it is discrete. And since you cannot undo knowing something, the jump to the wicked witch is always set in the context of once upon a time. The change to the wicked witch is not only discrete, it is irreversible. Construction events are irreversible; knowing something works like a ratchet. Thus, all stories are reversible and irreversible, as well as discrete and continuous. All this gives narratives their special power. They remain workable while being in contradiction. Telling stories appears to be the special talent of being human. Meanwhile, computers have difficulty dealing with the inconsistencies that appear naturally in narratives.

The local attractors do not have to be exactly consistent with the whole story. For instance, Kuhn (1970) tells the story of paradigm shifts with examples. Many historians of science are externalists, that is, they believe that there is a world that science can find out. They object to a social dimension to science, and they raise arguments presenting paradigm shifts that do not work as Kuhn says. The most energetic critics go to Kuhn's own examples and prove they did not work as Kuhn said. But those of us who have suffered the idiocy and jealousy in peer review know for sure that Kuhn is right. His grand narrative is correct, even if his examples fail. And given our new understanding of narratives, that should not be a surprise.

SOFT SYSTEMS METHODOLOGY FOR MANAGING
NATURAL SYSTEMS

The underlying physical dependence of ecological systems appears to have misled ecologists into thinking that if they only work hard enough, they will have hard systems and become hard scientists; not so! Invariant physical laws behind the scale-dependent parts of the discipline do make ecology a firmer science than some (compare social work), but it does not make ecology a hard science. Schneider and Kay had to restate the second law of thermodynamics in a way that would be pertinent to biology. They said that if a system is pressed away from equilibrium, it will use whatever it has available to it to resist that pressure. Their statement explains the physics of open systems far from equilibrium, systems that physics is almost impotent to measure and manage.

We have worked through a long and nuanced argument about what is a scientific narrative. A narrative works by rescaling processes down until they become events. Events are time dependent, but not rate dependent. Thus transformed, events can be strung together into a narrative that spans scales. Peter Checkland is an industrial engineer who has moved into problem solving, often in a business management effort. His method amounts to improving narratives. He has created an engine that drives situations into better narrative constructions (figure 8.12). He accepts that some systems cannot be made hard. Accordingly, his method invokes Soft Systems Methodology (SSM). It is the same message that Giampietro brings with his distinction between predicative and impredicative definitions and semiotics, as discussed at the beginning of this chapter.

By accepting that "point of view" as the very substance of the discourse, Checkland's SSM has given structure to what would otherwise be idle, capricious opinion. By now it should be clear that there are enough decision points in an ecological investigation to require some formalization of the decision-making protocol. We mostly know what we must do mechanically to solve ecological problems, but we fail because the social context of politics will always block the way. SSM should prove helpful in channeling basic research resources.

There are a number of protocols for dealing with soft systems, but SSM is particularly helpful for our purposes because it amounts to a translation of the central theme of this book into a series of steps for action. It is a scheme for problem solving in messy situations where there are too many competing points of view for simple trial and error to prevail. This is exactly the situation that pertains in writing environmental impact statements or in developing a plan for multiple use in an ecological management unit. It also will be useful in identifying questions of importance in the conduct of basic ecological research. The translation to basic science questions will require some modification of

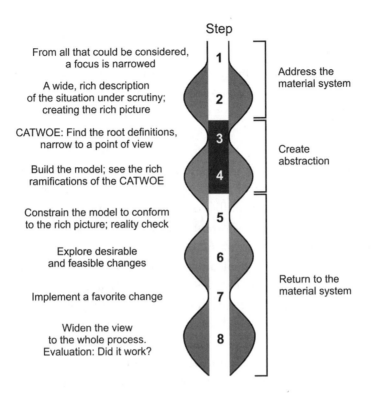

Step

1 — From all that could be considered, a focus is narrowed

2 — A wide, rich description of the situation under scrutiny; creating the rich picture

Address the material system

3 — CATWOE: Find the root definitions, narrow to a point of view

4 — Build the model; see the rich ramifications of the CATWOE

Create abstraction

5 — Constrain the model to conform to the rich picture; reality check

6 — Explore desirable and feasible changes

7 — Implement a favorite change

Return to the material system

8 — Widen the view to the whole process. Evaluation: Did it work?

FIGURE 8.12.   Checkland's problem-solving engine operates as a series of expansions and compressions that amount to function/relation switches and model/narrative switches. Of all the problems, step 1 focuses on a single problem. Step 2 expands the description of the problem. Step 3 bounds and scales the components of the description. Step 4 opens up the consequences of bounding and scaling in a model. Step 5 forces step four to be commensurate with the description in step 2. Step 6 expands the search for feasible and desirable actions in the light of the vetted model. Step 7 implements the option chosen. Step 8 expands the view to see how the process did in solving the problem.

the scheme, for it is distinctly a resource-use approach coming from business applications.

There are seven steps to the process of problem solving in the scheme. *First, is feeling the disequilibrium, recognizing that there is a problem, even if it is not yet expressed.* In ecology, this could be seeing that there is a problem in the sense that an environmental activist might raise as an issue. Something is changing, but the source of the problem is unknown. In ecological management, it might take the form of a major landowner who remembers that when he was a boy, the streams were clean and the lakes were clear. Apparently, there have been changes in land use without anyone noticing the fact, particularly the landowner.

At the early stage of Checkland's scheme, as it might be applied to ecology, the point of tension for an ecologist might only amount to a troubled feeling. A problem only felt intuitively could be one of joint production. Somehow, but in an as yet unspecified way, water, vegetation, and wildlife all occupy the same general area under joint production, but in ways that do not mesh conceptually. How can the basic scientist study these resources in a unified way?

"Mess" is a technical word here, coined by Russ Ackoff. It couches the situation in terms that recognize conflicting interests. Multiple use in a management unit is a mess, as would be a basic research question that mixed ecological subdisciplines. After intuiting as above that there is a mess, *the second stage is to generate actively as many points of view for the system as possible.* Checkland calls this stage "painting the rich picture," or the "problem situation expressed." The distinctive feature here is not the building of a model that has a particular point of view, but rather taking into account many explicitly conflicting perspectives. We have already seen this tension earlier in this chapter, as various stakeholders in a salmon fishery struggle to develop a mutually satisfying narrative (figure 8.5). This second stage generates the system as described after deliberation of stakeholders, rather than the system as given at the outset. It is the richness of the picture that is important at this stage, not the restricted mental categories one might create to deal with it. We are still considering the system on the side of the other, as it is directly observed, not the system as it is intellectualized later, on the side of the observer. Explicitly, a point of view is not the only legitimate one because many alternative points of view are employed. At this stage, it is important for the physiological ecologist to consider the view of other sorts of ecologists, or for other specialists, say community ecologists, to consider the physiological aspects of the problem. The danger at this stage is finding a particularly appealing angle and allowing that to curtail the generation of ideas. In dealing with spruce budworm outbreaks, Holling took the right posture in turning to his favorite explanation, bird predation, only after all other reasonable explanations had been rejected.

The third stage is the most critical, and involves the explicit development of abstractions. It puts restrictions on the rich picture in order to find leverage on a workable solution. Checkland calls this stage finding the "root definitions." There are certain aspects of the system that need to be identified at this point.

Root definitions are crucial to making the situation manageable. The point is to avoid talking at cross-purposes with respect to system attributes that produced the conflict and confusion that was felt at the outset. For instance, terrorist (Israel), freedom fighter (Palestinians), or both (U.S. Secretary of State) somehow need resolution. The particular physical parts of the system corresponding to the various system attributes will change as the scale of the system is changed, or as the point of view is altered. The point of view relates fairly directly to the sort of ecologist that the researcher or manager admits to being; a point of view is an admission to preferences and habits. So Checkland's third stage explicitly locks together a particular scale to a particular type of study. It amounts to using the abstract scheme we developed to separate conceptual levels from scale-dependent levels.

Client (or victim)   Who does or should
                            the system serve?

Actor   What or who is the
                            principal structure?

Transformation   What does the
                            system do?

Weltanschauung   The phenomenon
    (worldview)   of interest

Owner   Who or what can close
                            the system down?

Environment   What does the system
                            take for granted?

FIGURE 8.13.   The acronym CATWOE, for scaling and bounding, is unpacked.

Checkland's genius has been to identify the system attributes that link scale and structure with the phenomenon. He achieved that, not just in a tidy intellectual scheme like ours, but as a working, problem-solving engine. With these critical system attributes identified and linked, ambiguity disappears. The researchers or managers still have to find a workable solution, but at least by getting the root definitions they can avoid confusion and talking at cross-purposes. Explicitly, the root definitions—client, actor, transformation, worldview, owner, environment—can be remembered by the acronym CATWOE (figure 8.13). "C" is the client of the system; for whom does the system work? Sometimes the "client" is the person for whom the system does not work, namely the victim.

Ron McCormick was a consultant before his present position at the Bureau of Land Management. He was sometimes in conflict with his colleagues and clients in his management career because he always viewed his client as being the ecosystem.[36] The one called the client may not in fact be the functional client in the material ecological system; the client is usually the people hoping to use, or with responsibility for, the ecological materiality under investigation, but not necessarily so.

"A" refers to the actors in the system. These could be the client or victim as well, but often the actors are separate entities. In the scheme that we have used to this point, these are the critical structures. In human social problems, these structures are likely to be actual people whose scale depends on their scope of influence. However, the actor could be a forest in an ecological system. Implicitly, the actors set the scale with their size. All this is clear enough with hindsight, but in less obvious situations it is important to be open to new levels of analysis. Choosing an actor achieves that end.

"T" is the transformations or underlying processes. What does the system do? What are the critical changes? The actors generally perform these critical transformations. "W" identifies the implicit worldview invoked when the system is viewed in this particular manner. In the scheme that we raise in chapter 1, "T" identifies Rosen's "differentials," the naked, measured changes of state. "W" isolates the subjectivity embedded in the model. It is the values that are there anyway, but are explicitly recognized by a self-conscious worldview. Thus "A," "T," and "W" together identify whether it is a community study, an ecosystem study, or whatever else. They bring with them a set of scales.

"O" refers to the owners of the system, who can pull the plug on the whole thing. Like the actors, the owners could be the client or victim of the system, but usually the owner is someone else. The scaling issues of grain and extent emerge here. With power to terminate the system, the owner defines the extent aspects of the scaling of the study. By contrast, the actors will usually define the coarsest grain that can be involved in scaling the system. They have to be discernible at the level of resolution associated with the specification of the system. While owners fall out readily in an analysis of business or social management, it may be less important to identify them in strictly ecological settings. However, political owners and the public that elects them almost always stymie ecological action and remedies. Identifying owners in management can explain why even excellent programs are not enacted.

Even in a strictly ecological setting, it is helpful to know the ultimate limits to the functioning of an ecological system, and the concept of owner might be of service there. Ice ages are the owners of temperate communities, having pulled the plug on plant community associations in the past. This indicates the extent to which communities start as ad hoc entities. As an owner, global climatic shift puts limits on the evolved accommodation that is embodied in community structure.

As we said, in managed or restored ecological systems, the owner can apply to the situation in very literal terms of ownership. The owner might subscribe to the more physical view of sustainability, which stands in contrast to our view that asks first, "Sustainability of what, for whom, for how long, and at what cost?" Biological ecologists, as opposed to social ecologists, often put the physical limits on sustainability first, saying that "nothing can be sustained if the world cannot exist in that configuration." Our response is "true but trivial," because most of the decision making is in the social setting of livability, what the people living there (the owners) will tolerate. Even so, those physical limits that are the predilection of biological ecologists might be seen as the owner.

Finally, "E" identifies the environment, that is, what the system takes as given. The owner is a special case of the environment. Anything longer term and slower moving than the whole system is a context in which the system has to live. By default, the environment defines the scale of the system extent by being everything that matters that is too large to be differentiated. For instance, for a system that rescues people at sea, the environment is: "There will be storms, and people will get caught in them."

It is important to realize that the several different sets of root definitions are not only possible but also desirable. The actors in one set of definitions will be different from those in another. That presents no problem, but it is mandatory that the actors in question only act in the model for which they have been identified. They should not be mistaken for actors performing at some other scale on a different set of assumptions. Realism invites that error by suggesting that the actors are really there is every aspect of the situation. They may be there, but the situation is the one defined by looking at the actors in a particular way such that some or all may disappear. In fact, that error is exactly the sort of confusion that arises if the formal scheme recommended here is not followed. Mistakes are easy to make if there is not a formal framework to keep track of all the relationships. The logical error embodied in sliding the scale or change of worldview is a favorite device for vested interests to confuse the issue when they know that their own position is inconsistent. Lawyers representing the company in an environmental litigation, or an environmental action group bringing suit can confuse an issue in this way. It is a standard device if they are in danger of losing on rational grounds. Semantic arguments are confused as to which actors apply where.

Having defined the CATWOE, one is ready to *build the model, which is the fourth stage of Checkland's engine.* There will need to be a different model for each set of alternative root definitions. The model in stage four explores the consequences of the root definitions chosen. The *fifth stage returns to observation of the world, and the model is checked against what happened in the second stage.* If the actors are people, then one can ask them their opinion of the model and modify it to be consistent with their special knowledge. *At the sixth stage one identifies desirable and feasible changes for the system.* In an applied ecological setting, the plan for management is generated here. In basic research, the hypotheses and the protocol are married at this point. Note that desirable may not be feasible, or vice versa. If desirable and feasible changes cannot be matched, it is possible to return to the third stage and look for new root definitions. Sometimes there is no feasible and desirable course of action, in which case there is nothing to be done, but at least that fact is clearer than it was before (figure 8.14).

The seventh stage is implementation of changes. In the case of an ecologically managed system, that would be the implementation of the forest plan. In basic research, the tests are made, the data are collected, and the findings are studied. We should emphasize that this is not a fixed sequence, for one can jump from stage to stage, modifying as appropriate. Anyone who has done basic research knows that it is

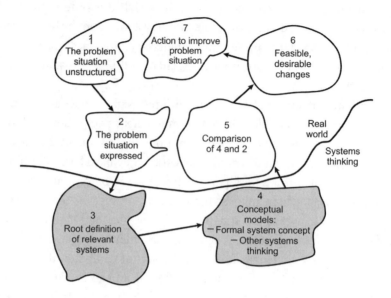

FIGURE 8.14. Checkland's (1981) scheme, laid out in his own terms. He emphasizes the distinction between systems thinking (gray-shaded components) and dealing with observations of the material system (clear components).

never done the way one writes it up for the journal article. Remember modernist distortions to make it look real, and the postmodern predilection to tell it like it is. By allowing steps to be repeated or taken out of order, the scheme acknowledges how researchers really operate. It does not try to force the process into the myth of the scientific method, which emerges in the modernist, sanitized account that is finally published in the journals.

Consider the difficulty that the USDA Forest Service has in managing to "maintain biological diversity," as they are indeed mandated to do by an act of Congress. Managing for diversity is clearly a mess, in Checkland's technical sense. The problem is that "maintaining biological diversity" is an undefined action leaving room for many points of view, each with its own agenda. Referring back to Giampietro, it requires soft semiotics and impredicative loops. The original congressional action might be the first step in Checkland's scheme, a general recognition that something is wrong. The second step, the problem situation expressed, in this case shows how complicated it is. There are many players, some of whom are: (1) hunters desiring diversity of animals for their sport are unhappy about closed growth that offers few resources for animals of choice; (2) lumberjacks may not care about diversity, but they change diversity (often increasing it) through logging old growth, therefore making a young growth more common than in the presettlement forest; (3) preservationists concerned with maintaining old growth in large blocks to provide for animals that require habitat that would otherwise become dissected (the irony is that preservation areas characteristically lose diversity); and (4) motorists who desire a diverse vista. And on to other stakeholders. Different CATWOEs can tease apart those conflicting views to allow development of pertinent models so that a plan can be made without ambiguity. With the scale, structure, and function of the forest defined, compromise can be sought in a meaningful way. An understanding of what is intended and how it is to be achieved will enlighten management action. Rhetoric and hidden agendas can be exposed and those that are disingenuous can be rejected.

In basic research, complex problems can be made explicit. They may not become simple, but at least the researchers are cognizant that they have a point of view and know what it is. Explicit entertaining of other scales and points of view, whether or not they are finally used in the investigation, highlights what has been done and sets it apart from what would have been interesting but has not been researched.

The literature of herbivory and over-compensation is a case in point that would be greatly improved by the application of the explicit scaling that Checkland recommends. We refer earlier to the work of Becky Brown in this regard. A CATWOE would tease these points of view apart.

CONCLUSION

In this chapter, we have unpacked the process whereby ecologists model their world. In the end, it all turns on improving the narrative with models. Checkland's scheme is really a way of creating narrative that is aware of all that a story entails. Stories are contradictory. For some, that is problematic, but we say, "Get used to it." Humans are remarkable problem solvers. They use narratives to do that. It seems that we have to take storytelling seriously enough that we see it even in the world of nature that may be coded (DNA, hormones, and mating dances) but is not sentient. So storytellers do not have to be thinking beings. There are competing pressures on organisms and their evolution. To give that an accounting requires something like the device of narrative to keep situations coherent as they slip slide between levels. Yes, there are physical limits, and sometimes we need to know what they are. But that is not where progress is to be made. There is action to be had in new ways of looking at what we encounter.

We have hinted that humanitarians are used to telling stories all the time, some explicitly so, but all to an extent. Applied humanitarians are lawyers. They exactly tell stories, and have to learn a new version of English as they enter law school. We have often given advice to concerned young, environmentally inclined students: "If you want to save the world, become a lawyer." It is through human devices like the law that we influence our context. Lawyers are far from being wilting flowers, but those in the more basic or pure humanities wilt all the time. In the expansion of the Northern Lakes Long-Term Ecological Research (LTER) project in Madison, Wisconsin, we were early to identify that there was an edge in grantsmanship to bringing in social scientists.[37] It worked, and the money rolled in. But in the beginning, the biophysical investigators gave only token attention to their social science colleagues. There was some private complaint from the social scientists, and it was clear to some of us biological scientists what was going on. The difference in culture was sharp. For instance, in the face of requirements to serve the whole project to keep it unified, coherent, and relevant, graduate students were to invest effort and time into doing just that. The social scientist took it very seriously, but the biogeophysical scientists said that simply doing your research as usual as a graduate student meets those criteria for contribution. But that was only at the outset. Having our social science colleagues around all the time, their culture began to rub off. In the end, biogeophysical management did its most creative work as it folded in the social part of management. Economic theory brought with it explicit statements of different levels with different relaxation times, thus generating abundant folded response surfaces. So the social scientists are now far from being tokens, and they keep Madison at the cutting edge. But we need to do the same for the humanities, or rather, let the humanities do the same for us.

Also on the Madison campus is the Center for the Humanities. They have wonderful sessions, sometimes with an environmental spin. But as the humanitarians look at environmental attitudes in the novels of this and that author, they admit that their respective scientists do not take them seriously. Someone was rude enough to ask the question explicitly, and the humanitarians all agreed that "No, environmental scientists would not be interested in what we are doing." The poise and nuance in the humanities is really impressive. We would recommend bringing humanitarians in on scientific proposals. At first they will be tokens, as were the social scientists. But in the end, they might even be taken seriously. They know the space wherein values and significance lie. They know about hidden places between levels.

A paper by John Needham in the Welsh *Poetry Nation Review* takes on board the ideas of physicist Michael Polanyi (table 8.1, bottom row). Needham took apart the Henry James 1879 novella *Daisy Miller*. Winterbourne is attracted to Daisy. Needham starts with a phrase, "small queer glance," that describes Daisy's response to Winterbourne's comment about her flirting with an Italian fortune hunter. "Small" is taken to capture some of Daisy's vulnerability. "Small" then tacitly influences what we think explicitly about "queer." It can mean upset, but it could also mean suspicious. A line or two later, "small queerer" comes up again, and both meanings are thus present rubbing up against each other. There is ambiguity pressed together. From this local ambiguity, Needham directs attention to other points of related tension. We are not sure if Winterbourne disapproves of Daisy. But at the level of several pages, Daisy is cold-shouldered by the hostess as she leaves the party. She leaves and Winterbourne goes to her defense. We are not certain about Daisy's feelings for the Italian. We are not sure what Winterbourne makes of her.

> After Daisy's death Winterbourne realizes that she liked him, respected him, and wanted his "esteem"—this is part of painful realization that he had misread her. (Needham 1988:38)

Needham goes up to the level of the whole novella, and finally back down to the echoes of "small" and "queer." He points to the ambiguities and uncertainty in Winterbourne and the reader of Daisy Miller. If the reader here wishes the whole story, Needham is well worth the read. But for our purposes, we show how the humanities are subtle and hierarchical with language in a way that might make ecologists feel a bit crass. We may or may not be so, but this chapter shows how ambiguity, uncertainty, and contradiction, all playing at multiple levels, are everywhere in ecological discourse. We are in trouble; the cavalry in the persons of our humanitarian colleagues might still ride in and save us with narratives. So we would recommend having a humanitarian as a Principal Investigator on the next grant, and not just as a token. They should be so respected that you will take them seriously enough to heed their words when they say, "No you cannot do that experiment because it does not fit with your narrative!" Perhaps all this can be summarized in the definition of a syndrome. A syndrome is a pattern waiting for a narrative. Ecology is significantly a set of syndromes. Our humanitarian colleagues may be abler in turning our syndromes into satisfying stories.

David Snowden's approaches to micronarratives are a promising point of purchase. Ecologists are often aware that laypeople on the periphery have deep insights; but how do we get them to communicate with comparable stories that might make a data set? Snowden has found that by collecting very small narratives, in large numbers and in equivalent terms, he is able to get statistically reliable numbers of stories. Those patterns get to the nub of the issue in a way that experts and controllers cannot. Snowden advises all sorts of power structures, including the Australian armed forces.[38] Unbeknownst to the commanders, a group of soldiers had a deep concern. They had to dig latrines. Soldiers have always dug latrines, so what was their complaint? Australia holds records against all comers with regard to how poisonous various types of venomous animals are, and their high numbers. And the animals like to live in shallow holes, something to worry about in a contemplative moment. The situation had been one of general excessive process control. This meant that getting forms filled out in time had obscured the issue. Immediately on hearing the ubiquity of the stories, the commander was on the telephone getting real portable toilets flown in. There is much shouting in the military everywhere, but a commander with something very high priority, like those latrines, will have a lower, more intense tone. Nobody had thought to tell him, and he did not think to ask.

Examples of using groundswell information are rare in ecology, although Shugart's FORET model (see chapter 4) for growing forests in computers is a counterpoint. FORET is not perfect, but it creates remarkably realistic vegetation. And there is a trick to it. Shugart calibrates his tree species with regard to shade tolerance and growth rates. But he does this by gleaning the common knowledge of local foresters sitting on a stump and looking and thinking as they learn their forest. Somehow he got hold of the stories that foresters remember and tell each other over their lifetimes. And the folk wisdom he collected had a rich background. That is probably the single feature of the model responsible for FORET's startling success. Fishery workers know a lot about the fish in their river. Beekeepers will know a lot about why their colonies are failing remarkably. So getting their stories may be the best data we can get, particularly in messy situations that need defining. We need to welcome Snowden's micronarratives into ecology. Why should the social sciences have a monopoly on him? We need to listen to our rank-and-file sea captains, just as Beaufort did.

APPENDIX 8.1

CRITERIA AND INDICATORS FOR ECOLOGICAL AND SOCIAL SYSTEMS SUSTAINABILITY WITH SYSTEM MANAGEMENT OBJECTIVES.

SOCIAL INDICATORS

Structure Indicators

Is the rank-size distribution of communities within the ecological system management unit changing in a direction unfavorable to smaller-sized communities?

Is the distribution of monetary incomes in the area of the ecological system management unit becoming more or less equitable?

Is the population age distribution among smaller communities within the forest management unit stable, becoming older, or becoming younger?

Is the number of locally owned businesses stable or increasing?

Is the number of businesses in smaller communities stable or increasing?

Is the level of economic activity among businesses in smaller communities stable or increasing?

Is the availability and feasibility of diverse economic pursuits stable or increasing?

Is the travel distance to basic goods and services stable or increasing?

Are cultural institutions (for example, theaters, museums, and churches) stable or increasing in variety and number across all communities within the ecological system management unit?

Is the physical infrastructure of the communities being maintained or extended?

Is there a full suite of government services within the area of the ecological system management unit?

Does local government have a land-use planning mechanism that helps to ensure community stability?

Does the local land-use planning mechanism operate cooperatively with the ecological system management planning mechanism?

Do nongovernment organizations provide a stable or expanding suite of services to the population of the ecological system management unit?

PROCESS INDICATORS

Are family monetary incomes within the ecological system management unit stable or increasing?

Is access to capital investment funds within the ecological system management unit stable or increasing?

Is the net capital monetary inflow stable or increasing?

Is the proportion of children choosing to remain in smaller communities stable or increasing?

Are people's expectations regarding the future of their communities being met?

Is access to vital forest resources stable or increasing (within the ecological systems capacity)?

Are human subsidies from outside the ecological system management unit decreasing (for example, redistribution of wealth by federal and state governments)?

Is the ratio of property value to local income stable?

Is the ratio of state and federal income tax to local tax stable?

Is the community sense of social/cultural identity stable or at a desired level?

Is consensual dispute resolution (for example, in ecological system management) stable or increasing?

Is the external institutional context of the ecological system management unit responsive to local needs?

Is there an increasing flow of information between local communities and their context beyond the ecological system management unit?

9

MANAGEMENT OF ECOLOGICAL SYSTEMS

THE IMPORTANCE of land to grow into was established in the very first years of the American Colonies and even before the colonies broke away from Great Britain. It is witnessed by both personal and national interest in acquiring as much territory as possible. George Washington demonstrated his sense of the value of land for expansion from his earliest years as a land surveyor, as a military leader protecting the western territories for the British general Braddock, and in his acquisition in the Ohio territory for his personal estate. In England, land was power. The land base significantly defined the great titled houses of England, and the ruling classes' fiat had land as their footing. Management of the English landscape was intense, not just for cropping but also for wildlands managed for producing wildlife for the hunt. The tradition of hunting in Britain is gentlemanly and aristocratic, far from what it is in the United States. English commoners hunt vermin, like rabbits. Only the gentry hunt deer and grouse. The dramatic reorganization of Britain in the twentieth century was caused by industrialization, making land a secondary source of resources. Industry brought land to a value where it could barely support itself, let alone be the power base of the aristocracy. The great families were caught between the cost of managing land and an inability to sell the estate. The cost of wages for farm managers, laborers, gardeners, verderers, and game keepers was significant. Many aristocratic families lost their grip completely.[1]

But at the founding of the nation, American leaders like Washington were functionally English gentlemen, with their values and attitudes. Along with Washington, other early leaders like Madison, Monroe, and Jefferson all had their personal wealth rooted in the size of their estates. This was the time before the Industrial

Revolution, when agrarian values associated with land were key to personal and national success. Jefferson, pushing some of his federalism aside when president, worked out a deal with France to purchase the Louisiana Territory, multiplying the new nation's land base several times. He then chartered the expedition of Lewis and Clark to the Northwest and the Pike expedition to the Southwest to learn more of what the new territory contained and offered the new nation. Lewis and Clark did not fully understand what they were going to find. For instance, the expedition bought dogs for food, while they were on the banks of the Columbia River, the greatest salmon run in the world. The nation's founders could only have sensed the value of land through an agrarian lens, with no way to anticipate the productivity that emanated from belief in the value of land.[2]

In his book *Seeing like a State*, Scott (1998) presents the conflicted story of governments' evolving policies toward the appropriation of the land base. Initially, the land base was largely in social and ecological equilibrium, but changed to one that is highly organized and far from equilibrium. Scott's account of scientific agricultural and forestry production has metaphorical value. It points to the dangers of dismembering an exceptionally complex and poorly understood set of existing social and ecological relations and processes. Such dislocation isolates attention on a single agriculture or forest of high value by its professional managers. From the purchase of the Louisiana Territory to the current time, land managers have evolved their practices one mistake after another as the land has taught them the error of their ways. It has been an unconscious, and so a fairly ineffective version of adaptive management. Whether the grand experiment will be successful is yet to be determined. We believe that the information in our book can begin to help land managers ask the right questions and seek the best scientific advice.

When George Washington routed the British, one of his key interests was to gain access to a much larger territory into which to grow the new country. Jefferson had the same motive when he negotiated the Louisiana Purchase from the French. They saw the land as a resource to grow a nation. They gave little consideration to the Native Americans who already fully occupied the land. We would do well to remember them better and take their advice now. Henry Lickers is a Seneca based in Akwesasne. He is a scientific leader in his community. He said to Allen, "We are still here and are waiting to help."

Management can invoke any of our criteria. But two very different criteria prevail, usually in parallel. Management raises human action to the fore, and we live and experience in the world of landscape. Things happen on landscapes to the point that we can see processes unfolding. The other criterion that dominates management is the ecosystem. But we have been at pains to emphasize that landscapes and ecosystems are more at odds than most criteria. In fact, they are put together at the beginning of our chapter-based treatments of criteria precisely so as to show how incompatible are landscapes and ecosystems. We act in the tangible spaces of landscapes, but the consequences of our action come to pass significantly through invisible ecosystem process. Thus, management action can work mysteriously and raise surprises. The surprises often appear in the other criteria, such as organism sickness or health, population increase or decline, and community collapse or biome shifts. Thus, management is the acid test of our whole approach. It forces integration of all our criteria; it is the unification process in unified ecology toward which we strive. Management is a perturbation of the land base and is likely to have an effect on many criteria. It is therefore important to use the tools of complexity, which function to link across criteria and scales. This is not going to be easy and will need an orderly plan.

Accordingly, we need some principles and guideposts so that management is more predictable and rational. These principles will touch most of the separate criteria, from which we have tried to construct a unified view and a unified plan for action.

1. Do as little management as possible. Apply management perturbations of natural systems only as necessary.
2. Do what the natural system does; then you will have fewer, smaller, and less surprising side effects.
3. Be careful to distinguish productivity that is integral to ecological function from productivity that can be taken and diverted toward human uses and consumption.
4. Realize that the criteria we have used for our chapters are not the types of ecological things that necessarily have privilege in nature, but they can still be useful organizers for multiple-use activities. The criteria are separate, but remain nevertheless connected.
5. Everything is not connected to everything else, but there are unexpected connections, precisely because it is the same physicality that emerges ordered in the criteria we choose.
6. Expect surprises anyway.
7. Expect an action under one criterion to have effects under another, to the point that you are always keeping an eye open for such effects.
8. The things that appear in ecosystem function will often have lead indicators on the landscape.
9. The capacity for narrative to deal with contradictions makes it a likely tool for assessing, describing, and understanding management action.

The intention in this chapter is to make what has gone before useful for the manager. There cannot be a recipe book for management because management prescriptions weave ecological narratives, not models. We can offer an intellectual framework that will help bridge the abstractions of basic science into application so that the manager can see where to translate narratives into management action by extending ecological science. It may appear a bit esoteric to traditional managers, but we can also try to persuade basic scientists that management issues are a wonderful place to test theory. Theorists working with managers may build the bridge to what managers can actually use. To get to the point where such recommendations make sense, we first develop a clear view of ecological management and its relationship to both pure and applied ecology. Given what we have learned about narrative, it is no surprise that we will use them to weave stratagems for integrating management.

We start with a particular narrative and associated models to show how the previously mentioned principles apply to particular management schemes, and how they turn up repeatedly. An example of a complex narrative is presented in chapter 1, and we use that again. Holling's work on modeling budworm outbreaks (figure 9.1) illustrates how to use several of the management criteria discussed earlier.[3]

The first principle is: *Do as little management as possible.* Holling's solution is to avoid spraying insecticide completely if possible, preferring to employ the second principle.

The second principle is: *Do what the natural system does; then you will have fewer, smaller, and less surprising side effects.* If there is to be spraying to kill the budworm, Holling recommends spraying not artificial chemical insecticides, but rather a fungus that makes the budworm sick. Chemical insecticides would bioaccumulate in the birds, eventually lowering their numbers, and making it easier for the insect to break out. But the fungi are part of nature and will not bioaccumulate, and the fungus is focused on the budworm with few side effects. The fungus does not eliminate the budworm wholesale, but it does slow them down.

In figure 9.1, one unstable equilibrium and two stable equilibria are presented. Between the unstable equilibrium (point B) and the low-density stable equilibrium (point A) is a trough in the graph wherein budworm density naturally declines through bird predation. Insect outbreaks occur as the trees get bigger and the budworms have more to eat and so can handle more predation. The effect of more food for budworms is to reduce the effectiveness of bird control. The effect of the fungus sprayed in management action is to suppress budworm populations to deepen the trough between A and B. This counters the effect of tree growth in increasing budworm populations, but in doing so only delays the eventual natural outcome of the outbreak. This delay in the outbreak gives more time for the trees to grow, making them more useful for forest products. A way to foil a pending outbreak is to harvest trees preemptively. In this way, the beetle food supplies are reduced. Harvesting need not be all or nothing. A clear-cut of trees takes all the food away, but if the manager only reduces tree density during the time interval between A and B, it will mitigate the impending epidemic. Without harvesting, the trees will be destroyed by the budworm anyway, so in a sense, the managers substitute their

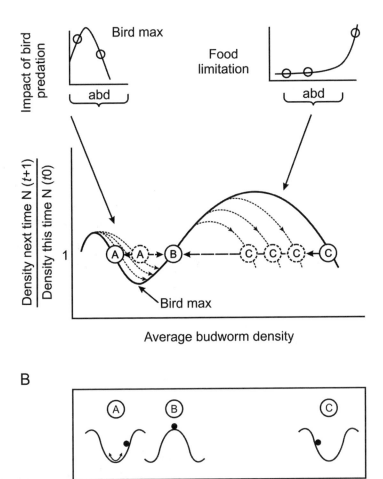

FIGURE 9.1. *A.* Reproduction of figure 1.13, but this time distinguishing between stable and unstable equilibria, as well as showing how the line of response deforms. The horizontal dashed line is the unitary equilibrial condition where the density this time equals next time. Therefore, any circle on the dashed line is at equilibrium. All the circles represent no change in the population at equilibrium, but two are stable equilibria while one is unstable. Circle A is a stable equilibrium point because local budworm densities move toward point A from a little below or from above in the trough between A and B. Circle B is an unstable equilibrium point because higher or lower densities of budworm tend to move away from B. C is another stable equilibrium point of sorts. The budworm densities at C stay on the unitary dashed line, where food supports high density. But at a higher level of analysis, C itself moves along the unitary line (arrows show where it will go). The line of the response deforms (see the sequence of dotted lines) as the food is eaten away. As trees get bigger before the outbreak, the line of response deforms at lower densities. The deforming line pushes stable equilibrium A toward B. Better-fed budworms can hold a higher population under bird predation. The stabilizing trough erodes away. Eventually, the trough between A and B disappears as A and B merge, and so lose the properties of equilibrium. The whole system then moves to C under the positive feedback of budworm growth with insufficient bird control to hold the budworm back. *B.* The lower panel shows equilibrium density points represented by balls on a surface. At A is a stable equilibrium ball in a cup, as is C. Move either ball from A or C and it returns to the bottom of the cup; the equilibria are stable. See the ball at equilibrium B, but poised on a peak, so that if anything moves the density (ball) either way from equilibrium, it keeps moving. B is therefore at an unstable equilibrium.

own strategic harvesting for budworm feeding. Harvesting denies the food source that budworms require to cause outbreaks. The manager does what nature would do in the form of budworms. This invokes the principle of "do what nature would do, and side effects will be minimized."

The third principle would note: *Productivity that is integral to ecological function must be carefully distinguished from productivity that can be taken and diverted toward human uses and consumption.* Holling recommends harvesting the trees at thirty to forty years, that is, before the budworm go into an epidemic outbreak. The populations of trees are to an extent adapted to outbreaks, and so harvesting trees at that age is what the budworms would do anyway. That is a harvest that the system expects and is ready to give up. With fungus spraying, the trees survive longer so as to increase tree growth as an integral part of ecosystem function. The manager separates integral growth from harvesting resources. The managers would prefer trees that are about seventy years old. Their preemptive cutting drives down regional budworm populations, allowing managers eventually to get closer to being able to harvest seventy-year-old trees.

It will take several cutting cycles to bring a managed forest to a stand rotation age that is consistent with the objective that has trees harvested on a rotation age closer to seventy years than forty years. All the while, the manger controls budworm populations and maintains other ecosystem components. We harvest trees on the budworm's schedule at first, so we can get the upper hand. As the regional budworm population comes under control, we can move the rotation age for tree harvest closer to our preferred longer harvest schedule. At first there is minimal tree value because of harvesting before the preferred rotation age. This is the cost of our previous failure to manage the forest with planned action rather than let the budworm manage the forest. However, that cost does underwrite a management plan that takes control back from the budworm. Implementing the management prescription over time continues iteratively to evolve the multiple-use management plan narrative and the management model.

The fourth principle is: *The criteria that we have used for our chapters are not absolutely the type of ecological things that necessarily have privilege in nature, but they can still be useful organizers for multiple-use management plan activities.* The criteria are separate but remain nevertheless connected. The ecological forest/budworm system from the managers' scientific point of view is a population issue of epidemic and predator-prey relationships. The controls on the system are bird predators and ecosystem primary production. Management perturbations substitute tree harvesting for beetle killing. A wrinkle in all this is the loss of organic matter returned to the soil. A solution to entertain is to leave a component of some downed trees during each cutting cycle, particularly the ones of little or no commercial value. Budworm beetle feeding infestation is on the leaves so there is little need to consider that whole downed trees will be a source of infestation. Downed trees might still provide a small reservoir of beetle pupae that emerge as adults to lay eggs; the magnitude of that implication deserves further attention. The expectation is never to rid the forest of the budworm, but rather to manage them at low population levels. The trade-off here is population considerations of beetles and ecosystem considerations of returning carbon to forest soils. All these factors need to be weighed against each other. Ecological activity under separate criteria is not physically separate; there is a crucial if somewhat mysterious material connection.

The fifth principal is: *Everything is not connected to everything else, but there are unexpected connections*. This is precisely because it is the same physicality that emerges when ordered under the criteria we choose. When managers normally spray chemical insecticides, they do so only when they can see a burgeoning insect population. At that point, it is too late and the outbreak is already in progress. One might imagine that careful spraying of powerful insecticides would knock out the pestilence, particularly since the spraying is exquisitely executed. Fixed-wing crop dusters can cover whole stands of trees, with less than a foot of overlap between passes. The bad news is that the insecticide simply holds the outbreak on the steep part of the insect population growth curve because the food supply is still abundant; it maintains the epidemic. This translates to endemic high populations that do not follow the natural rhythms of a thirty-to-forty-year cycle, but rather infest the landscape on a large scale. Preemptive harvesting might give managers more control at that regional scale. What was a western mountain problem has been able to move across the landscape such that spruce budworm now progressively threatens forests across northern North America when, before birds, mountains and prairie landscapes contained it.

The sixth principle is: *Expect surprises anyway*. The change from quiescent endemic to epidemic outbreak has sharp attack, and is a classic surprise in a technical sense. We know it is coming, but not exactly when; that is how surprises often come about. In fact, Holling introduced the model in the first place as an illustration of ecological surprise. As long as the spruce budworm/tree system is in the predator-prey cycle of birds and budworms, the long-term equilibrium changes little. An actual outbreak is a surprise in that little indicates it is coming immediately. The trigger for the outbreak is any of a large number of causes. It might be an adjacent outbreak. It might be a storm that brings in many insects synchronously in a pocket of fast-moving air. It might be a local bird flu that sets the birds back temporarily. It is a medium-number issue: any of a large number of causes could be the trigger. Of course, experience indicates that the attack is coming sometime, but we cannot say exactly when or where. Narrative has no trouble with the counterintuitive notion that surprises should be no surprise.

Petroski's work on bridge collapse makes the point exactly.[4] Any given bridge collapse is generally unknowable and an unexpected surprise. However, a dissertation by Paul Sibley in 1977 showed some unnerving patterns, revisited by Petroski (1993). Bridges that fall down are not generally the ones with distinctive daring design. They tend to be of a conventional design about thirty years after its original inception. Apparently, it takes engineers that long to become too confident about how to make incremental increases in size. We were due for a cable-stayed bridge to come down in 2005. Usually it is some recent new design that collapses, which explains Petroski's prediction. The exception here might be not a fairly new design, but a fix on an old design. So perhaps the collapse he anticipated was in 2007 on I-35 in Minneapolis. It was an old bridge, but one overhauled by adding a huge extra layer of roadbed.

The details of the collapse are not predictable, such that there can be planning so as to avoid it specifically. The narrative of the collapse of the Tacoma Narrows Bridge is harrowing, a gripping story (figure 9.2). Leonard Coatsworth, an editor at the Tacoma *News Tribune,* had the following eyewitness account from when he turned as he drove past the tollbooth of the bridge:

> I drove on the bridge and started across. In the car with me was my daughter's cocker spaniel, Tubby. The car was loaded with equipment from my beach home at Arletta.

FIGURE 9.2.   The Tacoma Narrows Bridge, moving like a bullwhip as it collapsed. (Photo courtesy of University of Washington, University Libraries Special Collections, UW20731.)

Just as I drove past the towers, the bridge began to sway violently from side to side. Before I realized it, the tilt became so violent that I lost control of the car. . . . I jammed on the brakes and got out, only to be thrown onto my face against the curb.

Around me I could hear concrete cracking. I started back to the car to get the dog, but was thrown before I could reach it. The car itself began to slide from side to side on the roadway. I decided the bridge was breaking up and my only hope was to get back to shore.

On hands and knees most of the time, I crawled 500 yards or more to the towers. . . . My breath was coming in gasps; my knees were raw and bleeding, my hands bruised and swollen from gripping the concrete curb. . . . Toward the last, I risked rising to my feet and running a few yards at a time. . . . Safely back at the toll plaza, I saw the bridge in its final collapse and saw my car plunge into the Narrows.

With real tragedy, disaster and blasted dreams all around me, I believe that right at this minute what appalls me most is that within a few hours I must tell my daughter that her dog is dead, when I might have saved him.[5]

In engineering, there is commonly overbuilding so the structure remains. Nevertheless, things can go wrong. In biology, systems are engineered to fail. Beavers expect dams to fail and so they use a design for easy replacement. If in biology we manage too closely and achieve a far-from-equilibrium solution, such as we do in agriculture, we can expect failure as something comes along and starts cheating. At one level, we manage agriculture so as to hold it too close to an emergent equilibrium of massive sustained yield. At another level of analysis, that massive sustained yield is very far from the equilibrium or homeostasis that would pertain in hunter-gatherer peoples. If there is large capital, something will start to game the system, and not just in human systems. For instance, *Atta* ants raise fungi in a highly organized scheme that goes to the point of focused genetic strains of fungus with seven times the normal potency of wild fungi. The ants, like humans, are managing very close to a high yield with a steady level of massive stable production. It is telling that there is a pest that steals the ants' resource base, and more to the point, the ants raise a species of bacterium that has an antibiotic effect on the cheating pest species (figure 9.3).[6] Only the most highly organized ants have to make pesticides, so there is a principle at work here for all highly organized biological production systems. Indications are that technical management will always look something like agriculture. But notice that the ants use a biological agent as their control device, and we should also. If the control is biological, management has evolution working as an ally as long as it sets the environment to select for better or at least stable control. We can expect surprises when we impose tight control far from equilibrium.

FIGURE 9.3.   *Atta* ants cut leaves that they take to their nests, where they grow fungi. (Photo courtesy of Martin Burd.)

Our seventh principle is: *Expect an action under one criterion to have effects under another.* That is often where the surprises come from. When an issue arises, it seems at first to be entirely a matter under the criterion where it first appears. But often better, more useful explanations arise under some other criterion. Remember the failure of the salmon.

Our eighth ecological management principle extends our seventh principle and argues how best to apply it. The eighth principle is: *We can expect the tangibility of the landscape criterion to give lead indicators on ecosystem management.* The general argument is that there will be lead indicators under the most tangible criteria, but the application for management is likely to be found elsewhere. The message is: in preliminary research, use signals that enter our sensory portals most easily. In other discourses, the same principle applies, but the lead indicators will be different from the landscapes of ecology. They will still be in the most tangible criterion for that discipline. For instance, the tangible entity in medicine is a human person who is compromised in some way.

Work at Oak Ridge investigated the influence of a heavy metal smelter on a forest and found landscape to be the primary indicator.[7] They were looking for lead to damage biota at large, but found no such thing. Cored trees showed the appearance of lead in the wood as the smelter started up, but when the smelting stopped, the wood reverted to the condition before the insult occurred. But the landscape told all. The forest toward the lead smelter had unusually deep leaf litter. The lead had killed the fungi (they are sensitive to heavy metals) so the mycorrhizae died. The litter showed failed decomposition. That told the researchers to look at nutrient loss, and there they found the heavy metal damage. Biota were mostly not influenced, but ecosystem function was destroyed. Landscapes tell the tale first.

In infectious diseases, the nature of the sickness might become understandable in spatial population terms, something we can see. The lead indicator is that there is illness in a person, and then some additional people. Cholera was not understood to be a waterborne disease; it was generally thought to be caused by "bad air." In 1854, there was an epidemic in Soho, London. Dr. John Snow mapped the incidences and that gave him the answer that it was waterborne. He wrote in a letter to the editor of the *Medical Times and Gazette*:

On proceeding to the spot, I found that nearly all the deaths had taken place within a short distance of the pump. There were only ten deaths in houses situated decidedly nearer to another street-pump. In five of these cases the families of the deceased persons informed me that they always sent to the pump in Broad Street, as they preferred the water to that of the pumps which were nearer. In three other cases, the deceased were children who went to school near the pump in Broad Street . . .

With regard to the deaths occurring in the locality belonging to the pump, there were 61 instances in which I was informed that the deceased persons used to drink the pump water from Broad Street, either constantly or occasionally . . .

The result of the inquiry, then, is that there has been no particular outbreak or prevalence of cholera in this part of London except among the persons who were in the habit of drinking the water of the above-mentioned pump well.

I had an interview with the Board of Guardians of St. James's parish, on the evening of the 7th inst and represented the above circumstances to them. In consequence of what I said, the handle of the pump was removed on the following day.

Our ninth and final management recommendation is: *Rely on narrative.* In any problem so large that we really want to fix it, there is almost always contradiction. Our pressing ecological issues are so large that we cannot expect to be able to create models to capture the whole issue. Chapter 8 focuses on narratives, showing how they are fully adept in the face of contradiction. In management, landscape and ecosystem are often bedfellows, although they are the most different of all the criteria. The two criteria do not map one onto another. This is bound to lead to inconsistency. Narrative is a likely way to straddle the divide.

An outbreak of budworm is most simply a process of impressive consumption focused on a local spot on the landscape every thirty to forty years. Tangible pattern appeared in tangible places. But Holling had difficulty making his model pulse when he put realistic movement on the landscape. It cycled just fine when he made the assumption of mass balance, common in physics, that as many budworms were entering a site as leaving it. But we know that epidemics are triggered by local influx. It took a holistic narrative approach to solve the issue. He calculated that he needed a certain pattern to slow the budworm, but could not find the empirical data to support that constraint. From his ecosystem/landscape pattern, he said he did not know what in nature he needed to create the pulses of epidemics, but it would have to have a particular signature of a process of certain intensity with a certain focus in space. In the end, he found the entity he needed. It was the combination of warblers and squirrels feeding on budworms. When he saw it, he knew that was what it had to be. Reductionists assert at the outset what they should model and investigate. Holists look for signatures and then only give them names when they find them. Warblers and like birds could not do it alone, but with squirrels, he got his thirty-to-forty-year outbreaks while still including realistic spatial movement. It yielded to a narrative approach that identified a critical unforeseen relationship.

Holling developed the budworm model in parallel with a general model called "panarchy."[8] With Lance Gunderson, Holling has written a whole book on it, and it is one of the more important narratives for management. We might say it is a general model, but would hasten to add that it is really a narrative. Narratives are improved by models, but have the added advantage over models that they do not have to be internally consistent, as we explain in the chapter 8. The internal inconsistency arises out of a change in level of analysis. It is not possible to get a consistent mapping across levels of analysis. In physics, their best shot at it is in statistical mechanics, where the contradiction is across determinate particles on the one hand but stochastic indeterminacy on the other.

Let us lay out panarchy and then explain the inconsistency within it. The chart that captures panarchy has two axes (figure 9.4A). The abscissa is complicatedness, organization, or otherwise degree of connectedness. The ordinate is capital in the system. The first two stations in the cycle are the bottom left and top right. The system spends most of its time moving between the r station, bottom left, and K, top right, in process of capital accumulation. The r station has low capital and low organization. The K station has high capital and intense organization. These terms are related to the r and K for growth rate and carrying capacity in the basic equations for population growth and limitation. At r, growth prevails. At K, constraints are great.

At K, the system is brittle. It is an accident waiting to happen. For the spruce budworm system, the K phase is thirty-year-old trees just before the epidemic outbreak. The next phase arises quickly in what Schumpeter, the mid-twentieth-century

A

Holling's Lazy 8 scheme is really a narrative and not a model.
Capital builds, is destroyed, and then reemerges as disorganized
capital (did it ever disappear?).

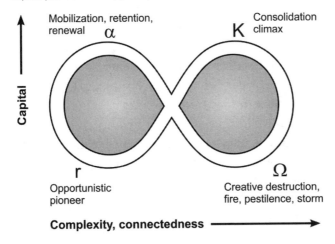

B

H. T. Odom pulsing

Odom runs Holling's concept over time but at
a slow, even pace. Odom's **fast** versus **slow** events
are graphed as **narrow** versus **broad** peaks. Where
Holling speeds up in some phases, Odom crowds
narrow, fast processes together so the time metric
stays constant.

Odom graphs **resource reserves** (Hollings capital)
in parallel with **assets**.

Efficiently conserved, organized
resource reserves

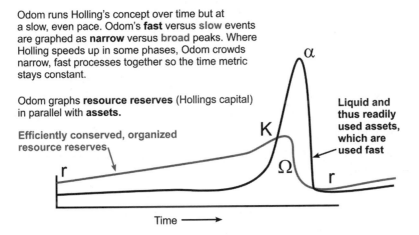

FIGURE 9.4. *A.* Holling's Lazy 8 narrative, which has come to be known as panarchy. The Lazy 8 scheme is really a narrative, not a model. Capital builds, is destroyed, and then reemerges as disorganized capital (did it ever disappear?). The track is from r to K, to Ω, to α, and back to r. At K, the system has much capital but is brittle and fragile. As in all fragile systems, collapse to Ω is fast. At α and K, capital is high. At α, H. T. Odom would say that capital is converted to liquid assets. For Holling, the distinction is that α is not organized. *B.* More like a model than figure 9.4A, Odom's maximum power principle plots capital on one line and liquid assets on the other. The two lines sidestep the inconsistency of panarchy, making it more of a model. In terms of high and low gain, K to Ω is the high-gain harvesting of standing crop. But the whole cycle for r to r is a low-gain scheme that keeps r to K growing with maximum power. From Ω to α to r to K is not just a passive phase of not harvesting; it is part of the low-gain strategy of maintaining maximum power. While the panarchy cycle has fast and slow transitions, Odom's scheme moves through time at the same pace, but the passage across the narrow (short time) α peak is where panarchy cycle appears to speed up. Without the Ω and α phases where there is lots of action, both systems would stagnate at K. Odum says maximum power always pertains.

economist, called creative destruction.[9] It is a creative destruction because it releases capital. Expressed in H. T. Odum's terms for the maximum power principle, capital is transformed into liquid assets (figure 9.4B).[10] This is not exactly destruction, but certainly a radical reorganization that removes capital as capital, converting it into something else. Holling labels the destructive phase Ω. In Odum's terms, Holling's creative destruction is just to the change from capital to liquid assets. In Holling's panarchy, the system moves from top right to bottom right on the diagram as capital is liquidated. Holling does not refer to liquidation, but only notices that capital recedes quickly. Liquid assets can be rapidly drawn down. Once the buying power is moved into liquid form under Odum, Holling moves his scheme up to the top left, his α phase. He is saying that something like capital has reappeared. But Odum emphasizes a critical difference that changes the name of buying power from capital to liquid assets.

Regarding the narrative nature of Holling's Lazy 8 chart (figure 9.4A), there is a contradiction. It comes in the disappearance of the capital and its creation anew. Capital reappears in the α phase, so the dilemma is, did it ever disappear? But it is a narrative, so dilemma and contradiction are all right. Odum's scheme is more model-like and is therefore internally consistent. He is simply graphing capital and liquid assets together over continuous time. He thus avoids the dilemma of one thing being two. The pulses of Holling's scheme are captured in the short time that Odum's graph collapses capital and shows only a narrow peak for liquid assets.

As happens with liquid assets, they are quickly spent. This moves the system briskly, but not instantly from top left, α, to bottom left, r. In Holling's budworm model the liquid assets in α would be dead trees after the budworm have killed them. Without the constraining organization that accompanies capital, the liquid assets are frittered away. In Odum's terms, there has been a transfer of high capital to high liquid assets, which are then spent on woody decomposition and disappear into the forest soil. After the budworm outbreak the dead trees simply fall and rot. If this were the end of the Roman Empire, the fungi consuming the dead trees might be Attila the Hun, who must have thought the collapse of Rome was a good idea. Attila takes the capital, but cannot hold on to it. This, of course, leads back to the r phase, where the system is open to opportunity, which unfolds as feudalism.

So a management question is how much of the liquid assets can be extracted for human use without removing necessary soil carbon resources to restart the capital accumulation in a new cycle. It is much easier to liquidate assets than it is to build capital. That is why K to Ω, to α, to r is fast, while r to K is slow.

With regard to management, we discuss the Tsembaga pig men in chapter 6 on population in terms of differential equations and basins of attraction.[11] As to the Lazy 8 diagram of panarchy, the Tsembaga build pig capital and then liquidate it. A slaughtered pig does not last long. The system pulses as it moves around the panarchy. Holling makes the point that the Tsembaga show resilience, like that of the host-parasite system of Holling and Ewing (see chapter 6, figure 6.9).[12] The pulses of pig growth and consumption also represent an example of Odum's maximum power principle.

Carpenter and Brock recently started to model management of lakes.[13] They modeled water quality from eutrophic to clear with a good fishery. At first they expected to have gradual change, with managers reaching and holding the desired condition. Immediately they discovered that folded-response surfaces were universal for almost all reasonable models. The models appeared to fit the data well. It seems that there was tension between a gradient toward a desired state and the opposing gradient of the expense of management. It gave a pattern of clear lakes becoming eutrophic, as effluent is cheaply dumped. The desired condition persists for a certain amount of time, and all appears well. But suddenly the lake passes to eutrophic, with attendant smells and green water. After the collapse of water quality, deep dissatisfaction leads to variously strenuous remedial measures. However, the lake usually stays eutrophic. There is a certain irreversibility to going over the fold in the surface. The shallow, clear prairie lakes, such as Clear Lake in Iowa, flipped in the early twentieth century to strongly eutrophic and have remained so despite a century of remedial action. The budworm outbreaks have a similar folded response surface, with a slow variable, the tree growth, and a fast variable, potential growth of unconstrained budworm. Some management does exist in zones of continuous change, but human reactions are generally so much faster than the variables that it is trying to control that folded pleats and surprises in behavior should be expected. Surprise occurs when a positive feedback is let loose. A case in point was the change in albedo with snow cover that can lead to rapid cooling into ice ages. The ice at the base of Greenland suggests that the change to permanent ice happened in just four years' time.

THE SUBCULTURES IN ECOLOGY

At professional meetings, groups of ecologists concentrate their attendance on particular sessions. The same faces are seen at either the landscape meetings or the population sessions, but not usually both. There are silos of interest, and only a minority moves between them. This political structure within the discipline seems innocuous enough. However, a price is paid when one group ignores the work of another as irrelevant when a cross-fertilization of ideas would, in fact, be helpful. A particularly wasteful schism is the one between pure and applied ecology. As we explain in following discussion, there is some contact across the divide, but there is also enough disjunction to waste opportunities for cooperation.

To meet this issue, the Ecological Society of America started a new journal for practical application of ecology. The British Ecological Society has published the

Journal of Applied Ecology since 1964, separately from its pure-science counterpart, the *Journal of Ecology*. Similar structure occurs in forestry with *Forest Science* and the *Journal of Forestry*. Thus, the division receives official sanction, albeit inherited from times when pure science was more confident and autonomous. The basic scientists' grants to applied agencies and programs start with: "To solve this applied issue we need basic science in such and such an arena," which is that of the proposer. The pressure to make basic science more obviously useful to the public that pays for it mostly gives rise to token window-dressing outreach, and not a move away from what basic scientists have always done.

One of the early big, important, and excellent works by Henry Horn is *The Adaptive Geometry of Trees*. It makes the distinction between monolayer trees and multilayer trees. The big oak trees and elms are monolayer, with leaves creating a sort of skin on the outside. The strategy of monolayers is to intercept all of the light in the vicinity of the canopy with no overlap. With such strong competition for light, self-shading must be avoided. Multilayer trees, by contrast, grow in relatively open habitats, where sidelong light hits the leaves up and down the tree early and late in the day. Multilayered trees capture the light of Gray's churchyard. "The curfew tolls the knell of parting day, the lowing herd wind slowly o'er the lea." We have never seen trees quite the same way after reading Horn.

We mention Horn because of an unwarranted, harsh review of Horn's book by John Harper, indicating dissatisfaction with the divide between basic and applied ecology. The review is a paradigm defense of crop and weed research. Paradigm defense and attack is often not fair, with mockery used here as a literary device.

> Much of this type of canopy analysis has been done before and in a much more sophisticated manner; reading this book is therefore rather like discovering a tribe lost to civilization that has quite independently discovered a primitive form of the internal combustion engine. Does one praise the originality or sympathize with the ignorance. There is a tragedy here, not just in Horn's book but in the failure of most ecologists to make the slightest attempt to follow the literature of agronomy and forestry. Ecologists must read this literature even if it shatters some of their conceit. For those many readers who will find that Horn's monograph opens a new vision on the nature of vegetation we append a brief bibliography to correct the perspective. (Harper 1972:662)

Harper's complaint, while unfair to Horn, was justified with regard to the separation of basic and applied research. Harper's influence over the mainstream of both applied and basic plant ecology persists today, more than thirty years after his retirement. That influence is captured, as Caccianiga and colleagues complained in 2006 (quoted in chapter 4 here), that the reductionist focus of theory comes from the fact that "the contemporary ecological mindset borrows heavily from agriculture." Harper's plants of choice were weeds and crop plants, though he was also in command of basic science ecology. With plants in cultivated fields, the larger picture of plants chronically and acutely limited is not seen as applicable. Finally, we are getting beyond the tight reductionist focus that yielded much but is now suffering diminishing returns. Zhu et al. (2000), working with rice, did indeed move to the wider view, their plant being a crop notwithstanding. They looked at their system in community terms and the whole system, which is an encouraging sign. Harper was right in the 1970s, in his general complaint about basic ecologists needing their "conceit" bruising. He would be less right today about that particular gripe because the better scholars do read applied and basic work, and it is time.

MANAGEMENT UNITS AS DEVICES FOR CONCEPTUAL UNITY

A significant part of the field manipulation conducted by the U.S. Forest Service (USFS) is imposed not on areas defined by community type or ecosystem function, but is applied to land management units that span community types and watersheds (figure 9.5). There is sometimes compromise between the definition of management units and terrain—a general community type, economic, or social factors—but management units are formally delimited to work largely as homogeneous production systems at the scale they represent. Management units may be defined on purely geographic convenience. Some cultural demarcation, such as a road, might cut across the middle of a homogeneous example of community or ecosystem, and yet for the management unit, the road might provide a very workable boundary (figure 9.5, right-hand photograph).

When it is to be seen through the eyes of the manager, a landscape falls into pieces whose identity turns on production of resources based on a complex of social, ecological, and economic considerations. Heterogeneous management units are common in initial management applications, becoming more homogeneous as management cycles play out over time. Only incidentally might a production unit map onto a homogeneous plant community or an area of homogeneous ecosystem function. The degree of homogeneity of the management unit is based on several ecological conditions: for example, geology, soils, topography, vegetation type, and, importantly, operability considerations. Tree age is a lesser consideration in early applications of management planning. It is an objective in longer-term management planning, as it will be the organizer for either a heterogeneous or homogeneous plant community at a particular successional stage. Land management units are a mixture of community, ecosystem, and landscape entities, ranging in scale from tens of acres to several hundred acres, depending on the heterogeneity in the major factors listed previously. Legislation requiring consideration of

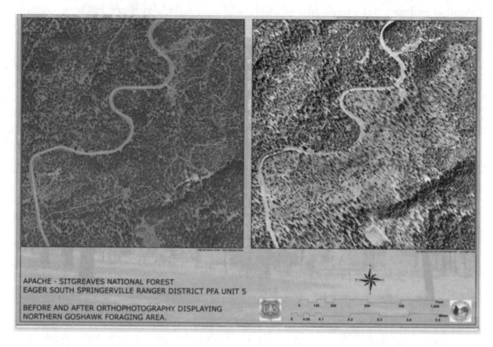

APACHE - SITGREAVES NATIONAL FOREST
EAGER SOUTH SPRINGERVILLE RANGER DISTRICT PFA UNIT 5

BEFORE AND AFTER ORTHOPHOTOGRAPHY DISPLAYING
NORTHERN GOSHAWK FORAGING AREA.

FIGURE 9.5.   Orthographic pictures of portions of adjacent land management units showing before (left photo) and after (right photo) management treatment. The area to the right of the
sigmoid-curved road (top to bottom left in the right photograph) shows the area treated to restore
the habitats of the plants and animals in the goshawk food web within a northern goshawk
foraging area. The management objectives were to create an open, uneven-aged, ponderosa pine
condition with a balance of age classes. Desired elements were groups of trees, scattered single
trees, open interspaces (grass, herbaceous), snags, logs, and woody debris. The cutting method
was individual tree selection. (Reynolds et al. 2013. Photo courtesy of USDA Forest Service,
Apache-Sitgreaves National Forest.)

joint production of public lands has increased the degree to which modern land
management units are multifaceted entities.

Once the land management units are established, silvicultural prescriptions,
such as a clear-cut or individual tree selection cut, are usually applied to the best
degree possible in a homogeneous way. Details of past influences such as fire or
management actions do introduce heterogeneity on the ground that can lead to
some variation in the application of a management prescription across the area at a
given time. A strategic forest plan is largely a narrative developed by the managers
through review of alternative options with the various public stakeholders. Originally, management alternatives were created with multiple-use optimization models. These optimization models received a wealth of information and, for example,
use minimum/maximum devices to find an optimal management scenario for a
given set of assumptions. In the series of books on complexity of ecological systems at Columbia University Press, in which this book is included, Hof and Bevers
have two books on the techniques of optimization for management of ecological

systems.[14] National forest planning today is less about optimization and more about public collaboration/involvement throughout the process. Today, the USFS seeks the public's involvement to jointly assess current conditions and trends of a national forest's ecological/social/economic resources. Also, they involve the public in identifying where the current, existing plan needs to change in order to better address the sustainability of the forest's ecological/social/economic resources. They solicit the public's involvement in developing and analyzing the components of the revised plan in response to the need for change that better address the forest's ecological/social/economic resources. And, finally, they involve the public in a predecisional objections process, instead of a postdecisional appeals process used originally to develop forest plans. The current forest planning process more closely follows the principles developed in chapter 8, especially Checkland's Soft Systems Methodology.

It is a mistake to dismiss studies arising from management goals as unnatural and therefore less likely to yield ecological understanding with general application. The power of multiple-use narratives for land management units as a basis for national forest plan revisions is crucial to understanding this messy situation. Resource use is no more arbitrary and anthropocentric than are the academic ecological entities; a nitrogen atom does not care if it is in the leaf of a community dominant, part of an ecosystemic nitrogen retention pathway, or located in a landscape entity like a hedgerow. Like land management units, communities, ecosystems, landscapes, and biomes are reflections of human ways of thinking, not reflections of ecological reality beyond perception and conception. Tansley's prescient comments on hypostatization (reification), quoted in chapter 1, are again pertinent. As a way to deal with the undefined fluxes of matter, energy, and information in ecology, land management units can be seen as just another conceptual tool to be used alongside academically defined ecological criteria.

There is a tradition of pragmatism in applied ecology. Academic ecological criteria are often applied in a manner that is just as ad hoc as management criteria. Therefore, it seems a pity that, since academic ecologists also pay the price of pragmatism, they often do not fully avail themselves of its utility. We might ask, "How do we get an additional 10,000 board feet out of here without negatively influencing water quality and elk habitat?" The power of using such management questions is that it forces a simultaneous application of narratives and models across conventional academic criteria.

Although nature functions simultaneously as communities, ecosystems, and landscapes, even dual, let alone tripartite, structural and process descriptions are rare in the literature. Since they introduce an ecological entity new to many students of natural systems, land management units may be used as a helpful device to pry open new intellectual possibilities. By considering management not only as a tool to achieve effective resource use but also as a tool for experimental manipulation in basic ecology, it is our intent to break old habits of using only one conception at a time.

Using land management units applies a perturbation across the major academic criteria. In management, we have a wealth of experience with many "experiments" already reported in the literature. Employing management action as an experimental manipulation, we can build a composite description using the power of the three conventional principles (community, ecosystems, and landscape) without being limited to any one of them. In fact, environmental impact statements legally require this rich conception of the manager. Adaptive management can use the wealth of reported management actions as superior starting places for their process of iteration.

Management plans for land management units are effective in satisfying narratives for ecological management. They establish a monitoring program that provides the basis for testing narrative outcomes of management in satisfying the narrative. In a North American (Canada, United States, and Mexico) test, Wright et al. (2002) put forth such a monitoring program in *Monitoring for Forest Management Unit Scale Sustainability*: the Local Unit Criteria and Indicators Development (LUCID) test was heavily based on and used the principles in the first edition of this book. The test was a technical test, and like many projects of this nature, the policy and political decision makers did not adequately understand the utility of the narrative-management-monitoring test. Such a monitoring process that interfaces land management and policy/political arenas could be important to the long-term social, economic, and ecological benefit of North American natural resources. One of the unique components of this effort was the structure and process criterion of social systems developed by Joseph Tainter (appendix 8.1).

If we are able to identify community criteria as well as ecosystem criteria that both map onto a given management practice, we may well be able to identify circumstances where a community entity is also a functional ecosystem flux. In tropical agriculture, the landscape mosaic of slash and burn allows community recovery. However, agricultural return time that is too short or fields that are too extensive both interfere with mycorrhizae and nutrient cycling. The degradation of communities therefore can be an ecosystem-related problem. We hope that our approach will uncover similar cross-links that were unsuspected until recently; it is designed to do so.

THE DISTINCTIVE CHARACTER OF THE MANAGED WORLD

At an autumn 1988 workshop held in Santa Fe, New Mexico, on the topic of ecology for a changing earth, one working group focused on the human component.[15] The group made three critical observations that can be woven together to give a broad picture of the world under human influence. It is on the parts of this world that managers focus their attention. The first point was that food webs containing humans have very indistinct boundaries. The second point was that the larger the human presence, the more leaky the ecological system. The third point was that, relative to the historical and prehistorical past, even the major ecological subsystems in the biosphere are now out of equilibrium.

FOOD WEBS HAVE DISTINCT BOUNDARIES

When the working group started to consider what was missing in the database to address their charge, Joel Cohen pointed out that very few published food webs have humans as one of the nodes. Human food webs are distinctive because of certain qualities pertaining to the system boundary. Putting humans in ecological systems does give a new perspective since we are so involved. New perspectives came out of comparing the University of Wisconsin, Madison campus, in terms of its watersheds and its sewersheds. When Cassandra Garcia presented her sewershed maps to water managers on campus, she could immediately see the excitement in her audience about a new set of insights.[16] In the same spirit, Bruce Milne is developing a science of foodshed analysis.

Africa is not well served by landline telephones.[17] The arrival of cell phones in Africa has changed networks considerably, to the advantage of local producers.[18] Groups of producers can now pitch in to buy one cell phone as a group, and in this way access more information about their network foodshed. The local buyer can no longer hold them hostage. There is much wealth at the bottom of the social pyramid, but it usually cannot be leveraged in a global system. Cell phones also provide a currency of cell phone minutes, which are valuable and readily transferrable great distances, allowing access to global markets with real capital.

THE LEAKINESS OF HUMAN SYSTEMS

The second critical observation of the 1988 working group was that pollution problems appear to be more deeply rooted than we first thought. Part of the problem is agricultural land use, where fertilizer gets into the aquatic system. Less expected, but very important, is the nutrient input that sheets off of suburban areas. This suburban runoff is not collected sewage or point discharge; rather, it is a reflection of the nutrient leakiness of suburbia as a whole. The points of nutrient enrichment on the Hudson River have been identified and mostly shut down; now unmasked, the full impact of suburban runoff is apparent. In the Lake Wingra project of the International Biological Program, the scientists found that the kick start of the eutrophic cycle in the spring came from the simultaneous melting of accumulated dog feces on suburban lawns, frozen through the winter.[19] By contrast, the forested lands on the opposite side of the lake contributed almost no nutrient

FIGURE 9.6.   An aerial photograph of Lake Wingra showing the suburban areas in contrast to the forested vegetation of the University of Wisconsin Arboretum just south of the lake.

load, and certainly not the pulse that was felt from the spring thaw of front yards (figure 9.6). Predominantly, natural ecological systems retain their nutrient material, and forests are masters at the game. It is by holding nutrients inside the forest through cycling that woodlands escape the pressing constraint of low nutrient input from the air. A nitrogen atom entering a forest system can be expected to be held for 1,810 years, even though it is likely to be mobilized for new growth at the beginning of most growing seasons.[20] The contrary characteristic of the suburban landscape applies to all other intensive human uses of the landscape. Human-dominated systems leak material.

BIOSPHERE OUT OF EQUILIBRIUM

The third observation of the working group was that human-dominated systems, even at the scale of the whole biosphere, are undergoing radical changes of state, such that the old equilibria are dysfunctional.

The changes of which we speak are so profound that wild and heavily managed systems alike are all casualties. Even situations that have already been fully impacted by humans, like the tall-grass prairie region that became the American corn belt appear not to be stabilizing in a new configuration. The continuous production of corn masks radical changes that are still happening on the farm. Prairie soil is still deep, but is eroding at a rapid rate by geological standards. Should the predicted, anthropogenic, global climate warming and attendant climatic shifts occur, the whole Corn Belt will move north to Canada. The trouble is that the Canadian Shield has no soil to grow corn, although the Canadian prairies do, and that is starting to happen. We note the concept of agrobiome in chapter 8, and it applies here as the description preferred over agroecosystem.

In forested areas, the changes are even more apparent. Vast areas of forest are being converted to grasslands across the tropics. Even areas that felt the heavy hand of humans long ago show this global pattern of change. The long-deforested lands of Atlantic Europe are accumulating organic material in bogs in a way that did not happen in the ancient forests. Monoliths of the first agriculturalists are found resting on soil not greatly different from that in the primeval forest, but whose stones are now buried under tons of peat.[21] Ironically, there are protests that the increasing mining of this peat for fuel is destroying the Irish landscape.

Certainly, very large changes have occurred in the biosphere before, but they have generally been so slow in coming that the major subsystems of the biosphere could accommodate them by moving to a new latitude. The speed of the change is presently so great that there is not enough time for natural systems to move with them without going out of equilibrium and losing integrity.

At the outset, the 1988 working group couched this observation in terms of an uncertain future. They were not prepared to predict the future of large-scale ecological systems because humans are changing the world so fast and so extensively. A view was expressed that we cannot predict because the system will be out of equilibrium for some time. However, it appears to us that the future envisaged by the working group is already here. There is no point in waiting for equilibrium because it will never come. At least it will be postponed indefinitely, until the human race is broken on the wheel of its own deeds.

The new entity, which we call the anthropogenic biosphere, is much smaller-scale than the entire global system, and is constrained by it (figure 9.7). Nevertheless, the anthropogenic biosphere is larger-scale than the major natural systems that it contains. The emergence of this new entity explains that (1) human food webs do not have discernible boundaries; (2) human-dominated systems are leaky, and (3) few, if any, major ecological systems can be adequately described with equilibrium models that would have applied before human intrusion.

The biosphere is large enough to leave plenty of slack for the old natural systems to relax close to some sort of equilibrium, but the anthropogenic biosphere works faster, imposing tighter constraints. That explains the apparent lack of equilibrium in the quasi-natural ecological systems in a human-impacted world. It is in the nature of the new anthropogenic upper-level structure that the parts will remain held out of equilibrium indefinitely. The anthropogenic biosphere is a local

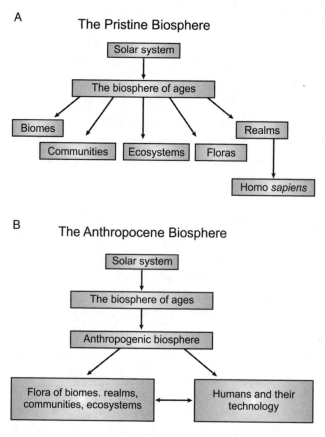

FIGURE 9.7.   *A.* The biosphere of ages has our species as a small part of a realm. We were then an uncommon, incidental omnivore. *B.* In the Anthropocene, humans are more significant, but certainly are not in control.

phenomenon relative to the long-term biosphere driven by volcanism, catastrophic meteors, and ice ages. Ice ages will come again, no matter what humans want or try to control. There have been much warmer epochs than this. Therefore, despite anthropogenic global climate change, we should not overestimate the significance of human activity at the scale of eons. Equally, we should not underestimate its effect on human systems, such as farming, fishing, or living close to water.

The anthropogenic biosphere is scaled much closer to the major ecological subsystems than was the primeval biosphere. Accordingly, the anthropogenic biosphere forces a much greater degree of integrity on the globe's biota, and not a good integrity. In the new biosphere, matter, energy, and biota are moving around the globe as never before. These fluxes are the connections that reflect the greater integrity.

In general, management is best achieved by playing to the ways of nature. The chinampas of Mexico take advantage of a natural process.[22] Tropical forest soil is nutrient poor because of high temperature mobilizing nutrients and high rainfall washing them all away. That water ends up in lakes. Thus, the nutrient status of many lakes in the tropics is high. If those lakes are shallow, there are possibilities for farming them with chinampas. The Aztecs farmed most of the lake that is now gone under development of Mexico City. They dug channels by scooping up mud and depositing it in long barrows. Thus, canoes could go in between these artificial islands to scoop up more mud and pondweed. Crops were then grown intensively.

Cortez reported the floating gardens of Montezuma. He was wrong. The barges with plants on them were for only small transplants in transit. The Spanish thought that the chinampas were sunken barges. A problem with this system is that the land for growing crops is highly derived and expensive. Another problem is that the tropical sun does not produce longer days in the summer. Normally, crops in north temperate regions suffer self-shading in June, but the extra summer sunlight compensates for the effects of crowding. Increased day length is not available in the tropics. The good news about the tropics is continuous cropping year-round. Transplanting allows valuable cropping space to be used intensively. Plants are grown very close together in seedbeds. When the plants get bigger, such that they would crowd badly, they are transplanted to a lower density. This is like extra light in the summer of temperate regions. It is a sort of time for space substitution. Plants grow all year, and in the extensive growing areas, the plants are always growing fast on the steep part of the growth curve with space to grow. Chinampas are one way of going with nature in the tropics. Pulsing allows for maximum growth all the time. This is the basis of H. T. Odum's maximum power principle to which we refer earlier. It would appear that harvesting is a high-gain activity of taking it all. But if the cropping system actively retreats, unconstrained growth is at its maximum. It is cropped again only after a significant period of release. This strategy allows for maximum growth in a low-gain, organized, long-term scheme.

Swidden agriculture is a pulse-and-rest system that has advantages over permacropping. The pre-Columbian natives of California lived in a pulsing environment of El Niño/La Niña, and the classic Mediterranean pulsing climate of cool, wet winters and hot, dry summers. That pulsing productivity creates a maximum power situation that in those times led to extraordinarily high human densities. With the prescription always to manage like nature, there comes the issue of "what is natural?" It appears that fields are natural. The native Californians had inherited chieftains, which hunters and gatherers cannot usually afford. Their population was about ten times the density of most hunter-gatherer systems. They dealt with the landscape by burning. Areas were burned and then left to regrow to an extent. The hunters would then return and hunt the edge of the open area. Game would concentrate there. The critical point is that when human density is high, open areas are a natural emergent property. The hunters were making fields; it is just that they were not doing agriculture on them.[23] So with a world of seven billion people, fields are a natural emergent. The point is that we now live in a world where steep gradients are presenting natural emergent entities. Whirlpools are strange but natural. Fields are a sort of human-induced whirlpool.

Fields follow the pattern of management units; that is, there is a homogenous application of human management. We might object to the way monoculture in industrial agriculture resembles a petri dish; industrial society can go too far. But, since our first edition, some remarkable achievements have been made public— yes, with fields, but with a new sort of homogeneous management strategy. Near Viola, in the poor, nonglaciated southwest of Wisconsin, Mark Shepard has taken a 106-acre farm that was in row crops.[24] He farms it with fields that are savannas. Our species evolved in savannas, and the New Forest Farm is an American savanna that feels like home.

Shepard uses natural selection to force his trees to produce early and in high densities. Those selected out are used for burning; others are for toy making in a local cottage industry. He farms the New Forest Farm very intensively, but with high diversity. He has bred hazelnut bushes to reach production very fast. They reproduce vegetatively, and he sells the stock. His hybrid chestnuts are resistant to the blight and produce early and heavily. The ground is hilly and is a poor part of the state. Shepard farms it with perennial crops that offer high added value. The produce of his apple trees is sold as hard cider. He uses grazing to recycle mineral nutrients. His system is complicated but highly organized, making it a complex emergent. While he does a lot to the land, he rides with nature. Shepard's book, *Restoration Agriculture: Real World Permaculture for Farmers,* tells how to do it.

One of the reasons that first world agriculture can be so damaging and inhumane is outside pressures on the farmer. Farmers often do what they do because they cannot afford to do anything else. Farmers pressed by bankers are an example of how important is context, but this time a bad context with sad outcomes. Shepard speaks of organic methods and multicropping, but most significantly, he has a chapter on making a profit. He has worked out how to create a fully functioning context for himself and his fields.

We have recommended to our students that they become lawyers, since humans act coherently with the law. We might also suggest that they go into business with the New Forest Farm model. Mainstream farming is a relatively high-pressure, high-gain system because of economic pressures. But what Shepard is recommending is a highly organized, low-gain way of increasing constraints and efficiencies. High gain is by definition short-lived and not sustainable. Low gain emerges when sustainability becomes an issue. Two years ago was a drought across the state; field crops from corn to hay were a disaster. But New Forest Farm had a great year, a bumper crop of hazelnuts, berries and apples, all species that are drought tolerant. Shepard has created a set of landscape berms that direct water across the slope and conserve it. It was a good year for "Shepard's Hard Cider." Shepard does impose homogeneity, as is characteristic of fields, but it is in the spirit of multiple-use management. The difference is that it is with multiple crops instead of multiple uses. Of course, multiple crops are common in tropical agriculture, but Shepard does it in a temperate region.

MANAGING ECOLOGICAL SYSTEMS AWAY FROM EQUILIBRIUM

Having recognized that the systems we are to manage will function away from equilibrium, we are now in a position to use an explicitly hierarchical approach to management. It is in the nature of hierarchical control that lower-level entities are held away from equilibrium by the constraints imposed by upper-level contexts. In this way, the second law of thermodynamics, the one about gradients, is not in contradiction of life but rather drives it. Life is held away from equilibrium by constraints at many levels, and the gradients thus created, drive organization (Schneider and Kay 1994).[25] Effective management needs to be particularly mindful of two points that follow from an explicitly hierarchical approach: (1) the system being managed will be out of equilibrium, all the more so because it is being managed, and (2) as a higher-level context, the management practice must offer a viable context for the system under its charge.

SUPPLY-SIDE SUSTAINABILITY

With Tainter, we wrote our book *Supply-Side Sustainability*.[26] The chapter so far has spoken in the spirit of that work, but let us be explicit about that scheme. The name, supply-side sustainability, comes from a set of principles laid out in that book. We have been talking around these principles heretofore, but now let us lay out those principles here.

The first principle says: *Do not manage for the resource, but rather manage for the health of the system that produces the resource.* If we manage for the resource in a fishery, we would harvest fish that give the greatest immediate weight of catch. That would take fish at the size when they are at their steepest part of their individual growth curve. When individuals are growing fastest, they are most efficient at converting the ecosystem into fish biomass. Older, bigger fish convert less, so the strategy would be to fish out the older fish. But the effect of that is to take out the context of the whole system. The old fish may not convert much ecosystem input to fish biomass directly, but they are stabilizers that offer the best-quality fingerlings to the next generation. Work to manage the whole system that produces coupled to the sun, and it will continue to produce.[27]

Thus, the second principle of supply-side sustainability is: *Manage from the context.* First, it is easier than managing the details, which will be different in each local setting. In fisheries that have been managed for maximum biomass of desired fish, one species at a time, the desired fish have been driven close to extinction. Apparently, backing off to allow recovery is action too late. The cod fishery is not coming back for reasons that are not transparent. The great fishery of the Georges Bank in Maine is now down to trash fish, whose growth rate is fast enough to survive intense human depredation. With such fast growth rates, the fishery now yields only to chaotic equations. Chaos cannot be managed even in the midterm. In other words, the fishery is not only being badly managed, it is unmanageable.

The lobster fisheries are doing well enough in Maine because they are managed from their context, the big old lobsters. Maine has legislated the use of lobster pots; bottom dredging is not allowed. The small lobsters can get out of the pots, while the big ones cannot get in. This preserves the young and gives the next generation a source of high-quality genetics. Any big lobsters that provide the context are not fished by lobster pots. The fisheries are also being heavily supported through hatcheries.

The third supply-side principle is: *If the system finds itself in a workable environment, it will support the sustainability effort.* The extra productivity can be taken. The manager has earned it by looking after the internal production needed for ecosystem function.

The fourth principle is: *If you want to shift the system, do it with positive feedbacks.* Simply pulling the system to where you want it is expensive and it will likely slide back once you stop paying for the engine that is doing the pulling. With positive feedbacks, the pressure to change takes the system into the new position and continues to hold it there with the feedback alone. It is best if the desired system state is an emergent property driven by feedback. Government regulation is important, but only as a catalyst.[28] Regulations are constraints.

The major positive feedback in the first world is at the heart of the mercantile system. Trade works with positive feedbacks. Without a healthy economy, there is no one to pay for environmental action. It is therefore appropriate to rely upon business to make the environmental changes. Business is often efficient in solving its problems. It can be directed to use that efficiency to the public good by government manipulation of its environment. Given regulation that tilts the landscape in a public-spirited direction, business can be trusted to make the best of that situation and act for profit in the public good. While government is well equipped to be a catalyst that tilts the landscape, it cannot work as the engine that actually delvers the change. Government constrains; it does not have positive feedbacks. It ratchets and does not come down in size when the pressure is off. So a mixed economy is probably the only way to put in place the feedbacks for change that we recommend.

Parts of business have discovered the principles of supply-side. There is a group of businesses applying supply-side principles, and is accordingly creating value much faster than that reflected in the major market indices such as the Dow Jones. The movement calls itself "Conscious Capitalism," and its proponents include: Whole Foods Market, Panera Bread, Southwest Airlines, The Container Store, Nordstrom, The Motley Fool, Stagen, Joie de Vivre hotels, and Trader Joe's. Their mantra is unleashing stakeholder value.

This view of the firm was presented thirty years ago by Russ Ackoff in his *Creating the Corporate Future*. It is not the usual top-down hierarchical control system view. The CEO is not the functioning head. The shareholders may have certain rights in the eye of the law, but they do not trump the other organization stakeholders because if they do, shares collapse in value. The company is beholden to its employees, creditors, customers, suppliers, and to the local government. If it does all that, it will deliver for its shareholders. Empower the stakeholder and the workers will be invested, the customers will be loyal, the government will be cooperative, and the creditors and suppliers will give good terms. And the outcome will be for the general good. It is a rational approach to collaboration that serves multiple self-interests. It works toward sustainability. We make similar points when we discuss commensurate experience in the chapter 8.

If management can achieve what we prescribe, then the managed unit should be serviced as if it were in context. Accordingly, well-managed units should behave as if the context were indeed there, even though it is not. Then the managed unit should be free to function without deprivations; what the extant context cannot offer, the human management system provides instead.

This leads to a principle of management that at present is only a hypothesis but may in time be verified: *If the management regime is effective, the managed unit will offer a maximum subsidy to the management effort.* If the managed unit is being provided with all it might expect from a natural context, then it can function to full effect. Humans get to choose the context, to an extent. Effective management would provide infusions of genetic diversity from a distant wild population or, if necessary, from a zoo. The Isle Royal wolf population is now down to eight, and inbreeding depression is suspected. Normally, the context would offer genetic diversity in the form of outcasts or strays from neighboring populations, but in their absence, human contrivance plays that role instead. The effect of these infusions should be to maintain the managed population as a vigorous unit, making it appear as close to self-sustaining as possible. Of course there is a cost to management, and the system is not self-sustaining in the normal sense. But if we cast humans and their management as being inside the system, then the larger scheme is self-sustaining. Part of the system feeds us, and we humans sustain what is not human in the system. The management cost must be sustainable, raising the questions of sustainability of what, at what cost, and for how long. Lower cost is likely to contribute time to the issue of how long. On the positive side, breeding in a managed population subsidizes the management effort. This feedback is what allows many management regimes to persist to the point where they are significantly sustainable with workable effort.

There is a difference between managing the system from outside, which is what we recommend, and forcing the system to perform in some focused, prescribed way aimed at the inside. Get the context right, and you do not need to understand the mechanisms of the managed system or the one it replaced. Do it right, and the system does what you want. See what works, and simply keep doing it. "Keep doing it" may itself invoke a sort of metalevel management that is the management of management.

In exactly that spirit, Eduardo Sousa makes foie gras. In 2006, his Extremeña Company, La Patería de Sousa, won first place at the International Food Products Exhibition in Paris, SIAL 2006. He is Spanish, and the French were incredulous and upset. Some of the outrage was that Sousa did not gavage his geese. Gavage involves force-feeding geese with a funnel and a tube into the stomach. In a few weeks of being involuntarily gorged, the goose liver expands to eight times normal. The livers make foie gras, a great delicacy. But fully deserved bad press has foie gras banned in Chicago, San Francisco, and New York. World-renowned chef Dan Barber went to visit Sousa, and gave a wonderful TED talk on the experience.[29] Sousa has his geese in a fenced-in area. The purpose of the fence is not to keep the geese in but to keep predators out—the electric wire is on the outside. He plants everything the geese want, a great diversity of plants. His geese are so happy that wild geese hear their calls and come, not just to visit but to stay. The geese know Sousa and appear to love him. Foie gras is usually bright yellow, but Sousa's foie gras, not involving corn, was gray. So he gathered seeds of a local yellow lupine. His geese loved the seeds and gobbled them down. His foie gras turned bright, bright yellow. The flavors in the foie gras come from his pepper plants and from plants he uses to control salinity. The geese are thus naturally salted. Sousa lets nature do it all for him.

Barber goes on to say that gavage is in fact a perversion, yet another example coming from a dysfunctional context. He reports that foie gras is an ancient Jewish device, raising the geese in the manner of Sousa. But in slavery, the story goes; the pharaoh demanded their foie gras from them in quantity. The only way to achieve such production was through gavage. The story may or may not be true, but it makes a point. Foie gras is made the Sousa way under low-gain constraints and complications. Gavage is a high-gain method that increases the flux on production. And so it is with mainstream agribusiness: focus on quantity lowers quality and generates instability. In a piece on the ecology and culture of Western European peasants, Estyn Evans makes exactly the same point. Abundance lowers quality. He says that if you want the best cheese, go to areas that produce little milk. There, they divert valuable milk into an even more valuable cheese. Heavy milk-producing regions that have excess characteristically dump the over-production into cheese as food preservation. Wisconsin, "America's Dairyland," as it says on vehicle number plates, does make some world-class cheese, but the bulk of it is for mass-produced pizzas and processed cheese.

In the management of the Great Lakes Ecosystem Basin, the International Joint Commission has adopted an approach that, like ours, explicitly includes humans and human activities. The "Ecosystem Approach," as it is called in the international agreements, argues that (1) ecological, (2) sociopolitical, and (3) economic systems all coexist in the functioning of the basin.[30] The "Ecosystem Approach" requires an integration of all three sectors in the search for management solutions. The emphasis is only slightly different from ours, for the absence of the natural context is just another way of pointing to the presence of the human sociopolitical and economic sectors. Ours is not a management scheme for a pristine world, but one for a world full of human activity, where even the fragments that resemble the primeval condition are artificial islands. If we manage them as pristine wilderness, all will be lost. If we acknowledge it is not nature that we manage, but managed units, there is a shot at competent intervention.

Ecosystem services have been a buzzword recently, but its very name indicates a fundamental misconception.[31] In the service sector of the economy, one party offers service as something good for some other party. The service provider works to satisfy the customer in return for money. Since ecosystems are not sentient, they are not in a position to offer service per se. They do have resources taken from them, but they do not offer services. In fact, it is the user of the so-called services who must offer service to the ecosystem. Normally, those receiving the service pay for those services. It is the ecosystem that pays, and it is the ecosystem that needs service as compensation for what it pays. Ecosystem services are simply backwards. Ecosystem service should be service offered by human consumers.

THE SOCIAL SIDE OF SUSTAINABILITY AND LIVABILITY

The manner of exploitation of ecosystems is generally not focused on sustainability. If a system is to reach a state of sustainability, it has to pass through a process that people living find acceptable. In other words, to reach sustainability, the system must meet criteria for livability. High and low gain again pertain. Livability is high gain in that there are no plans, and it is simply what the people living want.[32] Often the standards of livability are at odds with sustainability. The continued exploitation of fossil fuel is not sustainable, but its exploitation is left to the market. The market has no values; it just responds to flux and process. So exploitation will always follow the path of livability, which has little conscience and only addresses the short term.

The good news is that livability is labile. When resources are limited by depletion, the market dictates higher prices. In the end, they become prohibitive and a
new livability is imposed, like it or not. When the price of gasoline doubled in the
United States to reach four dollars per gallon for the first time, a new livability was
imposed. The Hummer production line was abandoned because the American
public had been awakened to the real cost of personal transportation.

There is always the danger of Jevons paradox blunting changes in livability. On
the face of it, efficiency should save fuel because less is used for a given amount of
work achieved. But increased efficiency usually leads to more, not less, consumption. Jevons in 1865 first wrote on "The Coal Question." He told of steam engines
and coal consumption. An efficiency of steam engines increased from 0.5 percent
(95.5 percent up the flue) to 2.5 percent around 1850. Efficiency of 2.5 percent may
not seem like much, but it was a 500 percent improvement that made steam engines
worthwhile. More steam engines came on line, so increased efficiency actually
increased coal consumption.[33] The same thing happened in the mid-1980s, when
internal combustion engines finally were made more efficient. The response was
the sudden popularity of sport-utility vehicles (SUVs) that drivers could afford to run,
even at higher fuel costs caused by scarcity. Increased efficiency in the face of declining resources allows rapacious consumption to persist. The villain in the coming
crisis of petroleum is the Prius, not the Hummer. The latter simply goes extinct, but
the hybrid car lets consumers continue to consume. As notions of livability change,
values shift. We can see the very same people with opposite values some time later.

An example is values in England with regard to enclosing spaces. With quite
opposite land ethics and values from today, the creation of British hedgerows
angered common folk. With the Elizabethan Acts of Enclosure in Shakespeare's
time, commons lands were fenced off, angering commoners. Later, parliamentary
acts enclosed larger parcels. Common folk were forced off the land and into cities
and industry. Their resentment is captured in an anonymous folk song of 1764:

> The law doth punish man or woman
> That steals the goose from off the commons,
> But lets the greater felon loose
> That steals the commons from the Goose.

Big landowners were the beneficiaries and, of course, approved of enclosure.
And it was not all a bad thing, as overgrazed land was brought under private ownership with its built-in self-regulation to avoid overexploitation. But over the centuries, the common folk have come to appreciate hedgerows as part of their heritage,
in a change of values as to what is livable. There is also a change in values of the
big landowners, who used to protect ancient sites. Now they are desecrating valuable old places, building shopping centers and the like, any place they can get

around regulations. As we mention at the beginning of this chapter, the land of big landowners was their power base, but it is no longer economically sustainable. Values for livability have flipped between aristocrats and common people. Values for livability are labile.

With the industrialization that followed enclosure, a change in fuel was forced on the populace. There was resistance because wood was free but coal had to be bought. Forest people in Epping Forest used to be entitled to lop trees at head height for fuel. Queen Victoria withdrew lopping rights, and in 1884, compensated the common people with Lopping Hall, a civic building. The trees still show the change in policy. Thick boles up to head height now have long, tall branches growing from them. That is the growth in the century and a quarter since lopping ceased (figure 9.8). The trees show the change in livability in the forest.

But resistance to changes in livability has produced a certain retrenchment in values, with some very strange patterns of land use. Regulations going back to 1790 give anyone with sizable land adjacent to Epping Forest open grazing rights

FIGURE 9.8.   Trees in Epping Forest showing a hundred years of growth from a lopped old trunk. (Photo courtesy of T. Allen.)

in the forest. Grazing practices more appropriate to Wyoming interdigitating with urban London make no sense! Allen lived on Whipps Cross Road, the North Circular Road, a main trunk route. But his mother would from time to time have to protect her flowers in the front garden by chasing cows out into the four-lane roadway traffic. Even when a motorcyclist was killed in 1977 from a collision with a steer in dense fog, it took another two decades for the grazing rights to be suspended. So livability is labile, but it takes pressure to change it.[34]

Another bizarre example is the entranceway to Whipps Cross Hospital, which serves the east quadrant of London with about a thousand beds. Allen's father, Frank Allen, was chief hospital pharmacist from the 1940s to the 1970s. The hospital is some thirty yards back from Whipps Cross Road, with what is technically part of Effing Forest between the hospital entrance and the road. It is a sad strip of trees, not really part of the forest in any functional terms. When Allen himself was a teenager, that entranceway was only a single lane. Ambulances would have to wait their turn to come out if another was coming in, and vice versa. Frank Allen reported that it took twenty years of negotiation with Epping Forest authorities to widen it to a rational two lanes so ambulances with the sick and dying could pass. In the end, livability will yield and values will change, although the examples above attest to how values can be entrenched until the last. We are not likely to move off fossil fuels until a decade or so too late.

Raw flux does not govern sustainability, planning does. It appears futile to hope to wrest control over careless resource use by populist politicians. Jimmy Carter had the solution to the gas crisis of his time. He proposed increasing tax on fossil fuel until use was diminished. Had he won out, this would be a different world now. But Congress members put the need to be reelected ahead of a rational energy policy, and they blocked him. To be fair to the politicians, the voters would probably have exacted their price in the cause of the livability they wanted. Even so, sustainability might be possible, at least in the midterm, because the present greedy livability will yield to shortages.

The danger is that running resources to too low a level might deny society enough energy to make the prudent shift. It takes resources to switch to low gain. But plans for sustainability do not go unnoticed. A more measured livability might emerge ahead of its time, before it is absolutely forced by lack of resources. The environmental moves made in the 1970s were not so much driven by actual poisoning of humanity at large as they were by wiser sentiments coming from seeing the whole planet from space, and feeling lost. Ironically, Richard Nixon was the last environmental president, forced there by popular sentiment. Some of the green parties of Europe do have enough seats in the government to be crucial players in some ruling coalitions. Even the U.S. presidential candidate of the Right in the 2012 election openly talked of energy independence in the short run. Whether or not it was a sincere statement, it was made under pressure of acceptance of the need for a sustainable policy perceived as a desire of the electorate. Sustainability must yield to livability, but livability is labile enough for it to be worth some low-tain planning toward sustainability.

CONCLUSION

To summarize this chapter, the world that includes modern humanity is held far from primeval equilibria. From this it follows that management is not only of systems that are out of balance, but management itself explicitly holds the system away from equilibrium. Out of this emerges a management strategy that is explicitly hierarchical, where management is a required substitute for defunct natural constraints. The central concept is subsidy: subsidy of the managed system in recompense for the destroyed context (absent forest, grasslands, and wetlands); subsidy of the human management activity by the managed unit.

The central theme of this whole book is a contrast between different ways of looking at ecological systems. In this chapter, we emphasize that the manager is forced to look at nature using several criteria simultaneously. The contrast of the manager with the restorationist is helpful because the latter generally deals with one ecological category at a time. The use of multiple criteria at one time, as is demanded of the manager, does present difficulty but also it exposes the richness of ecological material. We have analyzed what this means for management in contrast to basic research. Further implications for basic research of a multifaceted view of ecology are explored fully in chapter 10.

The different facets of ecology are not so much a matter of nature as they are a matter of divergence in human perception; the material system does not function discretely as a community or ecosystem, or any other conceptual categorized entity. By being forced to deal with several ecological categories at once, the manager comes face-to-face with the human subjectivity that makes ecology more a soft than a hard science. Lynton Caldwell once noted that when professionals manage an ecological system, they do not manage the ecological system itself; rather, they manage the people who live in and act upon the system.

Ecology is a fairly soft science. We have shown that the softness of a scientific endeavor is related to the changes in human value systems that occur when the object of study is raised. Hard science ecology would not only be impotent when it comes to management it would also be intellectually sterile. The essential beauty of ecological material can be seen with remarkable clarity through the eyes of the manager. The naturalist and the preservationist do not have a monopoly on the joy that is to be had from being an ecologist in the woods. Management is a very aesthetic matter. In the modern biosphere, human activity is part of the system in a new dynamic interplay. Our species has the next dance with nature, and it is the ecological managers who should be the dancing masters and the orchestra leaders.

10

A UNIFIED APPROACH TO
BASIC RESEARCH

A T THE outset, we used the Beaufort scale to characterize ecology. That device existed in a distinctly applied setting of mariners being able to read the situations they faced. Ecology involves being able to view a complicated situation and reduce it down to a set of observations or measurements, just as did Beaufort's sea captains. Spewing waves and tempest present an overwhelming set of inputs that must be codified so that observations can be repeated and compared. Thus, basic ecology can be set in applied situations, although in basic science, the line of investigation is more open and general, with only tangential connections to particular problems in particular settings.

FUZZY CRITERIA: A PROBLEM AND A SOLUTION

A tree can be a population member, a community member, or an ecosystem storage unit. Each conception emphasizes its own aspects of form and function. As a population member, the tree relates to other population members through reproduction and genetics. With the tree as an ecosystem part, reproduction and genetics are usually taken appropriately for granted. Ignoring reproduction and genetics might offend the sensitivities of population biologists, but population considerations have little relevance for most questions an ecosystem ecologist might ask. The ecosystem ecologist would see the tree as possessing biomass or a rhizosphere for nutrient cycling. Clearly, both the roots and seeds are crucial components, without which everything else about the tree could not exist. However, we are concerned here not with the tree itself, but with it as an object of study by an ecosystem ecologist, and that is a very different matter.

The aim of this chapter is to engender a more catholic worldview in ecological specialists. There is a formal algebra for dealing with situations that only belong in part to a defined criterion. It is fuzzy set theory.[1] The notion of fuzzy sets is not in any way vague, despite the name. Sets have been conventionally considered as crisp and quite concrete; that is, an entity is either a member of a set or not. In fuzzy sets, membership is not all or nothing, but is a matter of degree. The notion of fuzziness is helpful for sets such as the set of "fast cars," to which a vehicle belongs usually only to a degree. The archetype of the set might be a Formula One racing car, which would score 1.0, complete membership. A Model T Ford is not in the set. It is a slow car, scoring 0.0 membership. We already discussed our respective heights as fuzzy in previous chapters, but then the concern was predicative and impredicative definition.

Once there is a clear criterion for membership, there is a well-defined algebra of fuzzy sets, which is as particular as that which applies to discrete or crisp sets. Polar ordination of vegetation of Bray and Curtis is a rigorously defined technique, and is performed by the not-only-but-also operation of fuzzy sets.[2] It is a method of relative membership.

A given action or experimental stimulus may relate to several different criteria. The assignment of an ecological action to a given ecological criterion (or to a multiple of ecological criteria) is a fuzzy problem. The fuzziness comes about by virtue of a specified action having impacts that are variously split between ecological systems defined on alternative criteria. For example, removal of a tree or two, namely, the practice of selection cutting, is a community consideration, for it shifts the balance of dominance. Meanwhile, a clear-cut impacts the landscape. In practice, there is a continuum between the two extremes, such that forest management removes trees in various-sized mosaics, so as to impact both community and landscape to various degrees.

Thus, a linear monotonic increase in landscape effects, namely, an increase in size and completeness of the cut, might bear a complicated relationship to community effects (figure 10.1). A scale-of-action trajectory in a criterion-dimensional space is an instructive exercise for gaining insights into the relationships between the alternative criteria and ecological scale. Tools for thinking about multiple scales and multiple criteria are uncommon.

Ecologists make many types of measurements, but expend little effort relating across types, or even within related experiments. In communities we collect species lists, diversity estimates, relative size classes for major species, recruitment and regeneration estimates, seed rain, mortality, and wildlife foraging data. Detailed measures of ecosystems include fluxes in energy or matter: runoff, snow accumulation, evapotranspiration savings, nutrient and sediment flushing in streams,

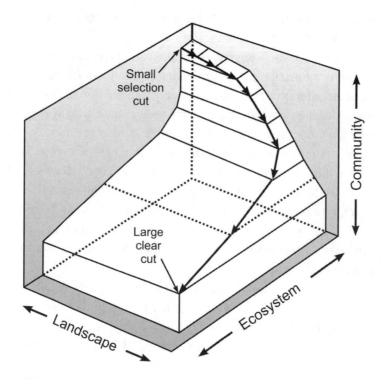

FIGURE 10.1.   A fuzzy assignment of ecosystem, landscape, and community criteria to cutting different-sized areas of trees. A simple example of complicated narratives are used in joint production; multiple criteria require more complex production interactions to meet multiple-use demands. A selection cut is high on community influence but low on ecosystem and landscape considerations. Progressively larger cuts have less community effect, and first increase in landscape considerations and later on the ecosystem criterion, as cuts follow stream corridors.

local temperature and humidity, and productivity. Finally, detailed measures of landscapes might be proportional area of land use types, composition of land use types, cover types, corridors, strips, intersections, edges and fragmentation effects, interiors, texture (evergreen-broadleaf mosaics), vistas, waterways, and so on. All of these categories of data were, in fact, collected in the Fraser Experimental Forest in Colorado (figure 10.2), and we found papers and reports published on them all.[3] Such a wealth of data from a particular forest should be intercollated. Surely, with big data capabilities, it is time for ecological information to be integrated. The ability to do this integration is recent.

In ecology, it is only just over a decade ago that Hans Rosling, to whom we refer in chapter 5, complained that data of human demographics existed but were not in an accessible universal format that allowed comparison.[4] And the demographic data are of a commensurate sort. He has given a series of talks on human populations using his Gapminder software from 2006 to 2014. He complained of "stupid passwords and boring data." on the internet from world population organizations. But

FIGURE 10.2.   The USDA Forest Service Frazer Experimental Forest is used to test a wide range of forest management practices that can provide data for various forest regeneration, watershed, and landscape modeling interactions. (Photo credit: Charles Rhoades, U.S. Forest Service, Rocky Mountain Research Station, Frazer Experimental Forest, Fool Creek, August 2008.)

now Rosling's transparent data and Gapminder software are available to all.[5] True statisticians might worry that it does not prove anything, but it is offered not as tests of hypotheses but for generation of hypotheses. He has generated some compelling ideas about human population dynamics, and exposed our ignorance. In his 2006 talk he told of testing his topflight students and the professors that select Nobel laureates. The test asked which country has the higher child mortality. The respective higher rate was twice that of the other in the paired nations, so the differences were significant. The pairs were: Sri Lanka–*Turkey*, *Poland*–South Korea, Malaysia–*Russia*, *Pakistan*–Vietnam, and Thailand–*South Africa*. The answer was that the respective country in italics had the higher child mortality. The top students in Sweden scored 1.8 correct out of five. The professors scored 2.5; that is, they might as well have picked at random. In the students Rosling had found systematic bias that lines up with "us and them." "Us" is the industrial world, and "them" is the third world.

Rosling's software lets him show countries moving over time on graphs of fertility and life expectation at birth. He started by looking at 1962, and discovered that the third world nations had the same demographics as the United States in 1900. By the year 2000, they had reached the same demographics as the United States in 1962. It was a remarkable change that was apparently unknown to the Swedish students and the professors, and probably the reader too, unless you have already heard Rosling's talks. Rosling recognized Africa as not being one place. Even so there was a general pattern that, in 2000, Africa lagged behind. Most African countries only reached the statistics of the United States in 1900. But Rosling observed that Africa had started in 1962 with the demographics of the Middle Ages. Rosling's optimistic scenario was that Africa was on schedule and in a few decades would reach the United States in 1962. Rosling went on to dissect improving health from improving economics. The United States was money first, then health. China was health first and then money, so they are separate issues. The United States still has lower health than their economic status would suggest. The recent strenuous debate on health care looks bizarre to those who are not U.S. citizens. Rosling found that different religions appear not to slow respective demographic improvement. While he showed how economic and health issues were good means to achieve improvement, he notes that economics and health are not the primary goals. The most important aspiration is culture. With strong, benevolent, and responsible cultures, almost anything is possible. Rosling swallowed a sword onstage to show that the impossible is indeed possible with the right cultural guidance.

The National Science Foundation (NSF) effort called the National Ecological Observatory Network (NEON) is trying to take big data seriously, but it will take complexity scientists who know some ecology to achieve even modest results. On its website, NSF says:

> The National Ecological Observatory Network (NEON) is a continental scale research instrument consisting of geographically distributed infrastructure, networked via cybertechnology into an integrated research platform for regional to continental scale ecological research. Cutting-edge sensor networks, instrumentation, experimental infrastructure, natural history archive facilities, and remote sensing will be linked via the internet to computational, analytical, and modeling capabilities to create NEON's integrated infrastructure.

The struggle will be away from reductionist tendencies of the discipline, and we can expect resistance. Reductionist instincts, for instance of John Harper, are still very influential, to the point of blunting progress:[6] "It is only from the work of many individuals working scattered over a variety of parts of the world, but concentrating their attention over long periods on the behaviour of individual plants, that development of ecology as a generalizing and predictive science may be possible" (Harper 1982).

NEON is the child of the Long-Term Ecological Research Program (LTER), and that makes it the grandchild of the International Biological Program (IBP) of the 1970s. From IBP, we learned in the most painful way that large-scale reductionism is not holistic, and that models need to be built always for a focused purpose, even if the issue is large.[7] The reductionist binge since then might indicate that those lessons have been forgotten. Normal science is slow to make fundamental shifts. For instance, the fanfare about genomics was at first empty. The preferred devices were arrays that could identify which of a large number of genes were turned on, as it were. Those that lit up were then subjected to the normal reductionist isolation, which had very little to do with the whole genome. So it will be in ecology, as the first attempts at big data are made. We might look to business models to get a leg up in ecology because those practitioners do know about complexity theory. Hans Rosling has much to teach ecology about big data and visionary thinking.

CYCLES AND THEIR POINTS OF CONTACT

If ecologists are to pull the loose ends of their discipline together, then they can seek natural structure, structure that is predictive and reliable. By natural, we do not mean somehow privileged in nature, although in the end we may see it as such. By natural, we mean patterns that coincide in many dimensions of variables. This is another way of saying that science might seek that which persists when measured in different ways, or viewed on several criteria. The several criteria each reveal a new aspect of the structure. Since the different criteria for observation and their respective emergent properties generally cannot be strictly connected, we can only pull them together with narrative. There is inconsistency, but the ecologist is still talking about the same thing. Look at a forest with photons and you get one boundary, which is where you literally see the forest. But if you "look" with a thermometer, you will "see" that temperature goes up as you enter the forest at night, and down when you enter the forest in the day. But the place where there is a change in temperature will not exactly correspond to where you literally see the edge of the forest to be. It may be close, but it is unlikely to be an exact correspondence.

Each new emergent property or phenomenon requires new explanatory principles. In terms of surfaces, predictive power is more general for entities defined inside surfaces that have high dimensionality. Surfaces are places where there is attenuation of one or more processes. The surface operates as a limit. Natural surfaces are where there is a coincident attenuation of many processes. Inside natural surfaces there is a closure of processes, a containment of cycles of behavior.[8] Sometimes the cycle might be a life cycle, in which various stages take their turn. Note that the tadpole in the life cycle is itself composed of a cycle of anabolism and catabolism, which itself can be broken down into smaller cyclical biochemical pathways. The cycle of interest could be a successional cycle or a nutrient cycle. Structure can be equated with the closure of cycles, like blood circulation or the cycling of the Krebs cycle.

We have discussed the importance of ecological entities, like trees, that are players in populations, communities, ecosystems, and landscapes. Remember that populations, communities, ecosystems, and landscapes can be considered as the embodiment of cycles. Actors, like trees, in more than one ecological structure occur in more than one ecological cycle. Trees play in many cycles. It is those structures in more than one game that embody the places where the cyclical processes that pertain to several criteria come together. Structures are the places where the various cycles of nature kiss. Those will be the instruments for generality that are central to the scheme that can pull ecology into a cohesive whole.

Much is already known about the behavior and evolution of cycles. Manfred Eigen has built a body of theory about cycles that identifies how upper-level structure emerges.[9] The new level results from a cycle of behavior that locks together a set of lower-level cycles. His ideas are related to Prigogine's notions of the emergence of upper-level structures through the amplification of fluctuations. In Prigogine's dissipative structures, when a system is away from equilibrium, a random fluctuation starts a set of coordinated functions. There are positive feedbacks driven by the gradient created when something is held away from equilibrium. The gradient is driven by the second law of thermodynamics as a series of most probable events occur. There is a strong tendency to move down the gradient. In the definition of the second law of thermodynamics in physics, moving down the gradient is what increases entropy in a closed system. Biological systems are so open that the biologist cannot get away with an assumption of closure. Inputs, perhaps food, keep the system away from equilibrium; inputs stoke the internal gradient at its top end. A biosocial structure is a set of negative feedbacks encountered by the expanding influence of the positive feedbacks driven by the gradient. Schneider and Kay offer a definition of the second law that does apply to biology and does not demand assumptions of closure. They say that if a system is pushed away from equilibrium, it will use whatever it has available to it to resist being pushed further away from equilibrium. Life is one of those things that are available.

The details of an emergent, new nonequilibrium system are very sensitive to initial conditions. For example, the form of a mass of crystals depends on the exact form of the seed crystal, and on when and where it was put into the supersaturated solution. With water in cold air, snowflakes appear, but no two are the same because the irregular detail of each seed particle is different. The importance of the unique initial conditions of Prigogine corresponds to the critical phase in the formation of Eigen's cycles.[10] The processes of the cycles are unlikely to yield to a strictly chaotic equation, but chaos theory is very useful in giving the minimal condition of infinite sensitivity. We should not be surprised about the ubiquitous presence of sensitivity to initial conditions. An accident may set up the first upper-level cycle, but once it has emerged, the new cycle outcompetes all other potential cycles. The first cycle to emerge at a given level gives the form of the large-scale, long-term dynamics thereafter, exactly as does the seed crystal in Prigoginian system reorganization. The seed crystal imprints itself on subsequent patterns of crystal formation. The seed crystal grows through cycles leading to patterned growth, which preempts the resources of other patterns of crystal growth. By the time the seed crystal has finished growing, there is not enough solute left to form any other crystalline pattern.

An example of the first cycle out-competing all later comers is the way that microbial life preempts the resources of any other system that otherwise might be the origins of a different form of life. As a result, all life that we know uses only a limited set of fundamental compounds. Life uses predominantly only one isomer of several major classes of molecule, namely amino acids and sugars. There is no mechanical thermodynamic reason for using one isomer over another. The isomers in life are historical accidents. There is an echo of early crystalline structure at the base of life. One cycle of organic reaction in the primeval soup locally out-competed another, thus fixing that cycle over the other. For example, the cycle of ATP to ADP and back again is what fuels all living systems. However, GTP could do the same job, and indeed it is part of the fueling system for protein synthesis. The "A" and the "G" above refer to adenosine and guanine. The cycle is from a phosphate chain of length two to a chain of length three, and back again: hence, diphosphate and triphosphate. At some stage, adenosine gained the ascendancy, and that fixed the rules for the future.

There are two sorts of memory: one is coded, for example, in DNA; the other is inertial. The two sorts of memory come together to create precedents in biology in cycles. A whirlpool remembers to keep turning the same way because of inertia of the gyre. In life, otherwise unstable emergence is coded so that it persists. As a result, in life there are levels of emergence that would normally be too close to each other in terms of scale to survive. We would expect them to compete and disappear. But the coding in life in symbolic terms of DNA, or hormones, or mating dances stabilizes separate degrees of emergence. As a result, patterns of emergence are stacked densely at closely adjacent levels. Living and social hierarchies are dense. The density of levels indicates a close stacking of cycles. First, there is a thermodynamic situation where some component of life lucks into being the most common. The memory for that pattern is inertial. It would give way in a sort of molecular mud wrestling match, except for coded memory that stabilizes the situation. Once ADP and ATP gain the ascendancy, they become coded in some sort of primitive genetics. The cycle then persists through all of life that follows.

In Eigen's scheme, hierarchies of cycles build ever higher, each stage depending on an accident outcompeting all later accidents. This ascendancy fixes a stable hierarchical structure. The mechanism of the competition is simple. Various separate structures become involved, each as one component in a cycle of behavior. There are other potential possible cycles. They could emerge, but they generally do not because the first cycle is already extant, and therefore uses its components with an efficiency that cannot be matched by an incipient cycle. The surprise is the survival of GTP as the energetic driver of protein synthesis. We do not know why ATP did not take over that job, but perhaps GTP got in first as that local environment appeared in the cell, and persists because in that special place it has precedence as the first cycle. In general, subsequent cycles never get started. Thus, in Eigen's scheme, a lower-level structure is held in just one upper-level cycle. Each component relates only to the next member in the chain. The track of the dominant cycle does not allow any other relationships to become as important.

At first sight, we would appear to be at odds with Eigen's evolution of cycles in our assignment of individual structures, say a tree, to several cycles under different ecological criteria. Each criterion implies some different separate cycle. In our scheme, the tree is part of a cycle of succession in the community. It is also a part of the carbon and nitrogen cycles under the ecosystem criterion. The cycles of the landscape are less obviously cyclical, but cycles are there all the same. In Georgia, the landscape is now in its fourth cycle from a forest to a field and back again. The coincidence of these cycles in involving a given tree is entirely intuitively reasonable, so we must be talking about something different from Eigen's primary cycles gaining hegemony, and indeed we are.

The competition between Eigen's cycles whereby one reaction outcompetes another only involves one criterion for both competitors, by our standards. Biochemical cycles work under the same set of biochemical constraints, really under the same criterion. By contrast, the cycle of succession compared to the nitrogen cycle operates in such different terms that there is no basis upon which they can compete. So ecosystem nutrient cycles do not readily compete with successional cycles. At the end of the nineteenth century, when the community lost a major stump-sprouting species, the American chestnut, the ecosystem nutrient cycling barely changed at all.[11]

Thus, we envisage a set of cycles that are free to share components because each cycle uses the entity as a part in such different ways that there is no competition. Once evolution has created a tree with its role in succession, the same death event in succession also releases the nitrogen in that dead tree for an ecosystemic cycle. It is the same material, doing the same thing, whether we view the death of a tree as an ecosystem event or a community event. The different cycles are only conceptually different, not physically different. Two biochemical cycles might indeed compete for a given molecule. Community and ecosystem do not compete. Quite to the contrary of competing, they coexist one with the other, and so form an accommodation to each other over time. At first, a commensurate ecosystem and community cycle might be independent of each other, but later they will form a structure of mutual reinforcement. The pin cherry is selected for quick establishment after deforestation, a community consideration. But in the long run, forests will hold on to their nutrients better in the ecosystem arena if pin cherry is selected as the critical pioneer.[12] Pin cherry can move succession and nutrient cycles forward, without competition between the cycles.

This discussion highlights what is missing in a conventional ecology divided along subdisciplinary lines. The conventional way to study ecology is to focus on one, maybe two, and hardly ever three criteria for those types of structures. It is reasonable to expect that there are critical structures that hold ecological structures together. These structures, mostly unidentified as yet in ecology, are the points of articulation inside a unified ecology. Ecologists, for the most part, study their material using only one of the types of cycles. The good stories in ecology turn on unsuspected points of articulation.

In our unified conception, lower-level ecological structures are shared by cycles of a very different nature, but all of them are ecological cycles. Imagine a given ecological structure, like a tree, as a bar magnet (figure 10.3). There is a magnetic

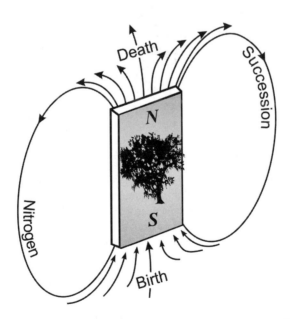

FIGURE 10.3.   A metaphor that sees a tree as a bar magnet. The lines of magnetic force correspond to the many ways and paths by which a tree is recycled: reproduction; seedling to death; as a mineral-recycling device; replacement in succession; and many other cycles.

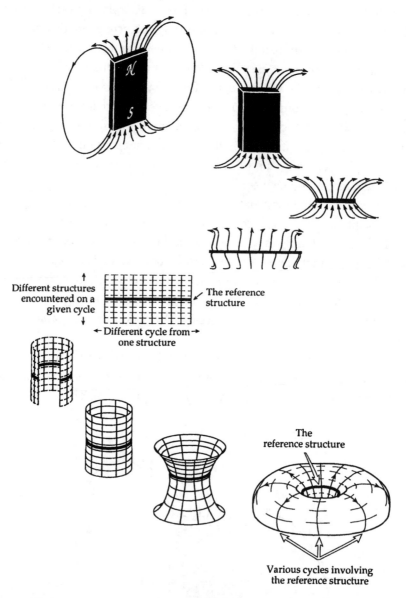

FIGURE 10.4.   An extension of figure 10.3, whereby the bar magnet is topologically transformed. If the bar magnet is the tree and the magnet has been transformed into the center of the torus, the thing that is the tree becomes the hole in a doughnut. This brings the cycles to the fore and places the structure of the tree in the background.

field of cycles around it in the analogy, each cycle being a line of magnetic force cycling from the North to the South Pole. An organism can be a part of many ecological cycles, each based on a different criterion. Furthermore, each of these cycles passes through many other structures, each one of them with its own field of cycles. The tree in the nitrogen cycle releases minerals in leaf drop. Before the local cycle returns the nitrogen to the roots, nitrogen has passed through many other structures. Some of them might be organismal, like earthworms, and others might involve whole guilds of species, as in the rhizosphere (see chapter 1, figure 1.15). At any given stage in this thought excursion, we could change criteria in one of the intermediate structures. At that point, we could head off into an evolutionary or genetic cycle, perhaps via one of the fungi in the rhizosphere. The physical material of which ecological systems are made does exactly this. It is not committed to being in a successional entity, or any other type of ecological structure. A carbon atom in a successional entity like a tree trunk could either pass next into the atmospheric part of an ecosystem or become incorporated in the DNA of a millipede.

It is important to recognize that it is not structures in the ecological cycles that are the focus of attention. They are only the devices that we need to follow the lines of connection in the ecological cycling of material. Perhaps the analogy of a magnet is inappropriate because it puts emphasis on the solid magnets instead of the lines of connection. More appropriate would be a torus, because there the hole in the middle is the ecological structure of reference (figure 10.5). Touching tori would be even better, so that the thought experiment can circle around from one torus to the next. In this sense, ecology can be imagined as a box of bagels or doughnuts, white with powdered sugar, which goes everywhere(figure 10.6). We would not normally point this out, particularly given the new fusion of cuisines around the world in the last two decades, except that a British reviewer of

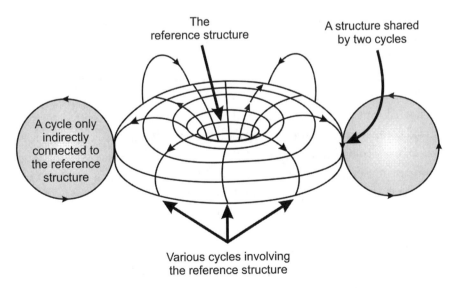

FIGURE 10.5.   Showing the torus at the end of figure 10.4 to emphasize that adjacent structures (perhaps other trees or animals) also have their cycles that will kiss the cycles of the torus for our tree.

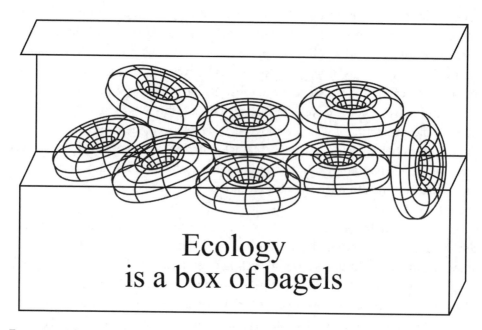

Ecology
is a box of bagels

FIGURE 10.6.   Showing that material in ecology cycles in lots of complicated ways, such that carbon in a tree can be moved to the DNA of a millipede, whereupon it becomes locked in the genetic cycle of a different sort of organism. The bagels have relationships, and are not just a jumble of cycles.

our first edition referred to "bangels," not bagels. He must have thought it was a typographic error.

The value of this approach is that suddenly our ecological concepts will be working for us, instead of us working for them. We are free to use an ecosystemic nutrient cycle as long as it serves our purpose. By emphasizing the arbitrary nature of the criteria that ecologists erect to further their ends, we can use the criteria and structures without being fettered by them. Only when ecologists have a clear distinction between what is physically necessary and what follows from conceptual devices can they avoid confusion and merely semantic argument. We can build large, powerful, and predictive ecological structures at a level that subsumes ecosystems, populations, organisms, or communities in any mixture of our choosing. In chapter 8, we are at pains to show how moving around cycles develops the narrative.

There is already an example of the approach we recommend, a study of prairies that involves three separate cycles focused on one ecological entity (figure 10.7). The tangible entity is big bluestem, the grass that dominated the American tall grass prairies before the coming of the Euro-Americans. Independent of our conception of coincident ecological cycles, David Wedin recognizes three cycles pertaining to three different system criteria, all focused on the tall grass prairie-dominant species.[13]

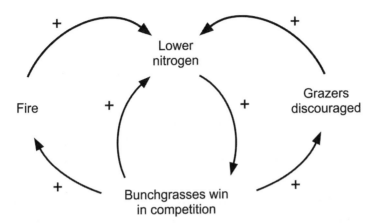

FIGURE 10.7.   The positive feedback cycles to drive the success of bunchgrasses, or might equally well drive the takeover of the site by Eurasian weeds after an agricultural interlude. The plus signs on the arrows mean that an increase in the donor compartment causes an increase in the recipient compartment. Decrease would cause decrease because of the plus sign on the arrows. Plus means "do the same as."

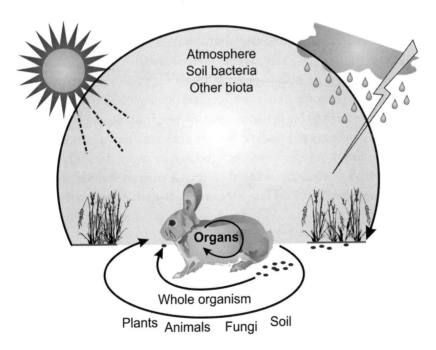

FIGURE 10.8.   The cycling of nitrogen occurs at many scales, from those inside an organism up to the scale of nitrifying and denitrifying bacteria interacting with biologically available nitrogen formed by electrical discharges in the atmosphere.

THE DIFFICULTY OF SEEKING ECOLOGICAL
SIZE OF NATURE

We never escape the measurement problem. Therefore, we cannot know what nature is, as opposed to what is being caused by observation itself. We do not say that nature itself has a certain scale or that simple measurement will reveal that scale. A scale of nature is meaningless if we cannot know fundamental units to which the scale applies; in fact, the concept of a unit at all is an arbitrary decision that the human mind needs as a crutch.

Photons may indeed be such fundamental units; perhaps we have to acknowledge the speed of light as a scale of nature. However, in ecology, we do not have anything approaching unequivocally fundamental units, not even in organisms, units of selection, ramets, or any other ecological entity. The scale of nitrogen recycling is not one scale, but a series of cycles of different scope and speed: inside organisms; a rabbit eating its own feces; feces into soil and back to plant food; cycling into and out of the atmosphere (figure 10.8).

SCALING INSIDE A CRITERION

Robert Rosen has put together some general principles of scaling, along with notions of similitude that should be helpful in relating different ecological situations.[14] He uses the van der Waals gas equation to advance the argument. Although, on the face of it, gas laws would appear to have little to do with ecology, Rosen is able to translate the conceptual devices for gases into biologically meaningful analogies. Gases have analogs for genomes, phenotypes, mutations, polygenes, pleiotropy, and environments once one reaches the powerful level of abstraction Rosen employs. Accordingly, we ask the reader's indulgence as we lay out Rosen's case.

The van der Waals equation is based on the equation of simple pressure, temperature, and volume equations for gases, the ideal gas law:

$$pV = rT.$$

The van der Waals equation is:

$$(p + a/V^2)(V\text{-}b) = rT.$$

The issue is that p is the pressure on the walls of the container, not the pressure on the particles, which is what we need. And V is not the volume we need. When we compress a gas, the space between the particles gets smaller, but the particles do not. The van der Waals equation has two new parameters, which work to respectively translate p and V into values we need. The quantity *a* pertains to the attraction between gas particles. That attraction increases the pressure on the particles. This is mediated by the volume and so we adjust p by adding to it a/V^2. That gives us the right p. To get the V we need, it is necessary to subtract the volume *b,* which relates to the volumes of the particles.

With little pressure, the ideal gas law is predictive, even though it is uncorrected for volume of the particles. Under pressures close to that sufficient to cause liquefaction, the volume of the gas particles themselves is a significant proportion of the measured volume. The proximity of the particles governs critical changes of state; therefore, close to liquefaction the volume of the particles themselves, *b,* is an essential part of a predictive equation. We need V-*b.* The ideal gas law already has a term associated with the temperature T. That parameter is *r* in the *r*T of the ideal gas law.

Rosen makes an elegant translation of the van der Waals equation to give it intuitive meaning for ecologists. He uses the conventional biological paradigm. The prevailing biological paradigm views biological form, or phenotype, as the result of the interaction between the genome of the organism and the environment in which it develops. Genome and environment are the primitives in the discourse (note that this is not so in Gaia). Life is seen conventionally as set in the inexorable physical environment in the prevailing Darwinist view. Mutation is viewed as a rare event that affects only a small part of the genome. Transfer of DNA between organisms is viewed as a special case, that is sex, or as a pathological departure from the reference conditions, viruses. The genetic material in the organism is taken to be fixed for most purposes.

In Rosen's scheme, each gas is a special case because, depending on the gas, it takes different pressures to cause liquefaction at a given temperature. The terms *a, b,* and *r* are different for each gas. They identify a given gas and make it distinct from all others. Rosen points out that these three parameters of the van der Waals equation are the genome of the gas that determines its response to its environment.

Now consider the basic variables in biological systems. They are the environment, on the one hand, and the phenotype, on the other. Of the three critical gas variables, pressure and temperature refer to the ambient conditions with which the gas is equilibrating. Accordingly, they are environmental variables. Volume is something the gas manifests once the environmental variables are given. So it follows that volume is the phenotype of the gas under the environmental conditions in question.

Note the three V terms in the equation, two in the V^2 and one more in the V of the volume posited in the van der Waals equation. Rosen points out: "The behavior of a gas at equilibrium depends on whether this cubic term has one or three real roots. The gas/liquid phase transition resides precisely here. [So does the transition to solid ice.]"

In geometric terms, phase transition between gas and liquid corresponds to a fold in the surface that relates volume to temperature and pressure. The fold is

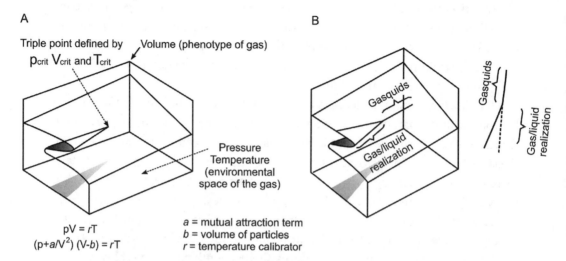

A.

Triple point defined by   Volume (phenotype of gas)
P$_{crit}$ V$_{crit}$ and T$_{crit}$

Pressure
Temperature
(environmental
space of the gas)

pV = rT
(p+a/V^2) (V-b) = rT

a = mutual attraction term
b = volume of particles
r = temperature calibrator

B.

Gasquids

Gas/liquid
realization

Gasquids

Gas/liquid
realization

FIGURE 10.9.   A. Rosen's schematic representation of the van der Waals response surface. The bottom plane is the environmental space of pressure against temperature. The space above comes from the phenotype of the gas, its volume. The pleat is where pressure and temperature are such that there is liquefaction or gasification. Behind the pleat, temperature and pressures are so great that water exists as a gasquid, a compressible liquid, as in deep ocean vents. The base of the pleat is the triple point where the equation has three real roots. It is the nub of the instability of changing state in a phase change. B. The schematic showing the line through the gasquid space where neither gas nor liquid can be realized separately. Beyond the triple point, that line bifurcates into liquid or gaseous manifestations of water. The side figure to the right shows how this all can be seen in hierarchical terms as a main stem splits.

positioned so that it corresponds to the range of pressures and temperatures at which liquefaction occurs (figure 10.9). The fold is what Thom has called a catastrophe cusp. As the environment moves across the region of the edge of the cusp, the system becomes unstable, and reorganizes by either liquefaction or volatilization into the other phase.

A model for the gaseous phase has nothing to do with the liquid phase. The behavior of gas molecules is almost independent, whereas in liquids the molecules are highly correlated in their behavior. When a gas becomes liquid, the constraints on the individual particles change. The new rules for being a liquid appear "as if from nowhere," Rosen says. We are talking here not of a continuous change that can be mapped dynamically. Rather, the change is to a new structure, and it is instantaneous. When a liquid becomes a gas, the constraints that hold particles in the highly correlated liquid state are broken. There is nothing about the high correlation of particles in a liquid that tells anything about this new, uncorrelated gaseous state, for the constraints that gave the old degree of correlation are instantaneously irrelevant. That is why we need a completely new model for liquids that has nothing to do with the model for gases.

Consider again the pressure, temperature, and volume space for a gas. There is the gaseous region and a liquid region, with transitional regions in between. Within a given region of the behavior space, behavior is continuous, such that a change in temperature and pressure always makes a change in volume. A small rescaling of the behavioral axis could accommodate this change in volume; we could rescale the units on the volume axis and remove the difference resulting from the small change in environmental variables. Why would one do that? It is because one cannot do that in certain special, interesting cases of small differences. A similar change in environmental variables that goes across into another region of the surface defined by the fold will give a radically different state on the behavioral axis. It is so different that a rescaling of the behavioral axis could not accommodate it. The shift is discontinuous. This is because, with a change in region giving a change of phase (liquid versus gas), there are new constraints in place. Therefore, the underlying rules of behavior are different. A simple rescaling can accommodate a small change, but only within a region because the same basic rules of behavior underlie what is found. In this sense, the entire gas phase is similar. The same applies within the liquid phase (figure 10.9A). But liquid and gas are not just different; they are fundamentally dissimilar.

Rosen goes further with his analysis of the van der Waals equation. There is a singular point where the cubic relation specified by the equation has three coincident real roots (otherwise called the "triple point"). It is at the base of the pleat in the behavior space (figure 10.9B). For each gas, that triple point is singular and different across all types of gas. The triple point for water corresponds to the one pressure, temperature, and volume point that lies adjacent to the gas, liquid, and solid phase. Behind the triple point that is on the flat surface behind the pleat, there is no clear distinction between gasses and liquids. That is the place of gasquids, compressible liquids. They pertain at the very high pressures and temperatures on deep ocean vents. The peculiar physics of water there has been argued as crucial for life origins.

According to Marc Brakken's master's thesis, there is something fundamental to hierarchies going on at the triple point. Observation along a single line in the gasquid phase cannot tell liquid water from gaseous water. On that line, neither gas nor liquid can be realized. But at the triple point, that line splits with one branch going to the liquid phase and the other going to the gaseous phase. Either gas or liquid as discrete considerations are realized (indeed, this is the same realization mentioned in chapter 8). A unified version of water bifurcates into liquid water and vapor with the move past the triple point (figure 10.8.B). We can see liquid water and water vapor as stems coming from a gasquid unity. In a sense, the gasquid is the essence of water. Mostly, we deal with one stem or another. Right at the triple point you see all the options—liquid water, vapor, and ice—but cannot realize any one of them. All this smacks of the slip-sliding with which models have problems that are no trouble for narratives.

The location of the critical triple point in the p, V, T space is determined by the genome a, b, r. The relationship to the critical points of the triple point is:

$$P_{critical} = a/27b^2; \; V_{critical} = 3b; \; T_{critical} = 8a/27rb.$$

It is possible to express the behavior of the gas in the three-dimensional space of p, V, and T, not on directly scaled axes of pressure, volume, or temperature, but in relation to the critical triple point. Thus, pressure, which might be expressed in pounds per square inch, can be expressed alternatively by a variable derived from $p/p_{critical}$. The values p and $p_{critical}$ cancel out to give a dimensionless number. This is a rescaling of the variables relative to the triple point. A given pressure, temperature, and volume will thus give different values for two different gases because the relativization terms $P_{critical}$, $V_{critical}$, and $T_{critical}$ are different for each gas. Such a rescaling of the behavior space allows direct comparison of the behavior of all gases (figure 10.10). We have rescaled the individual behavior spaces of all gases into comparable terms. Accordingly, it is possible to plot different gases as points in the genome space and then plot their dimensionless states as cross-sections of corresponding states above the genome space.

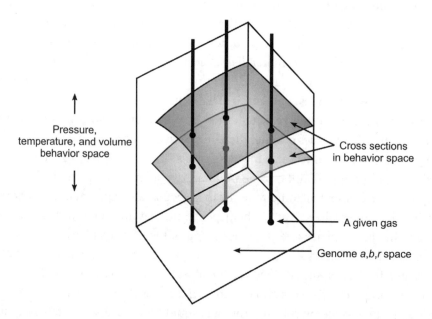

FIGURE 10.10.   An expression of all gasses, each one in its own position on an a, b, r plane. The space above is in itself three-dimensional in that the vertical lines are the pressures, temperatures, and volumes normalized relative to the respective triple points of the respective gasses. The figure is thus a six-dimensional expression, three for the plane below and three for the space above.

Since we can only draw in three dimensions, the three-dimensional space of the genomes is represented as only the two-dimensional plane on the bottom of the box. The three-dimensional variable space of rescaled pressure, temperature, and volume has to be collapsed into a single 3-D fiber above the point for the gas on the a, b, r plane. But all this is purely a matter of representation. Figure 10.10 is a compromised representation of a six-dimensional space, called a "fiber space." Note that the cross-sections of the genomic space corresponding to each behavioral state are not folded. This is because the folds in their state spaces have been normalized away in the relativization to their respective triple points at the base of each gas's respective fold. Remember, at the triple point you can see all the options, so all the options can be expressed in equivalent terms from that point. That is how the instability that creates a new phase of water can be normalized away.

Being able to express the behaviors without the folds in figure 10.10 is important because it makes all parts of the entire behavior space similar (as opposed to dissimilar across the fold). Remember how the discontinuity of the catastrophe cusp meant that all gaseous states were dissimilar to all liquid states, so that the change to or from a liquid phase could not be accommodated by a scale change. Having found a way to express the behaviors in terms of the genome of the gas without the manifestation of the instability around the cusp, a comparison between gases is now straightforward. This means that all gasses in Rosen's treatment are only mutants of each other, the mutations being in the a, the b, and the r. The gasses all can be cast in equivalent terms at this higher level of analysis. This is so even though at a lower level of analysis, in unnormalized terms, the individual gasses still show instability in between the liquid and gaseous phase because of their respective folds.

The whole discussion here focuses on problems of what it means to be similar despite differences. To summarize the above ideas, Rosen says, "Similarity means precisely that we can completely compensate for a change of control by means of a scale change (a coordinate transformation) of behavior alone," which would not be possible if the behavioral surfaces were folded.

Ecological concepts do not have a scale dependence of ecological actors. When the scale changes have been made, competition between bacteria can be compared to competition between whales. Of course there are differences between bacterial and balanid competition, but there had better be some relationships that translate between the two cases; otherwise, the term and the concept of competition are vacuous. It is no accident that terms like competition can be formulated into fairly straightforward algebraic statements, although there is fighting between the schools of students of competition as to what the exact formula should contain. As long as there is only a scale change, then the systems described are similar—different, perhaps, but still similar. Dissimilar situations, by contrast, involve changing the rules that constrain.

All this highlights the importance of having strict definitions of the major types of entities that ecologists study. This is so obvious as to be banal, except that by now it should be clear that even subtle differences in the nuance associated with a term can have radical consequences for scaling and what is meant to constrain what. Even slight changes of specification can introduce catastrophic consequences, in both human and strictly mathematical senses. A small change that brings with it a catastrophe cusp in the state space can make the situation dissimilar to what was specified before.

Differences in details of descriptions are one thing, for they can be accommodated by rescaling the system, but dissimilar situations are much more problematic. We need to pay enough attention to ecological descriptions so that we can tell the differences between ecological systems only rescaled as opposed to ecologically dissimilar situations. It should be clear by now that telling that difference is not easy.

Early in the twentieth century, well before the current interest in scaling biology, D'Arcy Thompson was explicit about the scaling of relationships in biology.[15] He suggested that closely related species are similar. Note that the closeness of the relationship is a matter of genomes, while the similarity is a matter of phenotypes, so the statement is not trivial. As we see in the ensuing discussion, Thompson's model is not always correct. He outlined a standard fish form on a grid, and then distorted the grid to transform the fish into a new form, an angelfish. He did the same for the face of a monkey and that of a human (figure10.11). He formally called these ideas the "principle of transformations." We now stretch these until they snap. Major ideas, like D'Arcy Thompson's principles, are most instructive when they fail. That is one of the ironies of science.

A stable space defines similar situations. In unstable parameter spaces with folds, the system passes into a new similarity class and manifests dissimilarity with the old behavior. Thus, two closely related species can be dissimilar. It is here that D'Arcy Thompson's principle of transformations fails. This would be because the small genetic metric distance between the species puts them into different classes of similarity; the change, which we could call a mutation, happens to cross the fold. Thus, some mutations may create a still closely related species, but one that is dissimilar. Inside a stable space, or inside a region, mutations generate similar species. Critical mutations fold the space to generate dissimilar species.

The human genome is very close to the ape genome, but a case can be made that we and apes are dissimilar. That dissimilarity can be profitably seen as coming from a rescaling of our development. Human babies are not simply inept relative to other animal babies; they are spectacularly hopeless at doing anything until three months old. We argue that the delay in an ability to act on the world gives time for baby cognition to come on line with protomathematics (more, less, many; one, two, many) and set theory (these things are equivalent). The difficulty for monkeys is that they learn about the world constructively before they can challenge the meaning of their understanding. Monkeys are tyrannized by reality.[16] We, however, discover a thesis and immediately erect an antithesis. We are then in a position to

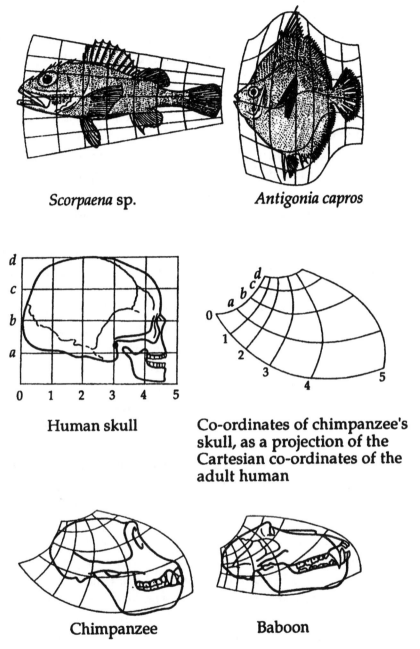

Scorpaena sp.

Antigonia capros

Human skull

Co-ordinates of chimpanzee's skull, as a projection of the Cartesian co-ordinates of the adult human

Chimpanzee

Baboon

FIGURE 10.11. A figure from D'Arcy Thompson, showing how he saw related species or genera as being simple rescalings of each other. As species become less related to each other, there is a successive rescaling. He is right until there is a fold in the rescaling, whereupon small differences in genes can lead to dissimilar morphology. Some mutations are critical; others are simply accumulated movement on the atomic clock.

move on to synthesis. Stand an object up (thesis). Knock it down (antithesis). Try to stand it on its corner (synthesis). The synthesis becomes a new hypothesis, and the baby is ready to go around again to reach a yet higher level. At more than three months old, human infants are little scientists, checking and rejecting hypotheses. Other primates get stuck on their original thesis, and so are dissimilar to humans.

This line of argument can be applied to the vociferous debates within ecological subdisciplines. Sometimes such discussions can degenerate into assertions about this or that definition being correct. More enlightened altercations are less concerned with the true definition of, say, competition, and focus on the meaning of choosing one definition as opposed to another. One might expect various dialects in competition theory to be only matters of style within a class of similar conceptions of competition. But the differences can be important, as when John Harper inserted the term interference in the competition debate. Harper could then dissect competition in new ways. His distinction had developed a new logical type in the domain of competition. In such cases, a simple rescaling can translate one dialect of competition into another by the inclusion or exclusion of some small conceptual device, or by changing the measurement of competition. The differences can be transformed away. Those speaking closely related dialects should form schools of thought. A trick here, when one finds oneself or someone else wrong, is to say, "All right, that is wrong, but what were they trying to say?" And then we tell a story.

All of this is a matter of degree. Rosen points out that while stability in his sense is all or nothing, there are degrees of bifurcation, that is, degrees of stability, degrees of dissimilarity. All this is reflected in the standard biological taxonomic order, where there is more similarity within a species than within a genus or a family. There are degrees of dissimilarity between different communities, or between different ecosystems, and all levels of difference are worthy of investigation. A less formal way to use Rosen's renormalization is to think of prairies in Colorado. Basically, all prairies in that state are short-grass.[17] All Colorado prairies are analog models of each other, as in the Rosen modeling relation. But at the Nebraska line near Julesburg, the vegetation changes color from blue-green to green and is taller. In crossing the line, there has been a change to mixed-grass prairie. The concept of short-grass prairie has gone unstable. The surface has folded. Short-grass prairies do not burn much because of insufficient biomass. Historically, vegetation was removed by buffalo. But if we relax the constraints for removal to be either bison or fire, we have normalized the difference away, and can keep modeling prairie all the way to the prairie-forest border region where Gleason grew up.

In an example from Bruce Milne, consider Tilman's notions of critical levels of resource determining who wins competition. Tilman has a term R*, which is the critical level for a resource, the lowest concentration that the species in question can take the resource and still break even. This is the argument we introduce earlier in the chapter to describe Wedin's work in figure 10.7. Again, we note that

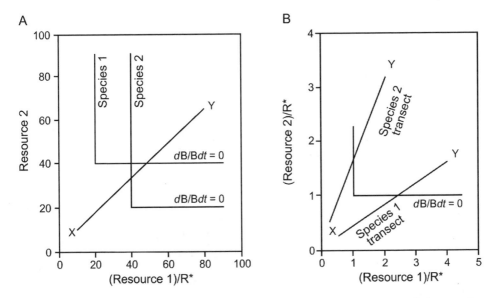

FIGURE 10.12.   *A.* Two species' breakeven resource levels (positive growth limits) plotted on a hypothetical two-dimensional resource space. A transect from resource level X to Y indicates that species 2 can grow closer to X than species 1 can. Even so, the relationship of the two species to the transect is less than transparent. *B.* The relationship of the two species to the XY transect in diagram *A* shown directly by plotting the resource levels on the transect relative to the breakeven resource levels (R*) of each species.

Grime has good arguments that challenge the concept of R*, but let us be tolerant once more. Two species might coexist because they each respectively corner one resource by taking it to a lower concentration than the respective other species can use. The two L-shaped limits show R* for the two species, with species 1 winning on resource 1, but species 2 winning on resource 2. In figure 10.12*A*, we can see a gradient, which is hard to interpret because it runs across both resources. Which species wins, and where, cannot be discerned. However, if we do a normalization of the space relative to each species' critical level of resource, we see the two breakeven points mapping onto each other. There are two spaces summarized as one relativized space. Note that the XY gradient splits into two, one each for the implications for the respective species (figure10.12*B*). Rosen's clarity of thought helped us here. It took the ambiguity out of two species reading an environment.

RESCALING ANIMAL FUNCTIONING FOR SIZE

In chapter 2, figure 2.13 and associated text, we look at landscapes and animal occupancy according to Holling's lump analysis. We are now in a position to explain the size of animals in light of the Rosen work on renormalization. Bruce Milne and

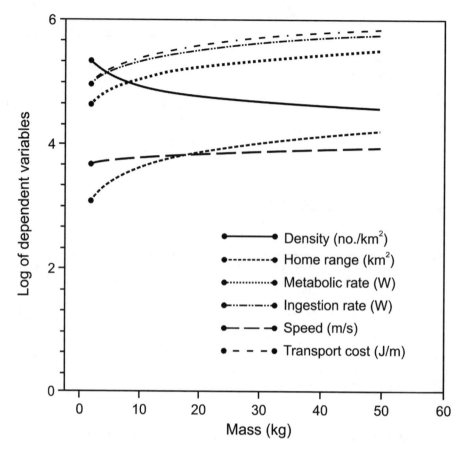

FIGURE 10.13. Allometric relationships used to regulate herbivory. The ordinate represents each of the following: density (Peters 1983:294), home range (Harestad and Bunnel 1979; cited by Peters 1983:296), metabolic rate (Stahl 1967), ingestion rate (Farlow 1976), speed (Buddenbrock 1934), and transportation cost (Taylor 1977). Notice that for increase in animal size, density decreases, while all other variables go up. Milne used these variables from the literature to calibrate the simulated animals in his model.

colleagues showed how differences in the size of animals can be rescaled to show equivalence with regard to size and resource use on landscapes.[18] Milne's work was a modeling experiment. Animals are scaled relative to mass with regard to several lifestyle characteristics (figure 10.13).

Bigger animals have a bigger home range. As in Holling's work, bigger animals move more mass and therefore have greater transportation costs, with a factor of increased speed of movement as part of that. Metabolic rate is also larger in larger animals, as is ingestion rate with quality of food as a factor in herbivores. Recent work by Warren Porter has made careful calculations with regard to body temperature and size.[19] His early work was with spherical and cylindrical animal approximations.[20] His new work is with shape and momentum included. His results show that big dinosaurs would have had to move slowly. They could not have moved as fast as they did in the movie *Jurassic Park*. Larger animals put greater demands on the local resource, so larger herbivores occur at lower densities on the landscape.

Milne took a real landscape remotely sensed and he filtered its image to give two states, one black and one white. The black state he asserted gave feeding areas. The feeding area pixels for year one (the year he used for the experiment) were scattered toward the northeast of the map, with almost no food pixels in the southwest corner. The other two years had more food pixels, as determined by the patterns captured in those years. He filtered the other two years of the landscape in the same fashion. He calibrated the movement, the density of their populations, and other resource-related factors for animals of four sizes between three and six kilograms. He had initial resource patterns for three different years. He parachuted the animals onto the resource area at random. These were theoretical animals and, therefore, they did not die when they ran out of sufficient resource; they simply moved at random to find a better spot. Each size of animal was simulated for each of the three starting conditions (three years). The critical simulation was what percentage of the animals of a given size were moving to find better spots over time.

The four animal sizes moved each year in different patterns. Initially, the general pattern was for a decrease in percentage of animals moving to a point, since increasing numbers of them had found enough food and did not have to move again. But animals accumulated on good spots and depleted the resources. At that point the percentage of animals moving increased, until all animals of all sizes were moving (figure 10.14). Milne was looking for those inflection points where animals changed from moving less over time to moving more. He could find these points unequivocally for all sizes of animal only in year one, so that is the year he used in his recalibration.

Milne's simulation told him the amount of joules in the feeding areas over time. The inset for figure 10.14*E* shows how animals of different sizes decreased the joules available over time. See how the different-sized animals behaved differently. But then Milne normalized the time relative to the time of the inflection points, where more animals started to move. He also normalized the joules available to the joules available for each animal at the inflection point. Rosen normalized against the triple point for his gasses. Milne normalized against the inflection points in his simulations. On the normalized graph (the main panel of figure 10.14*E*), it can be seen how all the animals used resources to the same degree over the same time, no matter what size they were. The critical points in his simulation rescaled the size differences out of his animals. This says that animals of different sizes are basically the same, except for size when you rescale for lifestyle parameters.

An interesting outcome was Milne's ability to see which size of animal feels more effect of normalization at what time. The smaller animals feel the rescaling of time less than the bigger animals, which feel the rescaling of energy more. This

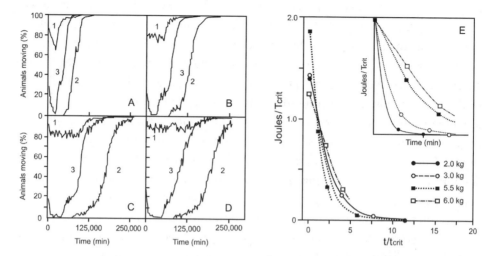

FIGURE 10.14.   Animals at first move less and less because they find the good spots. But these spots get eaten up, and more and more animals move until they are all moving on a depleted landscape. Panels *A–D* are for the four sizes of animals over a period of three years. Milne wanted the critical inflexion points between less and more movement, so he chose year 1 for all sizes of animals because in that year, there was a uniquely low number moving at a certain time. *E.* The inset panel plots the animals of different size on a time-to-energy space. The main panel shows what happens when the axes are normalized relative to the inflection points on *A–D.* The animals plot onto each other.

is like the result obtained by Porter and Gates (1969), as they scaled animals of different size in a space of air temperature and radiation input. From the outline of the climate space in which they could live (see chapter 5, figure 5.24), apparently smaller animals are more influenced by changes in air temperature, while bigger animals are more influenced by changes in wind speed. For Milne's animals, time is more critical for small animals. Energy limitation has a greater effect on larger animals that demand more resources.

CONCLUSION

In this chapter, we have looked at scaling in basic ecological science. There, the researcher floats the scale of the discourse so as to achieve convenient bounding of structures employed as a reference. For instance, ecosystem ecologists study whole watersheds, not fragments. But all this needs to consider the new theory in this book. Rescale to convenient boundaries, but consider many criteria instead of just one, such as ecosystem alone. We have shown in management how ecosystem and landscape need to be seen in basic scientific terms, but in parallel, as an accommodation. Our objection to mainstream population biology counting individuals via birth can be resolved by twinning populations and the energetics of the ecosystem flux approach. Of course, the energetics are different from the treatment of energy in ecosystems. Population biologists did not pick up on the full implications of Lotka denying privilege to birth because they are not used to energetic considerations. This chapter is an invitation to all types of ecologists to look toward the methods and postures of other criteria for their basic research.

Another idea is switching the object for its ground. We took the notion of an ecological structure and coupled it with processes and cycles that emanate from the structure. By morphing the shape into a torus of process, we were able to put the structure in the background, making it the hole, not the substance of the torus. This sort of reversal can be powerful, as when we identify that bilateral symmetry is best explained as the loss of symmetry in two polar dimensions. The separate identity of structures can get in the way of conceptualization. If we look at ecology as a set of kissing tori, it becomes easy to move between criteria in a measured way. A carbon atom moving through a material ecological system can move on in community succession, in ecosystem cycling, and through to genetic material in genetic recycling. It takes bold new conceptions to move away from stamp-collected ecological structures. The idea of fuzzy sets also helps to break down concrete structure that might otherwise get in the way.

Hans Rosling has shown us that we have plenty of data for most of our questions. Part of our emphasis on abstractions has been to distinguish between data problems in material issues and misconceptions in a semantic space. Before we look to new materiality, ecology would be advised to look creatively at the data we already have and conceive its implications in new ways that may be the solution. Unfortunately the granting process consistently favors new data that only make the confusion worse. In the end, creative use of extant data can lead us to things for which we have no data. In light of a data-driven background, first, we can first know that we do not have these data, and, second, we can understand that the new measurement is critical. Only do the heavy lifting with measurement if you know you really do need that number. Scientific measurement and control are so expensive; ecology simply cannot afford an unfocused search through a poorly defined data space. Students might be allowed to choose some natural historical setting, but experienced mentors should choose the discourse. Students entering graduate school often have a good natural historical knowledge, but they do not know enough theory to choose a critical approach to the natural history of their choice.

In particular, we have focused on Rosen's powerful notions on dissimilarity as opposed to mere difference. We did show how to use his methods in ecological research, but there is a useful general principle. Most ecological basic research calibrates local spaces so that they show mere differences. But if you want to find something new and important, take an idea and rescale it until it breaks. Many of the big new ideas were found that way. For instance, the notion of food chains is helpful and has a long history in ecology. But what to do about organisms that feed both one and two levels below? For instance, wrasse are fish that feed on benthic animals (level 2), but they also scrape algae (level 1). So, then, are wrasse at level 2.5? And is anything that eats wrasse at level 3.25? Food chain breaks down under close inspection (scale change) and from it comes the idea of food webs, a much later line of development. Notions of endosymbiosis stretch and collapse the idea of an organism. Rescale until your idea breaks down, and then build the patch. In this way, we can move between levels of analysis in a quantitative way.

CONCLUSION

WILLIAM SUTHERLAND gave the Tansley address at the Manchester summer meeting of the British Ecological Society in 2003.[1] It was refreshing, given in fine style. It has been published as "Predicting the Ecological Consequences of Environmental Change: A Review of the Methods." Predicting the ecological consequences of environmental change was really only the vehicle for the larger message, which was the remarkably tolerant review of the methods. Sutherland worked his way up from low-level local devices such as extrapolation, which can be very helpful when used cautiously; it is data based. But from there he moved to more general devices of complexity science with a more holistic approach. He was clear as to what each method was good for, and was critical in stating the respective limitations. Few have his understanding of what is going on across so many levels. While we have been using devices across a range of levels of analysis, Sutherland's was one of the first expositions coming from someone in the mainstream of ecology with such a catholic view. As to the methods, he writes in the abstract:

> The seven main approaches that are currently used are: extrapolation, experiments, phenomenological models, game-theory population models, expert opinion, outcome-driven modelling and scenarios. Each approach has different strengths and weaknesses. In practice, several approaches are often combined.

Sutherland notes that adaptive management meets with wide approval, but he thinks that the changes wrought by management have to be so large as to have an effect that political risk is considered to be too high. He claims it is used infrequently for that reason. But vigorous management action is performed every day by agencies responsible for forests. So the political resistance of which Sutherland speaks must come from ecological conservation. Actual ecological managers are in a position to use adaptive management.

He is more confident that evidence-based inquiry can do for ecology what it has done for medicine. This has the appeal of being a heuristic device, which means that you can have desired effects, but in general do not need to know how you did it. In the end, understanding what is going on does not help. A counterexample in evidence-based improvements in medicine is the widespread recommendation of statin drugs.[2] There is a statistically robust record that statin drugs lower cholesterol in the blood, and are even associated with better health outcomes by a small, statistically significant percentage. But, in general, medical researchers and doctors do not understand statistics.[3] In 2005, Ioannidis published a devastating argument whose title says it all: "Why Most Published Research Findings Are False."[4] His article refers almost exclusively to medical research. He reports that in medical research, probability values are often used inappropriately, with the result that most research results are simply false. Prasad et al (2013) performed an exhaustive study on a decade of papers in a well ranked journal and found reversals as almost the norm. Their devastating take home message is "Conclusion: The reversal of established medical practice is common and occurs across all classes of medical practice. This investigation sheds light on low-value practices and patterns of medical research." Ecologists are not much better; remember the sampling error in the big diversity studies of Hector and colleagues, and the mistaken criticism of Holling's lumps based on a sampling statistics instead of inventory statistics that actually should have applied. And these mistakes are from eminent ecologists, so we can expect less from the rank and file, who perform statistics as a voodoo dance to ward off evil reviewers.[5] A claim for statins improving heart attack outcomes by one-third is questionable because the group monitored was unlikely to have heart attacks in the first place. That greatly reduces the numbers actually affected, making the outcome marginal. Stephanie Seneff identifies that the outcome amounts to one in sixty avoiding one heart attack after 4.3 years of all receiving statin treatment. Seneff says the downside is that all sixty will have experienced frailty and mental decline. Statins poison mitochondria, a drastic thing to do, so muscles have to work harder. This raises lactic acid, which is good for heart function. So Seneff says avoid statins and eat more yogurt.

Gerd Gigerenzer is very critical of the statistics in medicine. He performs what he calls fast-and-frugal heuristics, and shows that heuristics work as well as, or better than, rational modeling. In a telling example, he reports that baseball players generally say a high fly ball goes up to its zenith and holds there; then the ball drops. They are wrong; the ball does indeed closely follow the path of a parabola. But if one tried to catch a fly ball using rational models of parabolas, the calculations would overwhelm most computation. The ball is hit; find the initial parabola. Then there is spin; change the parabola. Then at height is a wind; change the parabola. Then the wind changes direction with height; change the parabola. The spin slows, and so on, with humidity and other factors. So how do the players catch the ball, when they have a demonstrably bad model? They succeed anyway by focusing on the ball close to its zenith, and then they move as they must to hold that angle. The baseball drops into their hand.[6] Gigerenzer takes a hard line on science almost always being heuristics over understanding.[7]

One would expect that scientists having knowledge is a good thing, but knowing things comes at a price. Gigerenzer researched triage at Cook County Hospital as to how they make decisions on admitting possible heart attack patients; they cannot admit all comers. The doctors, with all their knowledge, get the triage decision right in the mid-eightieth percentiles. That is not bad, but there is a chart on the wall that asks only three questions in sequence, and it gets it right at about 10 percent better rates. The doctors pay for everything they know. The cost is that each piece of knowledge introduces a decision that they can get wrong. Their relative failure comes from them being "too clever by half."[8]

This suggests a utility to ignorance. Gigerenzer tests this when he asks Americans which is larger, Milwaukee or Detroit. The answer is Detroit, by about one hundred thousand people.[9] Americans get it right marginally above random at 60 percent. The surprise is that when he asks Germans, they get it right all the time, at 100 percent. The reason is that Germans have not heard of Milwaukee. It would appear that prediction in settings like ecology is better served by heuristics than even good models. Prediction is not about knowing what is right, so evidence-based research has a downside.

In his move up to more methods with a wider vision, Sutherland addressed expert opinion, output-driven modeling, and finally scenarios. Expert opinion is a path forward when there is no other information. It can be helpful, but there is no counter to it being mistaken. We have cited Becky Brown's research, which has a pattern of showing experts to be wrong quite often. Sutherland moves to output-driven modeling, which can set up outcomes and assess the probability of one versus another outcome. At a yet higher level of analysis, and the most open of Sutherland's devices, are scenarios. These are narratives told without any expectation that this or that story will play out. Scenarios are to open thinking as to what can happen, so plans may be made in the light of worst, best, or surprising outcomes. Most mechanistic ecologists will be disinclined to use scenarios. In his openness to the upper end of his options, Sutherland is rare and brave.

Sutherland does mention that the methods he reviews are often used in tandem. Even so, he does not much investigate the desperate problem of working at different levels of discourse and then making links across levels. In that light, this book has been an attempt to introduce critical points of tension that can only be resolved by a new way of thinking about ecological problems across levels. Business as usual in ecology has been distinctly incremental in recent years. Theory has been barely relevant to field practitioners and experimentalists, who have therefore been forced to take up the slack as best they can. As a remedy, let us find the critical issues in ecology and from that drive our thinking onto the knife edge of the dilemmas that the critical issues pose. Enough of patching over irritating inconsistencies; enough of poking ecological material to see what happens in some general way. Through fuller description, ecology needs to identify the critical points of tension, and then the empiricists can test predictions coming from explicit theory. Theory cobbled together by empiricists is unlikely to be adequate because that is not the specialty of empiricists, and we should not expect them to be experts outside the focus of their special talents.

So Sutherland addresses a lot that we do, which he does by looking across levels of discourse. Even so, we have laid out something more; two things. First, we show how ecologists can address the daunting task of moving between and linking levels of discourse. But, second, we acknowledge the different facets of ecology. As to linking across levels, it is not simple summation because in a change of level, the old and new entities are not additive. At first, small moves to new levels lead to nonlinearity, but quickly give way to folded spaces and catastrophe. Differences change from quantitative to qualitative. New constraints come at the scientist out of the sun. As to our second contribution, the multiple facets of ecology, they are the primary different points of view as to what matters. Meanwhile textbooks do often cover many of our criteria, they confuse scale with type, and they fail to compare the points of view. There is little appreciation of how scale in fact does enrich the discourses within the subdisciplines within the single criterion. That basic approach is in the first edition of this book.

In the update of this book there are more references to chaos and fractals. They are part of the same intellectual revolution; a new theory of complexity is emerging. The "Toward" in the title of this book is still important. The ideas in this book are incomplete, even in this second edition. As a whole, this volume represents some halting second steps, but now taken more critically. We move toward a more thoughtful conception based on observables rather than a reification of tangibles. Reification asserts what would happen in principle if we could only observe and take into account more detail. Increased computation encourages that misplaced logic. Bill Wimsatt refers disparagingly to "in principle arguments" that reify the system.[10] He mocks statements like, "If we could know everything about x through an infinity of infinitely fine observations." No, the premise is wrong. We cannot know everything because there is a discrete disconnect between the conception of parts and wholes, which show emergence. Observation collapses what infinities there are in the world down to discrete outcomes. Remember Schrödinger's cat. That is what observations do by definition. If it were possible to observe in a way that captured the infinities of the outside world, we would no longer be making observations. In the modern world, complexity is hard to manage, not because there are too many parts to track, but because of our limitations in conceiving the whole. It is a conception issue, not one of material difference. In practice, we find that as computers allow functionally infinite observation, what is predicted in principle often does not work out.

In the spirit of Beaufort, in the end, ecology wishes to organize experience to increase understanding, often to facilitate concrete action. "Understanding of *what?*" is a fair question. We refer to the understanding of the conceivable in light of the observable. While modernist holdovers insist it is an understanding of external reality, that position is wishful thinking; humans only have experience. N. Kathryn Hayles raises "constrained constructivism," where she acknowledges that experience is constructed in the observer through interaction.[11] Comforting to realists, she says it is still limited by possibilities dictated by externalities to experience. Experience may be limited by the unknown external reality; in fact, we think it is, despite our griping about excessive realism. However, that limitation by what is real in the external world is not the same as experience showing us the infinite richness of external reality inside the possibilities. Scientists can only model and infer, they cannot prove anything such that they know external reality. In all this, Hayles still holds to understanding as being constructed in the investigator by experience of interaction with influences that arise beyond our definitions. Meanwhile, she acknowledges that there is some correspondence to the external limits, and we think that too. It is easy to accept that there are some things that are impossible because they can never happen. What those limits are remains largely mysterious, although scientists do their best with randomization, experimentation, and other devices.

In American football, when forced to hand over the ball, the strategic play is kicking the ball downfield without the usual fail-safe precautions. The team relinquishing control of the ball must punt so as to drive the other team's field position back. When science has to admit that its controls have gone as far as they can, punting for reality is an abdication, like handing over control. But in chess, sacrificing a piece can open new lines of progress. Physics is masterful at sacrificing, even reason if it must. Although we eschew physics envy, ecology might learn on that point. It is a strategy of: "We can't do that which we want, but we can still do this." Such a posture is in the spirit of Hegel, when he said that "freedom is the recognition of necessity."[12]

The reason we wish to organize experience is so that ecological situations can be managed in what really does feel like very concrete ways. Even so, the reason for concrete action is from far behind many veils of human construction (models, narratives, inferences, recognition, experimental protocol, and so on, with regard to the presumed external reality). Accordingly, what we are really managing concretely is human experience. The feeling that action is concrete does not have to mean that we act on a concrete external world. We do believe there is (or at least may be) an external reality in there somewhere, but it cannot be known or accessed. In simple situations, scientists can get away with imagining a link to external reality. But that was ecology in the middle of the twentieth century. Now, in the face of complexity, deeply elaborate situations press themselves on contemporary ecological concerns. Often, the new complications are social. The intrusion of being a human observer is there in ways that make a difference. Concrete management is a contrivance; it may be the best contrivance science is able to achieve, but it is still a contrivance.

Even recently, in a retreat to realism, ecologists have muddled the material with the modeled system. This book develops a protocol to preempt that class of error. We have operationalized general systems theory for ecology with as little fanfare as possible, so as not to shy off the mainstream. Sometimes we may appear esoteric (our alliance with Kathryn Hayles and the like), but our bottom line is practice and action. Therefore, we sincerely hope that field ecologists and experimentalists have had the patience to bear with us. In the final analysis, we have filled a toolbox of concepts to help straighten out the unnecessary convolutions. That should free us all to tackle the necessary twists and turns of experience that were the intended objects of study all along.

We have not generated a set of explicit hypotheses that need to be tested. Our empirically talented colleagues can do that much better than us. We have not even offered a theory that might lead to hypotheses. However, we think that both better explicit theory and better testable hypotheses will follow from this work. If we have made a contribution, it is a metatheory, a way of looking at ecological material so that interesting theory about the material system can be pared away from semantic contention. Ecologists who have read this book and find it useful will continue to use standard methods of data collection and analysis. We do not intend otherwise. We just hope that empty contention about ontology and "correct" definitions is put in its place. More positively, we hope that there emerges a greater flexibility of thought and boldness of inquiry, leading to sounder practice in ecology.

There is some danger that modernists, still insistent on science addressing reality, will have become impatient with our attacks on reification. Bill Reiners with J. Lockwood have finished their splendid book on philosophy in the practice of ecology, being very kind to the work of Zellmer, Allen, and Koessebohmer.[13] They are pivotal for us, too, which earns them a place in closing our book. We relied heavily on the cycling scheme for building narratives in chapter 8. Reiners seems to buy the whole story more or less through his book, and gives us the anchor point on his very last page. But there is one reservation he has that we need to address. Reiners complains of our rejection of realism. Perhaps our ideas were not well stated in the publications a decade ago from which Reiners will have read. If we overstated our case then, it was because of our outrage at the unjustified use of realism in the literature, as well as the muddles and semantic arguments that come from naïve reification. So let us make a slightly softer statement here at the end of this book, so realists can come on board. As we might say to nervous swimmers, "Come on in, the water is fine."

We need to be clear that there is no way we can prove there is no external reality, and we do not wish to do so anyway. In our day-to-day life, we personally believe that there is an external world and that our experience relates to it in some way, albeit in a mysterious fashion. Allen really is here, in his real study, typing on what he thinks is his real computer. So why have we been so grumpy? It is that science observes, and raises first impressions of what it thinks is going on. But mature scientists know from experience that they are absolutely always demonstrably wrong in some way. It happens time and time again. In fact, the methods of science appear designed to show us that we all are wrong, all the time.

Science has to defy what we intuit is real because we humans are so clever at defending what we are prejudiced to think is true. It is not just lawyers and politicians who change the subject or change the level of analysis to win an argument spuriously. We have done exactly that in this book, many times, because the downside of narrative is that it invites such self-deception. But narrative makes us human, even if there is a price to pay. There is no way out. Get used to it! In fact, we have fixed a lot of those instances, as we did indeed catch ourselves in error. There still may be places in the book where we went wrong. If you find such mistakes, ask yourself, "What were they trying to say?" We are human and given to fooling ourselves, which is why we need science to keep us more honest than we would be otherwise. There are so many ways to win an argument spuriously, and even science cannot blunt them all. Science is simply the most worthy tool we have. And when we are challenging our beliefs with science, it is best not corrupted by using reality as the reference. It is hard enough to stay honest without inviting that temptation to take the easy way out.

We still insist that realism cannot be used validly to justify how we conduct science in practice. When the practice of it is all over for the moment, when we think we have it right, when all our friends agree, then suggesting that we are somehow closer to an external reality is perfectly acceptable. Reality is then a valid notion because, for the moment, the investigation is over. But start investigating again, and it is important to resist temptations to reify concepts. Rejecting realism in the practice of science while accepting it as part of the larger agenda is not in direct opposition. They are orthogonal, not opposite. It is a rotation of point of view.[14]

We stand in awe of Tansley in 1926, as he urged us not to mistake our concepts for reality. We quote the nub of his idea again. He said,

We must always be aware of hypostasizing abstractions, that is, of giving them an unreal substance, for it is one of the most dangerous and widespread of vices through the whole range of philosophical and scientific thought. (Tansley 1926:685)

He went on to say that we need to use the concept of climax community, but emphasized that it is only a set of species that persist together for a certain time, and an arbitrary time too. So our questioning of the pertinence of reality is in the spirit of Tansley.

We are grateful to Bill Reiners for his support at the end of his book. But there are many others who have helped us so much in the last quarter century; you can tell to a significant extent who they are from the way we clearly have favorite ecological scientists. Grime and Holling, for example, are well referenced in this work. We have returned to their work and that of others on several occasions. We have dissected the devices for ecological thought and practice. What we have debunked was others "hypostatizing abstraction." We admit that science is often most successful if its practitioners are realist. Newton was a realist, for instance. But ecology is a more tangled bank, where hypostatizing comes too easily. Our advice is mostly to keep doing what you always have done, but do it for somewhat new reasons. Be more philosophically cautious than ecology has been in recent decades.

NOTES

PREFACE

1. Schooler (2011); Schooler and Engstler-Schooler (1990).
2. Rowe, Repasky, and Palmer (1997); Polo and Carrascal (1999).
3. Pfeiffer and Hoffmann (2009).

INTRODUCTION

1. Huler (2004).
2. Huler (2004) reports that Beaufort was influenced by Alexander Dalrymple, who wrote a piece around 1790 that contained John Smeaton's 1759 scale for windmills. The 1790 piece was unpublished, but was preserved as a printer's copy.
3. Huler (2004).
4. Scott Huler, National Public Radio interview, September 7, 2004; http://www.npr.org /player/v2/mediaPlayer.html?action=1&t=1amp;islist=false&id=3893777&m=3893778 (accessed May 16, 2012).
5. Rosen (1991:44–66).
6. Kuhn (1970); the whole book is about paradigms.
7. Ahl and Allen (1996).
8. Knox (1998).
9. See Ahl and Allen (1996), and Allen, O'Neill, and Hoekstra (1984) for grain and extent.
10. See Ahl and Allen (1996) for levels of organization vs. observation.
11. Lotka (1956).
12. Rosen (1991:44–66).

13. Wainwright et al. (1982:355).

14. Chabot and Mooney (1985).

15. Salthe (1985) calls the three levels the "Triadic."

1. THE PRINCIPLES OF ECOLOGICAL INTEGRATION

1. Henry Horn, discussing an as-yet-unfinished 2010 draft of the preface for a monograph on "Social Butterflies," communicated to Allen in an email on December 17, 2013. In that same email, Horn confesses to the event of the water striders.

2. Tony Ives, in an email to Allen on December 16, 2013.

3. Box and Draper (1987:424).

4. Wald (1943).

5. Ernst Mayr, quoted in his obituary; Yoon (2005).

6. Tansley (1926).

7. Volk (2002).

8. Tainter and Lucas (1983).

9. Bishop (1995); Gurney (1997).

10. Wessman et al. (1988).

11. Krebs and Weitzman (1987).

12. Schneider and Kay (1994), most of which came from Kay's PhD dissertation.

13. Kelly and Mayo-Wilson (2010).

14. Personal communication with Kevin Kelly, Modes of Explanation Conference, Holiday Inn, Paris, April 2013.

15. Pattee (1979b).

16. Dublin Stationery Office (1946). "Defence Force Regulations: Musketry, Fire Control Training Regulations," no. 4, pt. 2. Image retrieved November 19, 2013, from http://michaelotoole.biz/wp-content/uploads/2012/11/MusketryCover.jpg.

17. Personal observation by T. Allen and communication from a ferryman, 1977.

18. Simon (2000:6): "Thus, biological organisms typically carry sizable inventories of food (for example, in the form of fat), but not of oxygen, which is used at too rapid a rate to be readily stored."

19. The observation of a dune system being cut off and constrained by an intermittent large river in Australia that always returned before dunes could build past it was a personal communication from Joe Walker of the Commonwealth Scientific and Industrial Research Organisation (CSIRO).

20. See note 18 herein.

21. Li et al. (2007).

22. Harper (1977).

23. Heslop-Harrison (1964) speaks of hybrid swarms in knapweed (*Centaurea*). Gurney et al. (2007) show *Primula* hybrid backcrossing.

24. *Larus* gulls form a circumpolar complex. Such situations are called ring species. Dawkins (2004:303) says they "are only showing us in the spatial dimension something that must always happen in the time dimension." Recent mitochondrial genetic

analysis suggests that the situation is not a simple ring (Liebers, de Knijff, and Helbig 2004), but the general principles still apply.

25. Hoekstra, Allen, and Flather (1991).
26. Bosserman (1979).
27. Tansley (1935).
28. Transeau (1926).
29. Harris, Kennerson, and Edwards (1977).
30. Clements (1905).
31. See, for example, Bormann and Likens (1979).
32. Brandner (2003).
33. Hetherington and Woodward (2003).
34. Van Voris et al. (1980).
35. Personal communication with R. V. O'Neill to Allen.
36. Schneider and Kay (1994).
37. Rosen (2000), chap. 1.

2. THE LANDSCAPE CRITERION

1. Humboldt and Bonpland (1807).
2. Clements (1905).
3. Wiens and Milne (1989).
4. Allen (1973).
5. Allen (1998).
6. Chander et al. (2004:2748).
7. Mandelbrot (1977).
8. Gleick (1987).
9. Rosen (1991).
10. Sugihara and May (1990). This work cites an unpublished mimeograph by Sugihara from the mid 1980s, which was circulated widely at Oak Ridge National Laboratory (ORNL) and beyond; it is probably one of the most influential unpublished documents in ecology.
11. Turner and Ruscher (1988).
12. Weinberg (1975). Small-, large-, and medium-number systems and their predictions.
13. Urban et al. (1988).
14. Stevens (1974:3).
15. Gardner et al. (1987).
16. References to neutral landscapes, percolation theory, and critical thresholds are all found in Gefen, Aharony and Alexander (1983); O'Neill and Gardner (1990); Milne (1992); Milne, Johnston, and Forman (1989); Gardner et al. (1987); Turner and Ruscher (1988) and Turner et al. (1989). The figure of the 30 percent threshold is not explicitly stated, but can be extracted from the graphs (personal communication with R. H. Gardner February 2013).
17. Puth and Allen (2005).

18. The U.S. Geological Survey (USGS) cites blue jay invasion in Nebraska in 1949 (Collister 1950) and more recent establishment along the Platte River (Johnsgard 1979, 1980); http://www.npwrc.usgs.gov/resource/birds/platte/species/cyancris.htm (accessed November 22, 2013).
19. Garcia (2003).
20. Therres (1993).
21. Allen and Holling (2008).
22. Heglund, Taylor, and McMahon (1974).
23. Brandner (2003).
24. Paine and Levin (1981).
25. Allen et al. (2009, 2010).
26. Allen et al. (2001, 2010).
27. Tainter and Allen (2014).
28. McCune and Allen (1985a, 1985b).
29. Prigogine has written extensively on the emergence of higher levels of order in Nicolis and Prigogine (1989); Prigogine (1978, 1982); Prigogine et al. (1969); and Prigogine and Nicolis (1971). Gleick (1987:311) discusses the emergence of patterns in snowflakes.
30. Haralick (1979); Haralick and Shanmugam (1974); Haralick, Shanmugam, and Dinstein (1973).
31. Musick and Grover (1990).
32. Prose and Wilshire (2000).
33. Kratz et al. (1991).
34. Dermott and Sagan (1995).
35. Gucinski, Miner, and Bittner (2004).
36. Eglash (1999).
37. McLuhan (1962).
38. Homer (2003).
39. Pausanias (1913) describes the sacred groves of Aesculapius at Epidaurus of Argus in Laconia and a sacred grove of plane trees at Lerna.

3. THE ECOSYSTEM CRITERION

1. O'Neill et al. (1986).
2. Sterner (2012).
3. Transeau (1926:5). Transeau goes back to Priestley's reference to "bad air," citing many of the early players from previous centuries who gradually worked out photosynthesis and the composition of air. Transeau cited early scholars in nutrient status.
4. Brown (1905), referenced in Transeau (1926). Brown was Transeau's anchor to something tangible, from which Transeau worked systematically upscale. Brown goes back to the work of scientists two hundred years earlier, leading Transeau into the same classical scientific literature.
5. Tansley (1935).
6. Odum (1969).
7. Lindeman (1942).

8. Shelford (1918) and Strøm (1928); both cited with figures by Sterner (2012).

9. Teal (1962) described a salt marsh estuary in ecosystem terms.

10. Through the 1970s, the International Biological Program (IBP) set up a series of biomes that were modeled elaborately with state-of-the-art computing. The faith was that a full enough account of each biome would allow questions to be answered by the model in a can. Coleman (2010), who was a player at the time at the center of the effort, wrote a history of *Big Ecology*. On page 35, he is explicit about the agenda of George van Dyne and George Innis being a model that could be used "in place of . . . field experimentation." The most important message learned from abundant failure was that models need to be conceived in the light of questions from the outset. The general model of Innis was not focused. Coleman (34) states it had "4,400 lines of code, 180 state variables, 500 parameters. It required roughly seven minutes to compile and run a two-year simulation with a two day step." Criticism has been overstated, forgetting we all learned a lot in a classic normal science refutation. Sadly, the cautionary tale appears forgotten in contemporary large-scale reductionism, for instance, in the big diversity experiments of Hector et al. (1999). Long-Term Ecological Research (LTER) is the child of IBP, focused largely on the same centers. The National Ecological Observatory Network (NEON) is the grandchild of IBP. The greatly increased funds of IBP have never gone away.

11. Emanuel, Shugart, and West (1978) and Emanuel, West, and Shugart (1978) under simulation showed that only minor effects in ecosystem function were caused by the loss of the American chestnut in eastern forests in the nineteenth century.

12. Hyatt and Stockner (1985) refer to a program of nutrient enhancement in lakes as a means of controlling the size of fish populations in Canada, which was conducted through the 1960s and 1970s. Explicit positive results are reported by Hyatt and Stockner (1985), Stockner (1981), and LeBrasseur and Parsons (1979).

13. Hocking and Reimchen (2002) and Reimchen (2001) report isotope ratios of nitrogen in western rivers and forests indicating salmon as a source of oceanic ratios in the terrestrial system. See Walters and Holling (1990) for an analysis of management strategies for Pacific salmon.

14. Thomforde and Allen (2010).

15. Some of the earliest work using radioactive traces to identify boundaries of ecosystems in the manner of biochemists was performed at Oak Ridge National Laboratory (Reichle and Crossley 1965).

16. Hubbard Brook was a center for ecosystem experimentation at the watershed scale through the 1960s and 1970s (Likens et al. 1967, 1977, 1978; Bormann et al. 1974).

17. Personal communication with James Kay on a car ride to a meeting in Ontario. Generally stated in Kay's Ph.D. thesis.

18. The forest simulation that calculated residence times for carbon at fifty-four years and nitrogen at 1,810 years was performed by O'Neill et al. (1977).

19. Simon (2000:6): "Thus, biological organisms typically carry sizable inventories of food (for example, in the form of fat), but not of oxygen, which is used at too rapid a rate to be readily stored."

20. Hoekstra, Allen, and Flather (1991).

21. The review of scaling effects in plant compensation is found in Brown and Allen (1989). The argument between McNaughton and Belsky first occurred in Belsky (1986) and McNaughton (1986).

22. Personal communication with Earl Aldon, a range scientist in the Albuquerque research lab of the U.S. Department of Agriculture Forest Service, around 1985 during a supervisory program review. Hoekstra was having a personal discussion about Aldon's work on the Rio Puerco watershed using Albuquerque sludge. Hoekstra was especially interested in it because of the cattle/sheep carbon drain part of the story. See also Aguilar et al. (1994); Finch and Tainter (1995).

23. The work on zooplankton correlation with phytoplankton being negative for a short period but positive for longer periods is found in Carpenter and Kitchell (1988).

24. Bradshaw and McNeilly (1981).

25. The work on turnover times of carbon and nitrogen on the grasslands is found in Parton et al. (1987), where the authors model organic matter and nitrogen in compartments that operate at different speeds. The model carbon flow has a hierarchy of carbon compartments with turnover times of 0.5, 1.5, 2.5, 3.0, and 1,000 years because of the different reaction rates. Grazing pressure does not much change the status of the faster-moving compartments because they merely draw on the resource base in the slow-moving compartments. See also Sala et al. (1988).

26. Margalef's pioneering work of broad scope in aquatic systems was published as Margalef (1968).

27. John Magnuson has published a series of papers that consider the relative size of a water body and the characteristics that can be used to predict the fish within it. He compares oceans and lakes, having collated the significant data set on small, temperate lakes (Magnuson 1988; Magnuson et al. 1989).

28. May (1981c:219) states, "This notion [complexity implies stability] has tended to become part of the folk wisdom of ecology."

29. Gardner and Ashby (1970); May (1972). In the context of diversity and stability relationships, various authors have identified that increase in diversity in a system that is connected at random will decrease stability in the final analysis (Levins 1974; Margalef 1972; MacArthur 1972; May 1974).

30. Levins (1974), MacArthur (1972).

31. A general discussion of community matrices and stability with respect to the loop structure of the subsystems is found in O'Neill et al. (1986). Other pertinent discussions occur in Tansky (1978), Levins (1974), McMurtrie (1975), Austin and Cook (1974), and DeAngelis (1975).

32. Van Voris et al. (1980:121).

33. Platt and Denman (1975) report extensive use of power spectra in ecological systems.

34. Ricotta (2005).

35. Fridley and Grime (2010). Experiments with genetic diversity from a local natural heath land with 2–8 species that grow together in nature. This is typical of Grime's measured approach with a lot all else equal.

36. Huston et al. (2000) shows the statistical errors. Such is the investment in big diversity experiments that it is hard for the protagonists of diversity and functionality to back down. There is a series of papers with most of the same authors of Hector et al. (1999) arising for more than a decade after the original publication (for instance, Spehn et al. 2005).

37. Patrick and Hohn (1956); Patrick, Hohn, and Wallace (1954).

38. Allen and Starr (1982) used Weinberg's (1975) notion of medium-number systems.
39. Allen (2010).
40. Simberloff and Wilson (1969).
41. Brown and Ewel (1987).
42. Fridley and Grime (2010).
43. Zimmerer (1998).
44. Zhu et al. (2000).
45. Allen, Havlicek, and Norman (2001). Also treated in Allen, Tainter, and Hoekstra (2003:344).
46. Schneider and Kay (1994) suggest that terrestrial vegetation works by using transpiration, itself driven by heat fed into the latent heat of vaporization. The cooler the canopy, the more work is being done. Far less energy is bound up in chemistry driven by photosynthates.
47. Tilman et al. (1997).
48. Resnik (2011). This paper deals with the pathology of the review process, but mistakes it for a data problem, and reports what proportion of scientists have what concerns. We already know about the dysfunction. We need to know what to do about it. Resnick only concludes with well-meaning urging to do better and be more ethical and rebuild trust. Ravetz (2006), a postnormal scientist, instead simply announces that we should accept that science is corrupt, and then indicates how to work within the context of it remaining corrupt. When science is driven by mercantile ethics at every level from individual scientists to funding agencies, it is best to accept the dysfunction and work from there. Mercantile ethics are unlikely to disappear.
49. In a series of three articles, J. D. Fridley—Fridley (2001, 2002); and Brown and Fridley (2003)—takes down the excesses of Hector et al. (1999).
50. Huston et al. (2000) criticized Hector et al. (1999).
51. There is a substantial literature on complexity, from which we can only indicate a few. Some of it is large-scale reductionism that is of no interest here. From business management, see Checkland (1981); Simon (1996); and Snowden and Boone. From biology, see Eigen and Schuster (1979); and Kauffman (1993). From ecology and resource management, see Giampietro (2003); Polimeni et al. (2008); Giampietro and Mayumi (2009); and Giampietro et al. (2014).

4. THE COMMUNITY CRITERION

1. Paine (1966).
2. Kimmerer and Allen (1982).
3. Connell (1978) pointed to intense or low levels of disturbance in tropical rain forests, and coral reefs led to lower diversity compared to intermediate levels of disturbance.
4. Hutchinson (1957, 1959) considers niche in terms of community and diversity. While he sets out the conventional view of increased diversity, increased stability (a notion that is now seen as at best incomplete, and mocked by May as the "folk wisdom of ecology"), Hutchinson looks validly at all the main issues of diversity and stability.
5. Loucks (1962) was concerned with nonlinear gradients, and black spruce occurred abundantly at opposite ends of the moisture gradient.

6. Bradshaw's work on the evolution of communities on toxic mine tailings is synthesized in Bradshaw and McNeilly (1981). Begon, Harper, and Townsend (1986:71) summarize the situation found in recently formed communities on toxic substrates: "When it [the pollutant] is newly arisen or is at extremely high concentrations, there will be fewer individuals of any species present (the exceptions being naturally tolerant variants or their immediate descendants). Subsequently, however, the polluted area is likely to support a much higher density of individuals, but these will be representative of a much smaller range of species than would be present in the absence of the pollutant. Such newly evolved, species-poor communities are now an established part of man's environment."

7. Bell (1971).

8. Examples of early biogeographical exploration are von Humboldt and Bonpland (1807), Grisebach (1838), and Unger (1836). The work of Unger is a detailed biogeographical study of the alpine valley in which he was the medical doctor. This work is distinctly ecological, identifying plants that grow on limestone as opposed to those on slate. Unger maps ninety-three species with dot maps in great detail, predating any other use of dot maps by a full twenty years. The Unger work had been forgotten until it was rediscovered by Hugh Iltis.

9. Allen, Mitman, and Hoekstra (1993).

10. The assertion that Clements died a Lamarckian is made by Worster (1977).

11. See Moorhead (1969) for background on Darwin. In his *The Descent of Man*, Charles Darwin acknowledges: "The conclusion that man is the co-descendant with other species of some ancient lower and extinct form is not new in any degree. Lamarck long ago came to this conclusion that has lately been maintained by several naturalists and philosophers" (p. 19). A full account of the German work on plant physiology that was the extension of Darwin's evolutionary ideas is found in Cittadino (1980, 1990).

12. Cittadino (1990) explicitly talks about imperialism as the driving force behind the late nineteenth-century German tropical biology. Examples of German tropical biologists mentioned by Cittadino are: A. F. W. Schimper, studying coastal vegetation in Indo-Malaysia; Georg Volkens, using Egyptian desert plants to test hypotheses developed in studies of native European vegetation; and Heinrich Schenk, who studied tropical climbing lianas. The Germans who most influenced the young Nebraskan ecologists were Warming (1896) and Schimper (1898). Of particular importance here is Drude (1890, 1896); direct evidence for Drude's influence is found in Pound (1896).

13. The Botanical Seminar at the University of Nebraska was an informal group begun by Bessey and his students in 1886. One of the principal activities of the seminar was the Botanical Survey of Nebraska. This work culminated in Pound and Clements (1900), a joint doctoral thesis. The first community work coming from that group was the thesis of MacMillan (1892, 1899).

14. Pound and Clements (1898) are explicit in their use of quantification of plants in quadrats.

15. Gleason (1920).

16. Cowles (1899) uses the term "plant society," while Pound and Clements (1900) used the word "association." All these workers used the term "formation" to imply something a little more inclusive than the association or plant society. Their prime influence appears

to be Drude, although the original use of formation goes back to Grisebach (1838). Smith (1898, 1899) talks about the plant association as being a community.

17. Clements (1916).

18. Levins and Lewontin (1985). Clark and York (2005) wrote a celebration of the twentieth anniversary of *The Dialectal Biologist*. It would not be unfair to say that the entire approach in this volume is already out there in Levins and Lewontin. The political baggage in the term "dialectical materialism" has blunted acceptance of their approach. We have developed our ideas in parallel, but without the political encumbrance. We too reject reductionism and idealism.

19. Gleason (1922) is not often cited in summary views of the Clements-Gleason debate. The concurrence of their views may have been lost since that paper was largely overlooked.

20. Greig-Smith (1964:133). But the quotation from Webb (1954) was important enough to still appear in Greig-Smith (1983:148).

21. Bird (1957) discusses catena gradients of soil down hillsides. He cites Milne from the 1930s in Africa, who first developed the idea. Curtis, a prime mover in vegetation analysis gradients, cites Bird. Catenas appear to have been a seed idea of gradients in the minds of vegetation ecologists.

22. Simberloff et al. (1992).

23. McIntosh (1967) reviews the idea. He was also in on the ground floor of gradient analysis, showing continuous variation of vegetation (Curtis and McIntosh 1951).

24. Greig-Smith (1983), chap. 8.

25. Principal component analysis (PCA) was first used in vegetation analysis remarkably early by Goodall (1954) under the name factor analysis, calculated with pencil and paper, and a gang of statisticians. New treatments were brought in by Orloci (1966).

26. Bishop (1995); Gurney (1997).

27. Personal communication to Allen in the refectory at University College of North Wales (UCNW) in Bangor.

28. Roberts (1984) drew on Sukachev and Dylis (1964); Roberts (1986, 1987a, 1987b, 1989).

29. Levins and Lewontin (1985); Clark and York (2005).

30. Watt (1926) identifies old stands on *Taxus*. For recent work on yew thriving in shade, see Perrin and Mitchell (2013).

31. Braun-Blanquet (1932).

32. Whittaker (1956) did direct gradient analysis. Curtis (1959) did indirect (vegetational) gradient analysis, but both took an environmental determinist position.

33. In popular works by Diamond (1999, 2005), he is a hard-line environmental determinist, to the disapproval of actual social scientists with an understanding of energy issues such as Tainter (2008) and Smil (2005). Forbes (1887) gives an account of balance and in a lake, where all the species play their role in the whole, even in sacrifice as prey. This reflects the maturing bourgeois revolution of Forbes's time, where balance and duty was important. Darwin's gradualism reflects the same issue of avoiding revolution.

34. See Allen, Mitman, and Hoekstra (1993) for an introduction to Roberts's maximal cliques.

35. See, for example, Knight and Loucks (1969).

36. Bakuzis and Hansen (1959).

37. Whittaker (1956).

38. Grime (1974, 1977).

39. Tilman (1988); the notion of R* is the breakeven point as the species in question forces a nutrient down. Tilman asserts that this determines the winner of competition.

40. Grime (2007): "But it is also evident that we must recognize that competition declines in importance under the impacts of reduced productivity and/or severe disturbance." Grime (2001:67–71) agrees with Caccianiga et al. (2006).

41. Caccianiga et al. (2006:19).

42. See the field research of Grime and Curtis (1976); the laboratory results of Mahmoud and Grime (1976) were supportive.

43. Grime et al. (1997).

44. Huston, DeAngelis, and Post (1988) used a similar fish example to explain individual-based modeling.

45. Shugart and West (1977) laid out the FORET model. Shugart, West, and Emanuel (1981) looked at implications.

46. Kuhn (1970). Normal science is the science that occurs to fill in the paradigm, the day-to-day science. It learns most from refutation of reasonable assertions that would otherwise be accepted.

47. Mooij and DeAngelis (1999) could not get the model criticized to behave as Ruckles-house, Hartway, and Karieva (1997) said it did. The modelers generously sent DeAngelis the explicit data as to the run of the model with noise inserted as a test, whereupon it became clear that the error noise Karieva and his colleagues had put in to perturb the model was seen to be not 16 percent but 16,000 percent. Ever the gentleman, DeAngelis was kinder in his rebuke than he might have been, a lesson to us all. Colleagues of DeAngelis know that he always gets technicalities right. This sweeps away all criticism of individual-based models from Karieva, Skelly, and Ruckleshouse (1997).

48. Watt (1947).

49. Roughgarden, Gaines, and Pacala (1987); see also Roughgarden and Pacala (1989).

50. Park (1948). Between two environments one species always won predictably; even in the intermediate environmental conditions one species still won, but which species was not determined.

51. West (1970).

52. Curtis (1959).

53. Potter et al. (2008).

54. Eickmeier, Adams, and Lester (1975).

5. THE ORGANISM CRITERION

1. Diamond (1999) reports that when a family is on the move in Papua, New Guinea, the women do all of the load bearing, while the men strut with a weapon. He interprets this not as protecting against snakes and large predators, but against other men. The selective advantage to this for women is that another man taking over, like a lion taking over a pride, will generally kill children of the previous husband.

2. Archibald (1976); Archibald and Mirande (1986); International Crane Foundation (2013), https://www.savingcranes.org/images/stories/pdf/about_icf/icf_strategic_plan_symbols _of_survival.pdf; Schoff (1991).

3. Grime, Hodgson, and Hunt (2007).

4. Schrodinger (1959).

5. Haldane (1963).

6. Lane (2011, 2012).

7. Atsatt (1988).

8. Heslop-Harrison (1964).

9. In an obituary to Ernst Mayr, Yoon (2005) quotes him as saying that he identified 136 bird species in a study in New Guinea, while natives were only two species off. He took this to mean that species are real, but as we discuss in chapter 1, he would not have achieved the same coincidence for lizards because major human input and output devices are the same as birds, but not lizards. We read the biology of speciation in nature with birds, so Mayr's observation for birds does not make species real.

10. Wright and Greengrass (1987).

11. Uexkull (1920, 1929); Uexkull and Kriszat (1934), all in German, but Uexkull (1957) is in English.

12. Hoekstra, Allen, and Flather (1991).

13. Neilsen (2006).

14. Tilman et al. (1994).

15. Ceier (2008). Online at Walton Outdoors; program manager, Lenny Fenimore.

16. Levin (1986). To reach his wonderful website, go to http://www.biotensegrity.com/.

17. Corner (1981).

18. Ibid.

19. Vogt and Riddifo (1981).

20. Corner (1981).

21. Haldane (1963).

22. Johanson and Edey (1981).

23. Porter and Gates (1969).

24. Dudley, Bonazza, and Porter (2013).

25. Mitman (1999).

26. Porter (2013).

27. Vogel (1988:116) gives a good general account of different Reynolds numbers. Using corn syrup instead of water allows models that work at large size, but as if they were small. Page 103 discusses the whirligig beetle screwing its way through water.

28. Vogel (1988:96–103); a wide discussion of the surface tension of water and capillary action.

29. Schmidt-Nielson (1984).

30. Allen and Starr (1982:256–58).

31. Kays and Harper (1974). Rely on a 2/3 scaling law. The 2 is thought to relate to the exponent of leaf area. The 3 is thought to pertain to the dimensions of the volume.

32. Brown et al. (2011).

33. Tainter (1988).

34. Giampietro (2003).

35. Rosling (2014).

36. Tainter (2008); a vigorous criticism of Diamond's *Collapse* (2005), with withering commentary on Diamond's *Guns, Germs, and Steel* (1999). Smil (2005) also has a short, very

hard review of *Collapse*; his 2008 book does the same, but with an extensive, dispassionate treatment of the worrying situation.

37. Tainter and Allen (2014).

6. THE POPULATION CRITERION

1. Bradshaw (1987).
2. Personal communication with James Kay on a car trip to a meeting in Ontario but also generally stated in Kay's Ph.D. thesis.
3. Lotka (1956).
4. Giampietro (2003:343).
5. Giampietro (2003:343) translated Lascaux (1921:330).
6. Hoekstra and Michael Wolfe, unpublished research.
7. Giampietro (2003:305), fig. 9.19, shows scatter diagrams where one can see nations that have passed the demographic transition on many parameters.
8. Malthus (1830). Shorter, revised version of the great work.
9. Malthus (1798); See Oxford's 1999 reprint, G. Gilbert, ed., chap. 7, p. 61.
10. Einstein, from "On the Method of Theoretical Physics," the Herbert Spencer Lecture, Oxford, June 10, 1933. This is the Oxford University Press version. The words "simple," "simplest," and "simplicity" recur throughout the lecture. The version reprinted in 1954 in *Ideas and Opinions*, 272, is a bit different. This sentence may be the origin of the much-quoted sentence that "everything should be as simple as possible, but not simpler," and its variants; http://quoteinvestigator.com/2011/05/13/einstein-simple/#more-2363.
11. Giampietro et al. (2014).
12. Patten and Auble (1980).
13. Giampietro (2003); see also Giampietro et al. (2014).
14. Mendelsson (1976).
15. MacArthur (1958).
16. May (1981).
17. Deevey (1947).
18. Harper (1967).
19. Park (1948).
20. Krejci and Dewey (n.d.). Good bibliography and thorough treatment; describes social systems of pronghorns and their relation to snow. See also Bleich et al. (2005).
21. Prince (1982). It was stated that the most southern population of the mopane tree in southern Africa was a vigorous small population in a favorable habitat with healthy adult individuals undergoing effective reproduction. Other favorable unoccupied sites were identified only two to three miles further south, but these were unoccupied by the species. This is an example of a distribution that can be explained by epidemic models for biogeographical limits (Carter and Prince 1981). The models are not to deal with disease per se, but liken occupied sites to infected individuals. In that infection that can be described as an all-or-nothing phenomenon, sites at the limit of a species range are either fully infected/occupied or not. Salisbury (1932, 1964) asserts that many species are abundant at the limits of their range.

22. Specifically, Harper, Williams, and Sagar (1965); but also Harper (1977).
23. Rejmánek, Sasser, and Gosselink (1987).
24. Leslie (1945).
25. Carosella (1978) for life stage analysis.
26. Rosen (2000) in his modeling chapter.
27. Gardner and Ashby (1970).
28. Johnson showed checkerboard patterns from stochastic automata. This represents the reverse direction in the relationship between stochasticity arising from determinism in chaotic equations. Much of the work is unpublished, but see Keymer, Marquet, and Johnson (1998). Also, similar patterns can be found in Regnault, Shanbanel, and Thierry (2010).
29. Southwood (1976) is explicit in comparing bacteria and whales.
30. Allen et al. (2010).
31. Canestrari et al. (2014).
32. Herr (1980).
33. DeAngelis, Post, and Travis (1986).
34. Volterra (1926).
35. Holling (1964).
36. Holling and Ewing (1971). Model was presented at Yale, 1969.
37. Rappaport (1968, 1972). Holling made the explicit connection between his equations and the Tsembaga in a 1973 talk, University of Wisconsin, Madison.
38. The American "ladybug" becomes "ladybird" in British parlance.

7. THE BIOME AND BIOSPHERE CRITERIA

1. Clements (1916).
2. Tansley (1926), published in 1929, is quoted at the beginning of this book on climax and succession.
3. Holling (1986).
4. Neilson (1986, 1987a, 1987b); Neilson and Wullstein (1983).
5. Personal communication with Ronald Neilson.
6. Brandner (2003).
7. Lovelock, Maggs and Rasmussen, R. A. (1972); Lovelock (1989).
8. Auton and Woolen (1991) could not show harmful effects of Freon, but there are records of death at high concentration in the order of 10 percent in the atmosphere. For general use, it appears remarkably safe, not affecting workers chronically exposed (Imbus and Adkins 1972). But of ninety-nine workers exposed from two to five years, three showed impaired psychomotor speed, impaired learning, and impaired long-term memory (Rasmussen, Jeppesen, and Arlien-Søborg 1988).
9. The half-life for diffusion out of the troposphere is twenty years (Hazardous Substances Data Bank 1994).
10. Green et al. (2013).
11. Saatchi et al. (2011) estimate around three gigatons of carbon fixed in tropical forests.
12. Wixon and Balser (2009). See also Wixon (2011).

13. "In 1972 I demonstrated that the summer time turbidity of the air over Southern England and Southern Ireland was man-made not natural by using the simultaneous presence of CFCs as a marker of air from urban sources. In the same year I measured during a ship voyage from the UK to Antarctica and back the abundance and the accumulation of CFCs in the global atmosphere." From Lovelock's own homepage: http://www.jameslovelock.org/page3.html.

14. Personal communication with Michael Corradini, Nuclear Engineering, University of Wisconsin, Madison.

8. NARRATIVES FOR COMPLEXITY

1. Personal communication with R. Rosen at the postmeeting session of the San Francisco Annual Meeting of the Society for General Systems Research, 1980.
2. Herr (1980).
3. Levin (1992) has covered much that we have about scale in narrative form, and before our first edition (it was a MacArthur lecture in 1989). Levin is a master of narrative, and field work, which is unusual for one with such mathematical strength.
4. Funtowicz and Ravetz (1992).
5. Ahl and Allen (1996).
6. Rosen (1991:54–55, 141–51).
7. Allen and Holling (2008).
8. Brown (1995).
9. Gerard (1969).
10. Giampietro (2003).
11. This is the notion of fuzzy sets. Membership is only to a degree, but that degree is unequivocal, so fuzzy has a particular meaning. Zadeh (1965a, 1965b).
12. Personal communication with Tainter.
13. Tainter (1988).
14. Working group session in Fort Collins, Colorado; Hoekstra et al. (1999).
15. Cohen (1971). Remarkably for us, our hero, Robert Rosen, was severely taken to task in Cohen's book review. Cohen does make some wonderful comments as to how "Lotka took the trouble to tabulate the average gains in weight of steers, and to relate these data to his models; Lotka *respected reality*, and the odor of that respect still rises from his still living pages" (our emphasis). Cohen claims to find many mathematical errors in Rosen's work. Even so, we can see the modernist realist in him, and his preference for predicative definition. There were few alternatives in 1971. The tension between the predicative and impredicative is stark in the difference between these two theorists, both of whom we respect mightily. We suspect Cohen might have more respect for Giampietro's reliance on data to calibrate complexity and impredicativity (Giampietro et al. 2014).
16. Rosen (2000), chap. 1.
17. Green et al. (2013).
18. Swetnam and Lynch (1993).
19. Wald (1943).

20. Cottam, Curtis, and Catana (1957); Cottam, Curtis, and Hale (1955). From the 1950s to the 1980s, Cottam taught ecological methods botany at the University of Wisconsin, and required students to write a paper on efficiency of methods of data collection. Not only data were collected on plants, but time sampling as well, so efficiency could be estimated. Cottam's work on vegetation sampling was so important that his methods are now ritual. Most have forgotten who did the foundational research, as important as it was in ecology. The emphasis on includers and excluders was a personal communication when Cottam and Allen taught the course together for a decade.

21. White (1967).

22. Allen, Tainter, and Hoekstra (2003).

23. Neilson et al. (1995).

24. Cronon (1992:1366–67). He quotes Plenty-coups from Lindeman (1930, reprinted 1962:311).

25. Pattee (1978, 1979a, 1979b).

26. For a summary that emphasizes the inconclusiveness of the experimental data as to the mechanisms of allelopathy, see Kershaw (1973:94–96).

27. Personal communication with Geoff Sagar at the Liverpool BES annual meeting.

28. Connell (1980).

29. Holling (1978).

30. Alfred Korzybski (1931) coined the expression "mistaking the model for what it models."

31. Norman (n.d.). Forthcoming from *Procedia Environmental Sciences*.

32. Brown and Allen (1989).

33. Simberloff (1983).

34. Minchin (1987).

35. Needham (1988).

36. Personal communication with R. McCormick.

37. T. Allen, personal observation.

38. A general citation for the approach is Snowden (2010).

9. MANAGEMENT OF ECOLOGICAL SYSTEMS

1. Cannadine (1990). This is a serious academic account of the decline; it assumes a working knowledge of twentieth-century British politics.

2. For Washington, see Ellis (2004); Chernow (2010). For Jefferson, see Jolly (n.d.); Ellis (1998).

3. Holling (1986).

4. Petroski (1993). Extends the work of Sibley (1977); and Sibley and Walker (1977).

5. http://www.wsdot.wa.gov/TNBhistory/People/eyewitness.htm#4.

6. Krakauer (2011). Reviews a book on super cooperators and captions a picture of *Atta* ants that raise bacteria "to keep their crop healthy."

7. For an account of the disruption of the forest ecosystem by the lead smelter, see Jackson and Watson (1977). For a more general account and hard data on the effects of heavy metal insults on ecosystems, see O'Neill et al. (1977).

8. Gunderson and Holling (2001).

9. Schumpeter (1950). In 1986, Holling explicitly borrows "creative destruction" as a term from Schumpeter.

10. Odum (1995); Odum and Pinkerton (1955); see also Odum and Brown (2007) for different statements of maximum power over time.

11. Rappaport (1968).

12. Holling and Ewing (1971).

13. Carpenter et al. (2011); see also Biggs, Carpenter, and Brock (2009); Carpenter and Lathrop (2008); Carpenter and Brock (2006).

14. Hof and Bevers (1998, 2002).

15. Working group on human affairs consisted of Timothy Allen, James Brown, Joel Cohen, Jared Diamond, Robert Howarth, Edward J. Rykiel Jr., and Stefan Somer. The meeting is reported in Brown and Roughgarden (1990).

16. Garcia (2003).

17. Personal observation by Allen, in Nigeria.

18. Prahalad (2004) is the anchor of reference to Peter Day, BBC World Business, July 4, 2005; http://news.bbc.co.uk/2/hi/business/4648049.stm.

19. Watson, Loucks, and Wojner (1981).

20. O'Neill et al. (1975).

21. Evans (1956).

22. For instructive contemporary photographs of making chinampas, see Gómez-Pompa, "Chinampas las Chinampas Tropicales (1975–1976): Antecedentes de los Proyectos de Construcción de Chinampas en el Tropic"; http://gomezpompa.blogspot.com/2011/04/17-chinampas-tropicales.html17. See also Jimenez-Osornio and Rorive (1991); Gómez-Pompa (1978).

23. Allen, Tainter, and Hoekstra (2003).

24. Shepard (2013). PhD research into working that farm is in progress by Peter C. Allen.

25. Schneider and Kay (1994).

26. Allen, Tainter, and Hoekstra (2003).

27. Appolonio (2002).

28. Gates (1998).

29. http://www.ted.com/talks/dan_barber_s_surprising_foie_gras_parable.html.

30. Allen, Bandurski, and King (1993).

31. Allen et al. (2013).

32. Allen (2010).

33. Jevons 1866; Polimemi et al. 2008.

34. Allen, T. 2010.

10. A UNIFIED APPROACH TO BASIC RESEARCH

1. Zadeh (1965a, 1965b).

2. Bray and Curtis (1957).

3. For a complete list of publications coming from the Fraser Forest, see Alexander (1978).

4. Rosling, H. (2014) This website has a series Hans Rosling TED talks on aspects of human populations from 2006 to 2014).

5. Gapminder (2014) web site. The software for Gapminder is publicly available.

6. Caccianiga et al. (2006) say that the present intellectual posture of reductionism comes from the use of crops as devices for testing theory of populations. Crops live in a peculiar environment of monoculture and abundant nutrients, making them suitable for reduction to one or a few-species investigations.

7. Coleman (2010:34–35) explicitly states that the big grassland model "was not terribly robust in predicting multi-year changes in organic matter." In a personal communication to Allen, W. S. Overton said that an oral presentation of a long simulation reported that the big grassland model, PWNEE, covered the western plains with feet of buffalo dung due to rounding error alone. The written publication of Bledsoe et al. (1971) does not offer long enough runs for that remarkable phenomenon to occur. Presumably, the rounding error problem was a manifestation of what Coleman means by "not terribly robust."

8. Allen and Starr (1982).

9. Eigen (1977).

10. Prigogine (1978, 1982); Prigogine et al. (1969); Prigogine and Nicolis (1971).

11. Shugart and West (1977).

12. Pin cherry appeared shortly after the clear-cut at Likens's Hubbard Brook experiment and appeared to amend nutrient loss.

13. Wedin (1990); Wedin and Tilman (1990).

14. Rosen (1989).

15. Rosen (1962); Thompson (1942).

16. Ahl (1993); Ahl and Allen (1996).

17. Allen et al. (1993).

18. Milne et al. (1992).

19. Personal communication, September 12, 2013, in a talk given to the Botany Department Colloquium, Madison, Wis.

20. Porter and Gates (1969).

CONCLUSION

1. Sutherland (2006).

2. Seneff (2014). Stephanie Seneff's home page includes several pertinent articles; http://people.csail.mit.edu/seneff/. She comments about the seriousness of the situation. "Should the disinformation [generated by the medical establishment] be overcome, there may be massive, chronic liability, akin to the mess left after asbestos."

3. All these papers document doctors failing to understand statistics. Ecologists are unlikely to be a lot better. Gigerenzer et al. (2007); Gigerenzer and Brighton (2009); Gigerenzer and Sturm (2012); Wegwarth et al. (2012); Gigerenzer and Wegwarth (2013); Gigerenzer and Galesic (2012); and Gigerenzer (2013).

4. Ioannidis (2005).

5. An example of voodoo dance statistics occurred in the review process of Carpenter et al. (1987). It is a classic paper, which is in itself a commentary on the review process in *Science*. Of the review process, Carpenter said in an email to Allen in 2007 (published with permission), how he was forced into the ritual dance: "The original version of the

paper was rejected by *Science* without review because the editors felt it was of insufficient interest. So we wrote a longer paper and sent it to *Ecology*. At the time, the discipline was strongly influenced by Hurlbert's 1984 paper on pseudoreplication. The reviewers were therefore quite attentive to statistical aspects of the paper. I would have preferred to write it without so many parenthetical statistics, but there was no real alternative. . . . [T]he subject-matter editor for the paper, did a masterful job of winnowing the review comments."

6. Talk on September 12, 2012, before the Institute for New Economic Thinking, Berlin; http://slightlytilted.wordpress.com/tag/gerd-gigerenzer/.

7. All these works emphasize heuristics over logic: Gigerenzer (2012); Gigerenzer and Brighton (2009); Gigerenzer and Sturm (2012); and Gigerenzer and Goldstein (1996).

8. Gigerenzer and Muir Gray (2011).

9. The same argument is raised with Milan and Moderna in Volz et al. (2006); Gigerenzer and Muir Gray (2011).

10. For the interview with Bill Wimsatt, see http://innovation.ucdavis.edu/people /publications/37%20Bechtel%20et%20al%202006%20BIOT-1.2-213-219%20 Wimsatt%20Interview.pdf/ See also Bechtel et al. (2006).

11. Hayles (1991) juggles the distinction of *true* versus *false* (which she says in unoccupied), as opposed to *not true* versus *not false* (which she says is occupied). Philosophers do that sort of thing. That distinction leads to a clearer view of truth in her "Constrained Constructivism."

12. Quoted by Engels (1947).

13. Reiners and Lockwood (2009).

14. Personal communication with Michael Lissack, April 2, 2014. It made sense once Lissack said it.

REFERENCES

Ackoff, Russell L. 1981. *Creating the Corporate Future: Plan or Be Planned For*. New York: Wiley.

Aguilar, R., S. L. Loftin, T. J. Ward, K. A. Stevens, and J. R. Gosz. 1994. *Sewage Sludge Application in Semiarid Grasslands: Technical Completion Report*. Effects on vegetation and water quality, p. 75. New Mexico State University, New Mexico Water Resources Research Institute, Las Crusas.

Ahl, V. 1993. Cognitive development in infants prenatally exposed to cocaine. PhD dissertation, University of California, Berkeley.

Ahl, V., and T. F. H. Allen. 1996. *Hierarchy Theory: A Vision, Vocabulary and Epistemology*. New York: Columbia University Press.

Alexander, Robert R., compiler. 1978. *Gen. Tech. Rep. Rm-40-A*. Fort Collins, Colo.: USDA Forest Service.

Allen, C. R., and C. S. Holling. 2008. *Discontinuities in Ecosystems and Other Complex Systems*. New York: University of Columbia Press.

Allen, P. 2009. Energy gain and transitions in social and ecological systems. Master's thesis, Gaylord Nelson Institute for Environmental Studies, University of Wisconsin, Madison.

Allen, T. F. H. 1973. A microscopic pattern analysis of an epiphyllous tropical alga. *Phycopeltis expansa* Jennings. *J. Ecol*. 61:887–99.

———. 1998. The landscape level is dead: Persuading the family to take it off the respirator. In D. Peterson and V. T. Parker, eds., *Scale Issues in Ecology*. New York: Columbia University Press.

———. 2010. Making liveable sustainable systems unremarkable. *Systems Research and Behavioral Science* 27(5):469–79.

Allen, T. F. H., B. Bandurski, and A. W. King. 1993. *The Ecosystem Approach: Theory and Ecosystem Integrity*. Washington, D.C.: International Joint Commission.

Allen, T. F. H., T. Havlicek, and J. Norman. 2001. Wind tunnel experiments to measure vegetation temperature to indicate complexity and functionality. In Sergio Ulgiati, ed., *Proceedings of the Second Biennial International Workshop on Advances in Energy Studies*, Porto Venere, Italy, May 23–27, 2000.

Allen, T. F. H., A. W. King, B. Milne, A. Johnson, and S. Turner. 1993. The problem of scaling in ecology. *Evol. Trends in Plants* 7:3–8.

Allen, T. F. H., G. Mitman, and T. W. Hoekstra. 1993. Synthesis mid-century: J. T. Curtis and the community concept. In J. Fralish, R. P. McIntosh, and O. L. Loucks, eds., *John T. Curtis: Fifty Years of Wisconsin Plant Ecology*, 339. Madison: Wisconsin Academy of Arts and Sciences.

Allen, T. F. H., R.V. O'Neill, and T. W. Hoekstra. 1984. Interlevel relations in ecological research and management: Some working principles from hierarchy theory. *General Technical Report RM 110*. Fort Collins, Colo.: USDA Forest Service.

Allen, T. F. H., D. Shaw, P. C. Allen, and J. Spohrer. 2013. Insights into the relationship between products and services coming from biology. *Systems Research and Behavioral Science* 30:570–79.

Allen, T. F. H., and Thomas B. Starr. 1982. *Hierarchy: Perspectives for Ecological Complexity*. Chicago: University of Chicago Press.

Allen, T. F. H., J. A. Tainter, P. Allen, A. M. Malek, J. Flynn, and M. Flynn. 2009. Confronting economic profit with hierarchy theory: The concept of gain in ecology. *Systems Research and Behavioral Science* 26(5):583–99.

Allen, T. F. H., J. A. Tainter, J. Flynn, R. Steller, E. Blenner, M. Pease, and K. Nielsen. 2010. Integrating economic gain in biosocial systems. *Systems Research and Behavioral Science* 27(5):537–52. doi:10.1002/sres.1060.

Allen, T. F. H., J. Tainter, and T. W. Hoekstra. 2003. *Supply-Side Sustainability*. New York: Columbia University Press.

Allen, T. F. H., J. A. Tainter, J. C. Pires, and T. W. Hoekstra. 2001. Dragnet ecology—"Just the facts, ma'am": The privilege of science in a postmodern world. *Bioscience* 51:475–85.

Appolonio, S. 2002. *Hierarchical Perspectives on Marine Complexities*. New York: Columbia University Press.

Archibald, G. W. 1976. *Crane Taxonomy as Revealed by the Unison Call*. Proceedings of the International Crane Workshop. Baraboo, Wis.: International Crane Foundation.

Archibald, G. W., and C. M. Mirande 1986. *Population Status and Management Efforts for Endangered Cranes. Transactions of the Fiftieth North American Wildlife and Natural Resources Conference*. Washington, D.C.: Wildlife Management Institute.

Atsatt, P. 1988. Are vascular plants "inside-out" lichens? *Ecology* 69:17–23.

Austin, M. P., and B. G. Cook. 1974. Ecosystem stability: A result from an abstract simulation. *Theor. Biol.* 45:435–58.

Auton, T. R., and B. H. Woolen. 1991. A physiologically based mathematical model for the human inhalation pharmacokinetics of 1,1,2-trichloro-1,2,2-trifluoroethane. *Int. Arch. Occup. Environ. Health* 63(2):133–38.

Bakuzis, E. V., and H. Hansen. 1959. A provisional assessment of species synecological requirements in Minnesota forests. Minnesota Forestry Notes 84. St. Paul: School of Forestry, University of Minnesota.

Bateson, G. 1980. *Mind and Nature: A Necessary Unity*. New York: Bantam Books.

Beals, E. W. 1973. Ordination: Mathematical elegance and ecological naïveté. *J. Ecol.* 61:23–35.

Bechtel, W., W. Callebaut, J. Griesemer, and J. Schank. 2006. Interview: Bill Wimsatt on multiple ways of getting at the complexity of nature. *Biological Theory* 1(2):213–19. https://smartsite.ucdavis.edu/access/content/user/00003578/Publications/Bechtel -etal-BT2006.pdf.

Begon, Michael, John L. Harper, and Colin R. Townsend. 1986. *Ecology: Individuals, Populations and Communities*. Sunderland, Mass.: Sinauer.

Bell, R. H. V. 1971. A grazing system in the Serengeti. *Sci. Am.* 225:86–94.

Belsky, A. J. 1986. Does herbivory benefit plants? A review of the evidence. *Am. Nat.* 127:870–92.

Biggs, R., S. R. Carpenter, and W. A. Brock. 2009. Turning back from the brink: Detecting an impending regime shift in time to avert it. *Proceedings of the National Academy of Sciences* 106(3):826–31.

Bird, E. C. F. 1957. The use of the soil catena concept in the study of the ecology of the Wormley Woods, Hertfordshire. *Journal of Ecology* 45:465–69.

Bishop, C. M. 1995. *Neural Networks for Pattern Recognition*. Oxford: Oxford University Press.

Blackman, F. F., and A. G. Tansley. 1905. Ecology in its physiological and phyto-topographical aspects. *NewPhytol* 4:199–203, 232–53.

Bledsoe, L. J., R. C. Francis, G. L. Swartzman, and J. D. Gustafson. 1971. PWNEE: A Grassland Ecosystems Model. *Technical Report* (US International Biology Program) 64. Fort Collins: Colorado State University Digital Collections. http://hdl.handle.net /10217/16080.

Bleich, V., J. Kie, E. Loft, T. Stephenson, M. Oehler Sr., and A. Medina. 2005. Managing rangeland for wildlife. In C. Braun, ed., *Techniques for Wildlife Investigations and Management*, 873–97. Bethesda, Md.: Wildlife Society.

Bormann, F. H., and G. E. Likens. 1979. *Pattern and Process in a Forested Ecosystem*. New York: Springer-Verlag.

Bormann, F. H., T. G. Siccama, R. S. Pierce, and J. S. Eaton. 1974. The export of nutrients and recovery of stable conditions following deforestation at Hubbard Brook. *Ecol. Monog.* 44:255–77.

Bosserman, R. W. 1979. The hierarchical integrity of *Utricularia* (bladderworts) periphyton microsystems. *Okefenokee Ecosystems Investigations,* Technical Report No. 4, p. 266, Department of Zoology and Institute of Ecology, University of Georgia.

Box, George E. P., and Norman R. Draper. 1987. *Empirical Model-Building and Response Surfaces*. New York: Wiley.

Bradshaw, A. D. 1987. Comparison: Its scope and limits. In I. H. Rorison, ed., *Frontiers of Comparative Plant Ecology*, 3–21. London: Academic Press.

Bradshaw, A. D., and T. McNeilly. 1981. *Evolution and Pollution*. London: Arnold.

Brandner, T. A. 2003. Reconceptualizing biome: A complex systems theoretical approach to understanding extinction events. PhD dissertation, University of Wisconsin, Madison.

Braun-Blanquet, J. 1932. *Plant Sociology: The Study of Plant Communities*. Translated by G. D. Fuller and H. S. Conard. New York: McGraw-Hill.

Bray, J. R., and J. T. Curtis. 1957. An ordination of the upland forest communities of southern Wisconsin. *Ecol. Monogr.* 27:325–49.

Brighton, H., and G. Gigerenzer. 2012. Are rational actor models "rational" outside small worlds? In S. Okasha and K. Binmore, eds., *Evolution and Rationality: Decisions, Co-operation and Strategic Behavior,* chap. 5, 84–109. Cambridge: Cambridge University Press. doi.org/10.1017/CBO9780511792601.006.

Brown, B. J., and T. F. H. Allen. 1989. The importance of scale in evaluating herbivory impacts. *Oikos* 54:189–94.

Brown, B. J., and J. J. Ewel. 1987. Herbivory in complex and simple tropical successional ecosystems. *Ecology* 68:108–16.

Brown, H. T. 1905. The reception and utilization of energy by a green leaf. *Nature* 71:522–26.

Brown, J. H. 1995. *Macroecology.* Chicago: University of Chicago Press.

Brown, J. H., W. R. Burnside, A. D. Davidson, J. P. DeLong, W. C. Dunn, M. J. Hamilton, J. C. Nekola, J. G. Okie, N. Mercado-Silva, W. H. Woodruff, and W. Zuo. 2011. Energetic limits to economic growth. *Bioscience* 61:19–26.

Brown, J. H., and J. Roughgarden. 1990. Ecology for a changing earth. Final Report of a workshop held in Santa Fe, New Mexico, December 1988. *Bull. Ecol. Soc. of Am.* 71:173–88.

Brown, R. L., and J. D. Fridley. 2003. Control of plant species diversity and community invisibility by species immigration: Seed richness versus seed density. *Oikos* 102:15–24.

Buddenbrock, W. 1934. Uber die kinetische and statische Leistung grosser und kleiner Tiere und ihre Bedeutung fur dem Gesamtstoffwechsel. *Naturwissenschaft* 22:675–80.

Caccianiga, M., A. Luzzaro, S. Pierce, R. M. Ceriani, and B. Cerabolini. 2006. The functional basis of a primary succession resolved by CSR classification. *Oikos* 112:10–20.

Canestrari, D., D. Bolopo, T. Turlings, G. Röder, J. Marcos, and V. Baglione. 2014. From parasitism to mutualism: Unexpected interactions between a cuckoo and its host. *Science* 343:1350–52.

Cannadine, D. 1990. *The Decline and Fall of the British Aristocracy.* New Haven, Conn.: Yale University Press.

Carosella, Tommi Lou. 1978. Population responses of *Opuntia compressa* (Salisb.) Macbr. in a southern Wisconsin sand prairie. Master's thesis, University of Wisconsin, Madison.

Carpenter, S. R., and W. A. Brock. 2006. Rising variance: A leading indicator of ecological transition. *Ecology Letters* 9:311–18.

Carpenter, S. R., J. J. Cole, M. L. Pace, R. Batt, W. A. Brock, T. Cline, J. Coloso, J. R. Hodgson, J. F. Kitchell, D. A. Seekell, L. Smith, and B. Weidel. 2011. Early warnings of regime shifts: A whole-ecosystem experiment. *Science* 332:1079–82.

Carpenter, S. R., and J. F. Kitchell. 1988. Strong manipulations and complex interactions: Consumer control of lake productivity. *Bioscience* 38:764–69.

Carpenter, S. R., J. F. Kitchell, J. R. Hodgson, P. A. Cochran, J. J. Elser, M. M. Elser, D. M. Lodge, D. Kretchmer, X. He, and C. N. von Ende. 1987. Regulation of lake primary productivity by food web structure. *Ecology* 66:1863–76.

Carpenter, S. R., and R. C. Lathrop. 2008. Probabilistic estimate of a threshold for eutrophication. *Ecosystems* 11:601–13.

Carter, R. N., and S. D. Prince. 1981. Epidemic models used to explain bio-geographical limits. *Nature* 293:664–65.

Ceier, L. 2008. Home on the "range': Burrowing owls call Eglin Air Force Base home. Walton Outdoors, http://www.waltonoutdoors.com/home-on-the-range/.

Center for Biotic Systems. 1974. Environmental analysis of the Kickapoo River impound-ment. Report to the U.S. Army Corps of Engineers. *IES Report 28*. Madison: University of Wisconsin.

Chabot, B. F., and H. A. Mooney. 1985. *Physiological Ecology of North American Plant Communities*. New York: Chapman and Hall.

Chander, G., D. L. Helder, B. Markham, J. D. Dewald, E. Kaita, K. J. Thome, E. Micijevic, and T. A. Ruggles. 2004. Landsat-5 TM reflective-band absolute radiometric calibration. *IEEE Transactions on Geoscience and Remote Sensing* 42(12):2747–60.

Chapelle, G., and L. S. Peck. 1999. Polar gigantism dictated by oxygen availability. *Nature* 399:114–15.

Checkland, P. 1981. *Systems Thinking, Systems Practice*. Chichester, U.K.: Wiley.

Chernow, R. 2010. *Washington: A Life*. New York: Penguin Press.

Cittadino, Eugene. 1980. Ecology and the professionalization of botany in America, 1890–1905. *Stud. Hist. Biol.* 4:171–98.

———. 1990. *Nature as the Laboratory: Darwinian Plant Ecology in the German Empire, 1880–1900*. Cambridge: Cambridge University Press.

Clark, B., and R. York. 2005. Dialectical nature: Reflections in honor of the twentieth anni-versary of Levins and Lewontin's *The Dialectical Biologist*. *Monthly Review* 57(1); www.monthlyreview.org/2005/05/01/dialectical-nature.

Clements, Frederic E. 1904. Developments and structure of vegetation. *Rep. Bot. Survey of Nebraska*, 7.

———. 1905. *Research Methods in Ecology*. Lincoln, Neb.: University Publishing.

———. 1916. Plant succession. *Carnegie Inst. Wash. Publ.* 242:1–512.

Cohen, Joel E. 1971. Mathematics as metaphor: A review of dynamical system theory in biology. Vol. 1, stability theory and its applications by Robert Rosen. *Science* 172(3984).

Coleman, D. C. 2010. *Big Ecology: The Emergence of Ecosystem Science*. Berkeley: University of California Press.

Colinvaux, P. 1979. *Why Big Fierce Animals Are Rare*. Princeton, N.J.: Princeton University Press.

Collister, C. N. 1950. Nesting birds on a Keith County farm. *Nebr. Bird Rev.* 18:1–4.

Connell, J. H. 1978. Diversity in tropical rain forests and coral reefs. *Science* 199(4335): 1302–10.

———. 1980. Diversity and the coevolution of competitors, or the ghost of competition past. *Oikos* 35:131–38.

Cooper, W. S. 1926. The fundamentals of vegetational change. *Ecology* 7:391–414.

Corner, E. J. H. 1981. *The Life of Plants*. Chicago: University of Chicago Press.

Cottam, G., J. Curtis, and A. Catana. 1957. Some sampling characteristics of a series of aggregated populations: Ecology populations. *Ecology* 8:610–22.

Cottam, G., J. Curtis, and B. Hale. 1955. Some sampling characteristics of a population of randomly dispersed individuals. *Ecology* 34:741–57.

Cowles, H. C. 1899. The ecological relations of the vegetation of the sand dunes of Lake Michigan. *Bot. Gaz.* 27:95–117,167–202, 281–308, 361–91.

Cronon, W. 1992. A place for stories: Nature history and narrative. *Journal of American History* 78:1347–76.

Curtis, J. T. 1959. *The Vegetation of Wisconsin*. Madison: University of Wisconsin Press.

Curtis, J. T., and R. P. McIntosh. 1951. An upland forest continuum of the prairie-forest border region of Wisconsin. *Ecology* 32:476–96.

Darwin, C. 1958. *The Origin of Species.* New York: New American Library.

Dawkins, R. 2004. *The Ancestor's Tale: A Pilgrimage to the Dawn of Evolution.*a New York: Houghton Mifflin.

DeAngelis, D. L. 1975. Stability and connectance in food web models. *Ecology* 56:238–43.

DeAngelis, D. L., W. M. Post, and C. C. Travis. 1986. *Positive Feedback in Natural Systems.* New York: Springer-Verlag.

Deevey, E. S. 1947. Life tables for natural populations of animals. *Q. Rev. Biol.* 22:283–314. Reprinted in W. E. Hazen, ed., *Readings in Population and Community Ecology.* Philadelphia: Saunders.

Dermott, S. F., and C. Sagan. 1995. Tidal effects of disconnected hydrocarbon seas on Titan. *Nature* 374(6519):238–40.

Diamond, J. 1999. *Guns, Germs, and Steel.* New York: Norton.

———. 2005. *Collapse: How Societies Choose to Fail or Succeed.* New York: Viking.

Dodson, S., T. Allen, S. Capenter, A. Ives, R. Jeanne, J. Kitchell, N. Langston, and M. Turner. 1998. *Ecology.* New York: Oxford University Press.

Drude, O. 1890. Handbuch der Pflanzengeographie. Stuttgart, Germany: Engelhorn.

———. 1896. Deutschlands Pflanzengeographie. Stuttgart, Germany: Engelhorn.

Dudley, P. N., R. Bonazza, and W. Porter. 2013. Consider a non-spherical elephant: Computational fluid dynamics simulations of heat transfer coefficients and drag verified using wind tunnel experiments. *J. Exp. Zool.* 319:319–27.

Dury, G. H. 1961. *The British Isles: A Systematic and Regional Geography.* New York: Norton.

Eglash, R. 1999. *African Fractals: Modern Computing and Indigenous Design.* Piscataway, N.J.: Rutgers University Press.

Eickmeier, W., M. S. Adams, and D. Lester. 1975. Two physiological races of *Tsuga canadensis* in Wisconsin. *Canadian Journal of Botany* 53:940–51.

Eigen, M. 1977. The hypercycle: Principle of natural self-organization. *Natur-wissenschaften* 64:541–65.

Eigen, M., and P. Schuster. 1979. *The Hypercycle: A Principle of Natural Self-Organization.* Berlin: Springer-Verlag.

Ellis, D., W. Sladen, W. Lishman, K. Clegg, J. Duff, G. Gee, and J. Lewis. 2003. Motorized migrations: The future or mere fantasy? *BioScience* 53(3):260–64.

Ellis, J. J. 1998. *American Sphinx: The Character of Thomas Jefferson.* New York: Vintage.

———. 2004. *His Excellency, George Washington.* New York: Alfred A. Knopf.

Emanuel, W. R., H. H. Shugart, and D. C. West. 1978. Spectral analysis and forest dynamics: Long term effects of environmental perturbations. In H. H. Shugart, ed., *Time Series and Ecological Processes.* Philadelphia: Society of Industrial and Applied Mathematics.

Emanuel, W. R., D. C. West, and H. H. Shugart. 1978. Spectral analysis of forest model time series. *Ecol. Modeling* 4:313–26.

Engels, F. 1947. *Anti-Dühring: Herr Eugen Dühring's Revolution in Science.* Moscow: Progress Publishers.

Evans, E. 1956. The ecology of peasant life in Western Europe. In W. L. Thomas Jr., ed., *Man's Role in Changing the Face of the Earth,* 264–86. Chicago: University of Chicago Press.

Falk, D. 1992. *Braindance: New Discoveries about Human Origins and Brain Evolution.* New York: Holt.

Farlow. J. O. 1976. A consideration of the trophic dynamics of a late-Cretaceous large-dinosaur community (Oldman Formation). *Ecology* 57:841–57.

Finch, D., and J. A. Tainter. 1995. Ecology, diversity, and sustainability of the middle Rio Grande basin. *General Technical Report RM-GTR-268.* Fort Collins, Colo.: USDA Forest Service.

Forbes, S. A. 1887. The lake as a microcosm. *Bulletin of the Peoria Scientific Association,* 77–87. Reprinted in 1925 in the *Bulletin of the Illinois State Natural History Survey* 15:537–50.

Forman, R. T. T., and M. Godron. 1986. *Landscape Ecology.* Chichester, U.K.: Wiley.

Fridley, J. D. 2001. The influence of species diversity on ecosystem productivity: How, where, and why? *Oikos* 93:514–26.

——. 2002. Resource availability dominates and alters the relationship between species diversity and ecosystem productivity in experimental plant communities. *Oecologia* 132:271–77.

Fridley, J. D., and J. P. Grime. 2010. Community and ecosystem effects of intraspecific genetic diversity in grassland microcosms of varying species diversity. *Ecology* 91:2272–83.

Funtowicz, S., and J. Ravetz. 1992. The good, the true and the post-modern. *Futures* 24:963–76.

Gapminder. 2014. www.gapminder.org.

Garcia, C. 2003. Development of a GIS based tool for stormwater management on the University of Wisconsin, Madison campus. Master's thesis, University of Wisconsin, Madison.

Gardner, M., and W. R. Ashby. 1970. Connectance of large dynamic (cybernetic) systems: Critical values for stability. *Nature* 228:784.

Gardner, Robert H., Virginia H. Dale, and R. V. O'Neill. 1990. Error propagation and uncertainty in process modeling. In R. K. Dixon, R. S. Meldahl, G. A. Ruark, and W. G. Warren, eds., *Process Modeling in Forest Growth Responses to Environmental Stress.* Portland, Ore.: Timber Press.

Gardner, R. H., B. T. Milne, M. G. Turner, and R. V. O'Neill. 1987. Neutral models for the analysis of broad-scale landscape pattern. *Landscape Ecology* 1:19–28.

Gates, J. 1998. *The Ownership Solution: Toward a Shared Capitalism for the 21st Century.* Reading, Mass.: Addison Wesley.

Gefen, Y., A. Aharony, and S. Alexander. 1983. Anomalous diffusion on percolating clusters. *Phys. Rev. Lett.* 50:77–80.

Gerard, R. W. 1969. Hierarchy, entitation and levels. In L. L. Whyte, A. G. Wilson, and D. Wilson, eds., *Hierarchical Structures,* 215–28. New York: American Elsevier.

Giampietro, M. 2003. *Multi-Scale Integrated Analysis of Agro-ecosystems.* Boca Raton, Fla.: CRC Press.

Giampietro, M., R. Aspinall, J. Ramos-Martin, and S. Bukkens. 2014. *Resource Accounting for Sustainability: The Nexus Between Energy, Food, Water and Land Use.* London: Routledge.

Giampietro, M., and K. Mayumi, 2009. *The Biofuel Delusion: The Fallacy Behind Large-Scale Agro-biofuels Production.* London: Earthscan Research Edition.

Gigerenzer, G. 2012. What can economists know? Rethinking the basis of economic understanding. Institute for New Economic Thinking, Berlin, September 12; http://slightlytilted.wordpress.com/tag/gerd-gigerenzer/.

——. 2013. HIV screening: Helping clinicians make sense of test results to patients. *BMJ* 347:f5151.

Gigerenzer, G., and H. Brighton. 2009. *Homo heuristicus*: Why biased minds make better inferences. *Topics in Cognitive Science* 1:107–43.

Gigerenzer, G., W. Gaissmaier, E. Kurz-Milcke, L. M. Schwartz, and S. Woloshin. 2007. Helping doctors and patients to make sense of health statistics. *Psychological Science in the Public Interest* 8:53–96.

Gigerenzer, G., and M. Galesic. 2012. Why do single event probabilities confuse patients? Statements of frequency are better for communicating risk. *BMJ* 344:e245. doi:10.1136/bmj.e245.

Gigerenzer, G., and D. G. Goldstein. 1996. Reasoning the fast and frugal way: Models of bounded rationality. *Psychological Review* 103:650–69.

Gigerenzer, G., and J. A. Muir Gray, eds. 2011. *Better Doctors, Better Patients, Better Decisions: Envisioning Health Care 2020*. Cambridge, Mass.: MIT Press.

Gigerenzer, G., and T. Sturm. 2012. How (far) can rationality be naturalized? *Synthese* 187:243–68.

Gigerenzer, G., and O. Wegwarth. 2013. Five year survival rates can mislead. *BMJ* 346:f548. doi:10.1136/bmj.f54.

Gleason, H. A. 1910. The vegetation of the inland sand deposits of Illinois. *Bull. Ill. St. Lab. Nat. Hist.* 9:21–174.

———. 1917. The structure and development of the association. *Bull. Torrey Bot. Club* 43:463–81.

———. 1920. Some applications of the quadrat method. *Bull. Torrey Bot. Club* 47:21–34.

———. 1922. The vegetational history of the Middle West. *Ann. Assoc. Amer. Geogr.* 12:39–85.

———. 1926. The individualistic concept of the plant association. *Contrib. NY Bot. Gard.* 279.

———. 1929. Plant associations and their classification: A reply to Dr. Nichols. *Proceedings of the International Congress of Plant Sciences*; vol. 1, 624–41. Banta, Menasha, Wis. Congress Meeting at Ithaca, N.Y., August 16–23, 1926.

———. 1939. The individualistic concept of the plant association (with discussion). *Am. Midland Nat.* 21:92–110.

———. 1953. Autobiographical letter. Bull. Ecol. Soc. Am. 34:40–42.

Gleick, J. 1987. *Chaos: Making a New Science*. New York: Viking Penguin.

Gómez-Pompa, A. 1978. *Vino Nuevo en Odre Vieja: Mazingira 5*. Oxford: Pergamon Press.

Goodall, D. 1954. Objective methods for the classification of vegetation. III. An essay in the use of factor analysis. *Austral. J. Bot.* 2:304–24.

Green, M.B., A. S. Bailey, S. W. Bailey, J. J. Battles, J. L. Campbell, C. T. Driscoll, T. J. Fahey, L. C. Lepine, G. E. Likens, S. V. Ollinger, and P. G. Schaberg. 2013. Decreased water flowing from a forest amended with calcium silicate. *Proc. Natl. Acad. Sci.* 110(15):5999–6003. doi:10.1073/pnas.1302445110.

Greig-Smith, P. 1964. *Quantitative Plant Ecology*. 2nd ed. London: Butterworths.

———. 1983. *Quantitative Plant Ecology*. 3rd ed. London: Blackwell.

Griggs, Robert F. 1914. Observations on the behavior of some species at the edges of their ranges. *Bull. Torrey Club* 41:25–49.

Grime, J. P. 1974. Vegetation classification by reference to strategies. *Nature* 250:26–31. doi:10.1038/250026a0.

———. 1977. Evidence for the existence of three primary strategies in plants and its relevance to ecological and evolutionary theory. *American Naturalist* 111:1169–94.

———. 1985. Towards a functional classification of vegetation. In J. White, ed., *The Population Structure of Vegetation*, 503–14. Dordrecht, Netherlands: Dr. W. Junk.

———. 2001. *Plant Strategies, Vegetation Processes and Ecosystem Properties*. 2nd ed. Chichester, U.K.: Wiley.

———. 2007. Plant strategy theories: A comment on Craine (2005). *Journal of Ecology* 95(2):227–230. doi:10.1111/j.1365–2745.2006.01163.x.

Grime, J. P., and A. V. Curtis. 1976. The interaction of drought and mineral nutrient stress in calcareous grassland. *Journal of Ecology* 64:976–98.

Grime, J. P., J. G. Hodgson, and R. Hunt, eds. 2007. *Comparative Plant Ecology: A Functional Approach to Common British Species*. Dalbeattie, U.K.: Castlepoint Press.

Grime, J. P., K. Thompson, R. Hunt, J. G. Hodgson, J. H. C. Comelissen, I. H. Rorison, G. A. F. Hendry, T. W. Ashenden, A. P. Askew, S. R. Band, R. E. Booth, C. C. Bossard, B. D. Campbell, J. E. L. Cooper, A. W. Davison, P. L. Gupta, W. Hall, D. W. Hand, M. A. Hannah, S. H. Hillier, D. J. Hodkinson, A. Jalili, Z. Liu, J. M. L. Mackey, N. Matthews, M. A. Mowforth, A. M. Neal, R. J. Reader, K. Reiling, W. Ross-Fraser, R. E. Spencer, F. Sutton, D. E. Tasker, P. C. Thorpe, and J. Whitehouse. 1997. Integrated screening validates primary axes of specialisation in plants. *Oikos* 79:259–81.

Grisebach, August. 1838. Ueber den Einfluss des climas auf die Begranzung der naturlichen Flora. *Linnaea* 52:159–200.

Gucinski, H., C. Miner, and B. Bittner, eds. 2004. *Proceedings, Views from the Ridge: Considerations for Planning at the Landscape Scale*. U.S. Forest Service, Pacific Northwest Research Station, General Technical Report PNW-GTR-596 USDA FS, Portland, Oregon.

Gunderson, L. H., and C. S. Holling, eds. 2001. *Panarchy: Understanding Transformations in Human and Natural Systems*. Washington, D.C.: Island Press.

Gurney, K. 1997. *An Introduction to Neural Networks*. London: Routledge.

Gurney. M., C. D. Preston, J. Barrett, and D. Briggs. 2007. Hybridisation between Oxlip *Primula elatior* (L.) Hill and Primrose *P. vulgaris* Hudson, and the identification of their variable hybrid *P. ×digenea* A. Kerner. *Watsonia* 26:239–51.

Haldane, J. B. S. 1963. On being the right size. In R. M. Hutchins, M. J. Adler, and C. Fadiman, eds., *Gateway to the Great Books, No. 8, Natural Science*. Chicago: Encyclopedia Britannica.

Haralick, R. M. 1979. Statistical and structural approaches to texture. *Proceedings of the IEEE* 67:786–804.

Haralick, R. M., and K. S. Shanmugam. 1974. Combined spectral and spatial processing of ERTS imagery data. *Remote Sensing of Environment* 3:3–13.

Haralick, R. M., K. S. Shanmugam, and I. Dinstein. 1973. Textural features for image classification. *IEEE. Transactions on Systems, Man and Cybernetics SMC* 3:610–21.

Harper, J. L. 1967. A Darwinian approach to plant ecology. *Journal of Ecology* 55:247–70.

———. 1972. Ecological study of form. *Science* 176:662.

———. 1977. *Population Biology of Plants*. Chicago: Academic Press.

———. 1982. After description. In E. I. Newman, ed., *The Plant Community as a Working Mechanism*, 11–25. Oxford: Blackwell.

Harper, J. L., and I. H. McNaughton. 1962. The comparative biology of closely related species living in the same area: VII. Interference between individuals in pure and mixed populations of *Papaver* species. *New Phytologist* 61:175–88.

Harper, J. L., J. T. Williams, and G. R. Sagar. 1965. The behavior of seeds in soil. Part 1. The heterogeneity of soil surfaces and its role in determining the establishment of plants from seed. *J. Ecol.* 53:273–86.

Harris, W. F., R. S. Kennerson, and N. T. Edwards. 1977. Comparison of below ground biomass of natural deciduous forest and Loblolly Pine plantations. *Pedobiologia* 17:369–81.

Harestad, A., and F. Bunnell. 1979. Home range and body weight - a reevaluation. *Ecology* 60: 389–402.

Hayes, M. A., A. E. Lacy, J. Barzen, S. E. Zimorski, K. A. L. Hall, and K. Suzuki. 2007. An unusual journey of non-migratory whooping cranes. *Southeastern Naturalist* 6(3):551–58.

Hayles, N. K. 1991. Constrained constructivism: Locating scientific inquiry in the theater of representation. *New Orleans Review* 18:76–85.

Hazardous Substances Data Bank (HSDB). 1994. MEDLARS Online Information Retrieval System, National Library of Medicine.

Hector, A., B. Schmid, C. Beierkuhnlein, M. C. Caldeira, M. Diemer, P. G. Dimitrakopoulos, J. A. Finn, H. Freitas, P. S. Giller, J. Good, R. Harris, P. Hogberg, K. Huss-Danell, J. Joshi, A. Jumpponen, C. Körner, P. W. Leadley, M. Loreau, A. Minn, C. P. H. Mulder, G. O'Donovan, S. J. Otway, J. S. Pereira, A. Prinz, D. J. Read, M. Scherer-Lorenzen, E. D. Schulze, A. S. D. Siamantziouras, E. M. Spehn, A. C. Terry, A. Y. Troumbis, F. I. Woodward, S. Yachi, and J. H. Lawton. 1999. Plant diversity and productivity experiments in European grasslands. *Science* 286:1123–27.

Heglund, N. C., C. R. Taylor, and T. A. McMahon. 1974. Scaling stride frequency and gait to animal size: Mice to horses. *Science* 186(4169):1112–13.

Herr, D. G. 1980. On the history of the use of geometry in the general linear model. *American Statistician* 34:43–47. doi:10.1080/00031305.1980.10482710.

Heslop-Harrison, J. 1964. *New Concepts in Flowering-Plant Taxonomy*. Cambridge, Mass.: Harvard University Press.

Hetherington, A. H., and F. I. Woodward. 2003. The role of stomata in sensing and driving environmental change. *Nature* 424:903–8.

Hocking, M. D., and T. E. Reimchen. 2002. Salmon-derived nitrogen in terrestrial invertebrates from coniferous forests of the Pacific Northwest. *BMC Ecology* 2, article 4.

Hoekstra, T. W., T. F. H. Allen, and C. H. Flather. 1991. The implicit scaling in ecological research: On when to make studies of mice and men. *Bioscience* 41:148–54.

Hoekstra, T. W., T. F. H. Allen, J. Kay, and J. A. Tainter. 1999. Appendix H: Criteria and indicators for ecological and social systems sustainability with system management objectives. In *North American Test of Criteria and Indicators of Sustainable Forestry*, vol. 1, compiled by S. Woodley, G. Alward, L. Iglesias Gutierrez, T. Hoekstra, B. Holt, L. Livingston, J. Loo, A. Skibicki, C. Williams, and P. Wright. USDA Forest Service Inventory and Monitoring Institute, Report No. 3; http://www.fs.fed.us/institute/cifor/cifor_1.html.

Hof, J., and M. Bevers. 1998. *Spatial Optimization for Managed Ecosystems*. New York: Columbia University Press.

———. 2002. *Spatial Optimization in Ecological Applications*. New York: Columbia University Press.

Holling, C. S. 1964. The analysis of complex population processes. *Canadian Entomologist* 96:335–47.

———, ed. 1978. *Adaptive Environmental Assessment and Management.* Chichester, U.K.: Wiley.

———. 1986. The resilience of terrestrial ecosystems: Local surprise and global change. In William C. Clark and R. E. Munn, eds., *Sustainable Development of the Biosphere.* Cambridge: Cambridge University Press.

Holling C. S., and S. Ewing. 1971. Blind man's bluff: Exploring the response space generated by realistic ecological simulation models. In G. P. Patil, E. C. Pielou, and W. E. Waters, eds., *Statistical Ecology. Vol. 2. Proceedings of the International Symposium on Statistical Ecology.* University Park: Pennsylvania State University Press.

Homer. 2003. *The Odyssey.* Translated by G. H. Palmer. New York: Barnes and Noble Classics.

Horn, H. 1971. *The Adaptive Geometry of Trees.* Princeton Monographs in Population Biology 3. Princeton, N.J.: Princeton University Press.

Huler, Scott. 2004. *Defining the Wind: The Beaufort Scale and How a 19th Century Admiral Turned Science Into Poetry.* New York: Three Rivers Press.

Humboldt, A. von, and A. Bonpland. 1807. *Essai sur la geographie des plants.* Paris: F. Schoell; reprint New York: Arno Press, 1977.

Hurlbert, S. H. 1984. Pseudoreplication and the design of ecological field experiments. Ecological Society of America. *Ecological Monographs* 54(2):187–211.

Huston, M. A., L. W. Aarssen, M. P. Austin, B. S. Cade, J. D. Fridley, E. Garnier, J. P. Grime, J. Hodgson, W. K. Lauenroth, K. Thompson, J. H. Vandermeer, and D. A. Wardle. 2000. No consistent effect of plant diversity on productivity. *Science* 289:1255a.

Huston, M., D. DeAngelis, and W. Post. 1988. New computer models unify ecological theory. *Bioscience* 38:682–91.

Hutchinson, G. E. 1957. Concluding remarks: Cold Spring Harbor. *Symp. Quant. Biol.* 22:414–27.

———. 1959. Homage to Santa Rosalia or Why are there so many kinds of animals? *American Naturalist* 93:145–59.

Hyatt, Kim D., and John D. Stockner. 1985. Responses of sockeye salmon *Onchrhynchus nerka* to fertilization of British Columbian coastal lakes. *Can. J. Fish. Aqu. Sci.* 42:320–31.

Imbus, H. R., and C. Adkins. 1972. Physical examinations of workers exposed to trichlorotrifluoroethane. *Arch. Environ. Health* 24:257–61.

International Crane Foundation. 2013. *Cranes, Symbols of Survival: Ten Year Strategic Vision of the International Crane Foundation.* Baraboo, Wis.: ICF. https://www.savingcranes.org/images/stories/pdf/about_icf/icf_strategic_plan_symbols_of_survival.pdf.

Ioannidis J. P. A. 2005. Why most published research findings are false. PLoS Med 2(8): e124. doi:10.1371/journal.pmed.0020124.

Jackson, D. R., and A. P. Watson. 1977. Disruption of nutrient pools and transport of heavy metals in a forested watershed near a lead smelter. *J. Env. Qual.* 6:331–38.

Jevons, W. S. 1866. *The Coal Question* (2nd ed.). London: Macmillan and Company.

Jimenez-Osornio, J. J., and N. Rorive, eds. 1991. *Memorias del Simposio Internacional Sobre Camellones y Chinampas Tropicales.* Villahermosa, Tabasco: Publicaciones de la Universidad Autónoma de Yucatán.

Johanson, D., and M. Edey. 1981. *Lucy, the Beginnings of Humankind.* New York: Simon and Schuster.

Johnsgard, P. A. 1979. *Birds of the Great Plains: The Breeding Species and Their Distribution.* Lincoln: University of Nebraska Press.

———. 1980. *A Preliminary List of the Birds of Nebraska and Adjacent Plains States*. Lincoln: University of Nebraska Press, School of Life Sciences.

Jolly, S. K. (n.d.) *Jeffersonian Federalism: State Rights and Federal Power*. Syracuse, N.Y.: Maxwell School of Syracuse University; faculty.maxwell.syr.edu/skjolly/jeffersonianfederalism .pdf, pp. 13–16.

Karieva, P. M., D. Skelly, and M. Ruckleshouse. 1997. Reevaluating the use of models to predict the consequences of habitat loss and fragmentation. In S. T. A. Pickett, R. S. Ostfeld, H. Schachak, and G. E. Liens, eds., *The Ecological Basis of Conservation*. New York: Chapman Hall.

Kauffman, S. 1993. *The Origins of Order*. New York: Oxford University Press.

Kays, S., and J. L. Harper. 1974. The regulation of plant and tiller density in a grass sward. *J. Ecol.* 62:97–105.

Kelly, Kevin T., and Conor Mayo-Wilson. 2010. Ockham efficiency theorem for stochastic empirical methods. *Journal of Philosophical Logic* 39:679–712.

Kershaw, K. A. 1973. *Quantitative and Dynamic Plant Ecology*. 2d ed. New York: Elsevier.

Keymer, J. E., P. Marquet, and A. R. Johnson. 1998. Pattern formation in a patch occupancy metapopulation model: A cellular automata approach. *J. Theor. Biol.* 194:79–90.

Kimmerer, R. W., and T. F. H. Allen. 1982. The role of disturbance in the pattern of riparian bryophyte community. *American Midland Naturalist* 107:37.

Kleiber, M. 1961. *The Fire of Life: An Introduction to Animal Energetics*. Chichester, U.K.: Wiley.

Knight, D., and O. L. Loucks. 1969. Quantitative analysis of Wisconsin forest vegetation on the basis of plant function and gross morphology. *Ecology* 50:219–34.

Knox, E. B. 1998. The use of hierarchies as organizational models in systematics. *Biological Journal of the Linnean Soc.* 63:1–49.

Korzybski, A. 1931. A non-Aristotelian system and its necessity for rigour in mathematics and physics. Paper presented before the American Mathematical Society, meeting of the American Association for the Advancement of Science, New Orleans, Louisiana, December 28. Reprinted in *Science and Sanity*, 1933, 747–61.

Krakauer, D. 2011. Laws of cooperation. *Science* 332:538–39.

Kratz, T. K., B. J. Benson, E. Blood, G. L. Cunningham, and R. A. Dahlgren. 1991. The influence of landscape position on temporal variability in four North American ecosystems. *Am. Nat.* 138:355–78.

Krebs, H. A., and P. D. J. Weitzman. 1987. *Krebs' Citric Acid Cycle: Half a Century and Still Turning*. London: Biochemical Society.

Krejci, K., and T. Dewey. (n.d.) *Antilocapra Americana* pronghorn. http://animaldiversity .ummz.umich.edu/accounts/Antilocapra_americana/#geographic_range.

Krummel, J. R., R. H. Gardner, G. Sugihara, and R. V. O'Neill. 1987. Landscape patterns in a disturbed environment. *Oikos* 48:321–24.

Kuhn, T. 1970. *Structure of Scientific Revolutions*. Chicago: University of Chicago Press.

Lane, N. 2011. Energetics and genetics across the prokaryote-eukaryote divide. *Biol. Direct.* 6:35. doi:10.1186/1745–6150-6-35.

———. 2012. Is complex life a freak accident? UCL lunchtime lecture; http://nick-lane.net /Nick%20Lane%20Talks.htm.

Lascaux, R. 1921. *La Production et la Population*. Paris: Payot.

LeBrasseur, R. J., and T. R. Parsons. 1979. Addition of nutrients to a lake leads to greatly increased catch of salmon. *Environ. Conserv.* 6:187–90.

Leslie, P. H. 1945. On the use of matrices in certain population mathematics. *Biometrika* 33:183–212.

Levin, Simon. 1992. The problem of pattern and scale in ecology. *Ecology* 73:1943–67.

Levin, Steve. 1986. The icosahedron as the three-dimensional finite element in biomechanical support. In J. Dillon, ed., *Proceedings International Conference on Mental Images, Values and Reality.* Salinas, Calif.: Intersystems Publication.

Levins, R. 1974. Discussion paper: The qualitative analysis of partially specified systems. *Ann. NY Acad. Sci.* 123:38.

Levins, R., and R. Lewontin. 1985. *The Dialectical Biologist.* Cambridge, Mass.: Harvard University Press.

Li, M., K. Xu, M. Watanabe, and Z. Chen. 2007. Long-term variations in dissolved silicate, nitrogen, and phosphorus flux from the Yangtze River into the East China Sea and impacts on estuarine ecosystem, coastal and shelf. *Science* 71:3–12.

Liebers, D., P. de Knijff, and A. J. Helbig. 2004. The herring gull complex is not a ring species. *Proceedings of the Royal Society* B 271(1542):893–901.

Likens, G. E., F. H. Bormann, N. M. Johnson, and R. S. Pierce. 1967. The calcium, magnesium and potassium budgets for a small forested ecosystem. *Ecology* 48:772–85.

Likens, G. E., F. H. Bormann, R. S. Pierce, and N. M. Johnson. 1977. *Bio-geochemistry of a Forested Ecosystem.* New York: Springer-Verlag.

Likens, G. E., F. H. Bormann, R. S. Pierce, and W. A. Reiners. 1978. Recovery of a deforested ecosystem. *Science* 199:492–96.

Lindeman, R. L. 1942. The trophic-dynamic aspect of ecology. *Ecology* 23:399–418.

Lindeman, F. 1930. Plenty-coups, Chief of the Crows. Reprint 1962. Lincoln: University of Nebraska Press.

Liu, J., A. Li, and S. Seneff. 2011. Automatic drug side effect discovery from online patient-submitted reviews: Focus on statin drugs. 91–6. 1st International Conference on Advances in Information Mining and Management, Barcelona, Spain, October, 2011; http://people.csail.mit.edu/seneff/.

Liu, Z., J. M. L. Mackey, N. Matthews, M. A. Mowforth, A. M. Neal, R. J. Reader, K. Reiling, W. Ross-Fraser, R. E. Spencer, F. Sutton, D. E. Tasker, P. C. Thorpe, and J. Whitehouse. 1997. Integrated screening validates primary axes of specialisation in plants. *Oikos* 79: 259–81.

Lotka, A. J. 1956. *Elements of Mathematical Biology.* New York: Dover.

Lovelock, J. E. 1989. *The Ages of Gaia.* Oxford: Oxford University Press, UK.

Lovelock, J. E., R. J. Maggs, and R. J. Wade. 1973. Halogenated hydrocarbons in and over the Atlantic. *Nature* 241 (5386): 194. doi:10.1038/241194a0.

Loucks, O. L. 1962. Ordinating forest communities by means of environmental scalars and phytosociological indices. *Ecol. Monogr.* 32:137–66.

Lougheed, L. W., and D. J. Anderson. 1999. Parent blue-footed boobies suppress siblicidal behavior of offspring. *Behav. Ecol. Sociobiol.* 45:11–18.

MacArthur, R. 1958. Population of ecology of some warblers of northeastern coniferous forests. *Ecology* 39:599–619.

———. 1972. Strong, or weak, interactions? In E. S. Deevey, ed., *Growth by Intussusception: Ecological Essays in Honor of G. Evelyn Hutchinson. Trans. Conn. Acad. Arts Sci.* 44:177–88.

MacMillan, Conway. 1892. *The Metaspermae of the Minnesota Valley: A List of the Higher Seed-Producing Plants Indigenous to the Drainage Basin of the Minnesota River.* Minneapolis: Harrison and Smith.

———. 1899. *Minnesota Plant Life: Report of the Survey, Botany Series III.* Saint Paul: University of Minnesota.

Magnuson, John J. 1988. Two worlds for fish recruitment: Lakes and oceans. *Am. Fish. Soc. Sym.* 5:1–6.

———. 1991. Fish and fisheries ecology. *Ecol. Applications* 1:13–26.

Magnuson, John J., Cynthia A. Paszkowski, Frank J. Rahel, and William Tonn. 1989. Fish ecology in severe environments of small isolated lakes in northern Wisconsin. In R. R. Sharitz and J. W. Gibbons, eds., *Freshwater Wetlands and Wildlife,* Conf-8603101, DOE Symposium Series No. 61. Oak Ridge: USDOE Office of Scientific and Technical Information.

Mahmoud, A., and J. P. Grime. 1976. An analysis of competitive ability in three perennial grasses. *New Phytologist* 77:431–35.

Malthus, T. 1798. *An Essay on the Principle of Population.* Edited by Goeffrey Gilbert. Reissued 1999. Oxford World's Classics. Oxford: Oxford University Press.

Malthus, T. 1830. A summary view of the principle of population. In *Three Essays on Population,* 13–59. Reprint 1958. New American Library. New York: Mentor.

Mandelbrot, Benoît. 1967. How long is the coast of Britain? Statistical self-similarity and fractional dimension. *Science* 156(3775):636–38.

———. 1977. *Fractals: Form, Chance and Dimension.* San Francisco: W. H. Freeman.

Margalef, R. 1968. *Perspectives in Ecological Theory.* Chicago: University of Chicago Press.

———. 1972. Homage to Evelyn Hutchison, or why there is an upper limit to diversity. In E. Deevey, ed. *Growth by Intussusception: Ecological Essays in Honor of Evelyn Hutchinson. Trans. Conn Acad. Arts and Sci.* 44: 213–35.

May, R. M. 1972. Will a large complex system be stable? *Nature* 238:413–14.

———. 1974. Stability and complexity in model ecosystems. *Mono. Pop. Biol.* 6:1–265.

———. 1981a. Models for single populations. In R. M. May, ed., *Theoretical Ecology: Principles and Applications.* 2d ed. Sunderland, Mass.: Sinauer.

———. 1981b. Patterns in multi-species communities. In R. M. May, ed., *Theoretical Ecology: Principles and Applications.* 2d ed., 197–227. Sunderland, Mass.: Sinauer.

———, ed. 1981c. *Theoretical Ecology: Principles and Applications.* 2d ed. Sunderland, Mass.: Sinauer.

McCune, B., and T. F. H. Allen. 1985a. Forest dynamics in the Bitterroot Canyons, Montana. *Can. J. Bot.* 63:377–83.

———. 1985b. Will similar forests develop on similar sites? *Can. J. Bot.* 63:367–76.

McIntosh, R. P. 1967. The continuum concept of vegetation. *Bot. Rev.* 33:130–87.

McLuhan, M. 1962. *The Gutenberg Galaxy: The Making of Typographic Man.* Toronto: University of Toronto Press, Scholarly Publishing Division.

McMahon, T. A., and J. T. Bonner. 1983. *On Size and Life.* New York: Scientific American Library.

McMurtrie, R. F. 1975. Determinant of stability of large randomly connected systems. *Theor. Biol.* 50:1–11.

McNaughton, S. J. 1986. On plants and herbivores. *Am. Nat.* 128:765–70.

Mendelssonn, K. 1976. *The Riddle of the Pyramids*. London: Thames and Hudson, 1974; Sphere Cardinal edition.

Miller, James G. 1978. *Living Systems*. New York: McGraw-Hill.

Milne, B. T. 1988. Measuring the fractal geometry of landscapes. *Applied Mathematics and Computation* 27:67–79.

———. 1990. Lessons from applying fractal models to landscape patterns. In M. G. Turner and R. H. Gardner, eds., *Quantitative Methods in Landscape Ecology*. New York: Springer-Verlag.

———. 1991. The utility of fractal geometry in landscape design. *Landscape and Urban Planning* 21:81–90.

———. 1992. Spatial aggregation and neutral models in fractal landscapes. *American Naturalist* 139:32–57.

Milne, B. T., Kevin M. Johnston, and Richard T. T. Forman. 1989. Scale-dependent proximity of wildlife habitat in a spatially neutral Bayesian model. *Landscape Ecology* 2:101–10.

Milne, Bruce, Monica G. Turner, John A. Wiens, and Alan R. Johnson. 1990. Interactions between fractal geometry of landscapes and allometric herbivory. *Bull. Ecol. Soc. Am.* 71:257.

Milne, B., M. G. Turner, J. A. Wiens, and A R. Johnson. 1992. Interactions between fractal geometry of landscapes and allometric herbivory. *Theoretical Population Biology* 41:337–53.

Milne, G. 1935. Some suggested units for classification and mapping, particularly for East African soils. *Soil Res.* 4:1–27.

Minchin, P. 1987. An evaluation of the relative robustness of techniques for ecological ordination. *Vegetatio* 69:89–107.

Mitman, G. 1999. *Reel Nature: America's Romance with Wildlife on Film*. Cambridge, Mass.: Harvard University Press.

Montgomery, Ruth Ann. n.d. History of Union Township. http://www.evansvillehistory.net/files/Union_Township_History.html.

Mooij, W. M., and D. DeAngelis. 1999. Error propagation in spatially explicit models: A reassessment. *Conservation Biology* 13:930–33.

Moorhead, Alan. 1969. *Darwin and the Beagle*. New York: Harper and Row.

Musick, B., and H. D. Grover. 1990. Image textural measures as indices of landscape pattern. In M. G. Turner and R. H. Gardner, eds., *Quantitative Methods in Landscape Ecology*. New York: Springer-Verlag.

Needham, J. 1988. The limits of analysis. *Poetry Nation Review* 14(6):35–38.

Neilson, M., S. L'Ialien, V. Glumac, D. Williams, and P. Bertram. 1995. State of the lakes conference paper. Nutrients: Trends and system response. U.S. Environmental Protection Agency, EPA 905-R-05=95–015.

Neilson, R. P. 1986. High-resolution climatic analysis and southwest biogeography. *Science* 232:27–34.

———. 1987a. Biotic regionalization and climatic controls in western North America. *Vegetatio* 70:135–47.

———. 1987b. On the interface between current ecological studies and the paleobotany of pinyon-juniper woodlands. In proceedings, Pinyon-Juniper Conference, USDA Forest Service, General Technical Report INT-215.

Neilson, R. P., and L. H. Wullstein. 1983. Biogeography of two American oaks in relation to atmospheric dynamics. *J. Biogeogr.* 10:275–97.

Nichols, G. E. 1923. A working basis for ecological basis of plant communities. *Ecology* 4:11–23.

———. 1929. Plant associations and their classification. *Proceedings of the International Congress of Plant Sciences*, Ithaca, New York, August 16–23 1926, vol. 1, 624–41. (published 1929. Banta, Menasha, Wis.)

Nicolis, G., and I. Prigogine. 1989. *Exploring Complexity: An Introduction.* New York: W. H. Freeman.

Nielsen, J. 2006. *Condor: To the Brink and Back: the Life and Times of One Giant.* New York: Harper Perennial.

NOAA Photo Library. 1957. NMFSPlate 216. NOAA Central Library Call Number: SH91. R9 1957.). http://www.photolib.noaa.gov/brs/fsind78.htmO'Donoghue, M., S. Boutin, J. Krebs, Z. Gustavo, D. L. Murray, and E. J. Hofer. 1998. Functional responses of coyotes and lynx to the snowshoe hare cycle. *Ecology* 79:1193–1208.

Odum, E. P. 1969. The strategy of ecosystem development. *Science* 164:325–49.

Odum, H. T. 1995. Self-organization and maximum empower. In C. A. S. Hall, ed., *Maximum Power: The Ideas and Applications of H. T. Odum.* Boulder: University Press of Colorado.

Odum, H. T., and M. T. Brown. 2007. *Environment, Power and Society for the Twenty-First Century: The Hierarchy of Energy.* New York: Columbia University Press.

Odum, H. T., and R. C. Pinkerton. 1955. Time's speed regulator: The optimum efficiency for maximum output in physical and biological systems. *Am. Sci.* 43:331–43.

O'Neill, R. V., B. S. Ausmus, D. R. Jackson, R. van Hook, P. van Voris, C. Washburne, and A. P. Watson. 1977. Monitoring terrestrial ecosystems by analysis of nutrient export. *Water, Air and Soil Pollution* 8:271–77.

O'Neill, R. V., D. L. DeAngelis, J. B. Waide, and T. F. H. Allen. 1986. A hierarchical concept of ecosystems. *Monographs in Population Biology* 23:1–272.

O'Neill, R. V., and R. H. Gardner. 1990. Pattern, process, and predictability: The use of neutral models for landscape analysis. In M. G. Turner and R. H. Gardner, eds., *Quantitative Methods in Landscape Ecology. The Analysis and Interpretation of Landscape Heterogeneity.* Ecological Studies Series. New York: Springer-Verlag.

O'Neill R. V., W. F. Harris, B. S. Ausmus, and D. E. Reichle. 1975. A theoretical basis for ecosystem analysis with particular reference to element recycling. In F. G. Howell, J. B. Gentry, and M. A. Smith, eds., *Mineral Cycling in South Eastern Ecosystems.* Department of Energy Symposium Series, Conf. 740513. Oak Ridge, Tenn.: Oak Ridge National Laboratory, Technical Information Center.

O' Neill, R. V., A. R. Johnson, and A. W. King. 1989. A hierarchical framework for the analysis of scale. *Landscape Ecology* 3:193–205.

O'Neill, R. V., J. R. Krummel, R. H. Gardner, G. Sugihara, B. Jackson, B. T. Milne, M. G. Turner, B. Zymunt, S. W. Christensen, V. H. Dale, and R. L. Graham. 1988. Indices of landscape pattern. *Landscape Ecology* 1:153–62.

O'Neill, R. V., B. T. Milne, M. G. Turner, and R. H. Gardner. 1988. Resource utilization scales and landscape pattern. *Landscape Ecology* 2:63–69.

Orloci, L. 1966. Geometric models in ecology. I. The theory and application of some ordination methods. *J. Ecol.* 54:193–215.

Paine, R. T. 1966. Food web complexity and species diversity. *Am. Nat.* 100:65–75.

Paine, R. T., and S. Levin. 1981. Intertidal landscapes: Disturbance and the dynamics of pattern. *Ecological Monographs* 51:145–78.

Park, T. 1948. Experimental studies of interspecies competition. I. Competition between populations of flour beetles: Tribolium confusum duvall and tribolium castaneum Herbst. *Ecol Mono.* 18:267–307.

Parton, W. J., D. S. Schimel, C. V. Cole, and D. S. Ojima. 1987. Analysis of factors controlling soil organic matter levels in the Great Plains grasslands. *Soil Sci. Soc. Am. J.* 51:1173–79.

Patrick, R., and M. H. Hohn. 1956. The diatometer: A method for indicating the conditions of aquatic life. *Amer. Petroleum Inst.*, Proc. III, Refining 36(3):332–39.

Patrick, R., M. H. Hohn, and J. H. Wallace. 1954. A new method for determining the patterns of the diatom flora. *Notulae Naturae of the Academy of Natural Sciences of Philadelphia* 256:1–12.

Pattee, H. H. 1972. The evolution of self-simplifying systems. In E. Lazlo, ed., *The Relevance of General Systems Theory.* New York: Braziller.

———. 1978. The complementarity principle in biological and social structures. *J. Soc. Biol. Structures* 1:191–200.

———. 1979a. The complementarity principle and the origin of macromolecular information. *Biosystems* 11:217–26.

———. 1979b. Complementarity vs. reduction as an explanation of biological complexity. *Am. J. Physiol.* 236(5):12241–46.

Patten, B., and G. T. Auble. 1980. Systems approach to the concept of niche. *Synthese* 43:155–81.

Pausanias. 1913. *Pausanias's Description of Greece.* Translated with a commentary by J. G. Frazer, 6 vols. London: Macmillan.

Perrin, P. M., and F. J. G. Mitchell. 2013. European effects of shade on growth, biomass allocation and leaf morphology in European yew (*Taxus baccata* L.). *Journal of Forest Research* 132:211–18.

Peters, R. 1983. *The implications of body size.* New York: Cambridge University Press.

Petroski, H. 1993. Predicting disaster. *American Scientist* 81:110–13.

Pfeiffer, T., and R. Hoffmann. 2009. Large-scale assessment of the effect of popularity on the reliability of research. *PLoS ONE* 4(6): e5996. doi:10.1371/journal.pone.0005996.

Platt, T., and K. L. Denman. 1975. Spectral analysis in ecology. *Ann. Rev. Ecol. Syst.* 6:189–210.

Polimeni, J., K. Mayumi, M. Giampietro, and B. Alcott. 2008. *Jevons' Paradox: the Myth of Resource Efficiency Improvements.* London: Earthscan, Research Edition.

Polo, V., and L. M. Carrascal. 1999. Shaping the body mass distribution of Passeriformes: habitat use and body mass are evolutionarily and ecologically related. *Journal of Animal Ecology*, 68: 324–337. doi: 10.1046/j.1365-2656.1999.00282.x.

Porter, W. 2013. Connecting art and science to determine climate change effects on sea turtle nesting and oceanic distributions. Botany Department Colloquium, University of Wisconsin, Madison, September 9.

Porter, W., and D. M. Gates. 1969. Thermodynamic equilibria of animals with environment. *Ecological Monographs* 39:227–244.

Potter, K. M., W. S. Dvorak, B. S. Crane, V. D. Hipkins, R. M. Jetton, W. A. Whittier, and R. Rhea. 2008. Allozyme variation and recent evolutionary history of eastern hemlock (*Tsuga canadensis*) in the southeastern United States. *New Forests* 35:131–45. doi:10.1007/s11056-007-9067-2.

Pound, Roscoe. 1896. The plant-geography of Germany. *Am. Nat.* 30:465–68.

Pound, Roscoe, and Frederick E. Clements. 1898. A method of determining the abundance of secondary species. *Minn. Bot. Studies*, 2d series, Part 1:19–24.

———. 1900. *The Phytogeography of Nebraska*. 2d ed. Lincoln, Neb.: Seminar.

Prahalad, C. K. 2004. *Fortune at the Bottom of the Pyramid: Eradicating Poverty Through Profits*. Upper Saddle River, N.J.: Prentice Hall.

Prasad, P., A. Vandross, C. Toomey, M. Cheung, J. Rho, S. Quinn, S. J. Chacko, D. Borkar, V. Gall, S. Selvaraj, N. Ho, and A. Cifu. 2013. A decade of reversal: An analysis of 146 contradicted medical practices. *Mayo Clin. Proc.*;88(8):790–798. http://dx.doi.org/10.1016/j.mayocp.2013.05.012 www.mayoclinicproceedings.org.

Prigogine, I. 1978. Time, structure, and fluctuations. *Science* 201:777–85.

———. 1982. Order out of chaos. In W. J. Mitsch, R. K. Ragade, R. W. Bosserman, and J. A. Dillon Jr., eds., *Energetics and Systems*. Ann Arbor, Mich.: Ann Arbor Science.

Prigogine, I., R. Lefever, A. Goldbeter, and M. Herschkowitz-Kaufman. 1969. Symmetry breaking instabilities in biological systems. *Nature* 223:913–16.

Prigogine, I., and G. Nicolis. 1971. Biological order, structure, and instabilities. *Quart. Rev. Biophys.* 4:107–48.

Prince, S. D. 1982. The southern boundary of the mopane tree *Colophospermum mopane* in Botswana. *Brit. Ecol. Soc. Bull.* 13:187.

Prose, D. V., and H. G. Wilshire. 2000. The lasting effects of tank maneuvers on desert soils and intershrub flora: A product of the recoverability and vulnerability of desert ecosystems project. U.S. Geological Survey, Open-File Report OF 00–512.

Puth, L., and T. F. H. Allen. 2005. Potential corridors for the rusty crayfish, *Orconectes rusticus*, in northern Wisconsin lakes: Lessons for exotic invasions. *Landscape Ecology* 20(5):567–77.

Rappaport, Roy A. 1968. *Pigs for the Ancestors: Ritual in the Ecology of a New Guinea People*. New Haven, Conn.: Yale University Press.

———. 1972. The flow of energy in an agricultural society. *Sci Am.* 225:116–132.

Rasmussen, K., H. J. Jeppesen, and P. Arlien-Søborg. 1988. Psycho-organic syndrome from exposure to fluorocarbon 113: An occupational disease? *Eur. Neurol.* 28:205–7.

Ravetz, J. 2006. When communication fails. In A. G Pereira, S. Guedes Vaz, and S. Tognetti, eds., *Interfaces between Science and Society*, 16–34 Sheffield, U.K.: Greenleaf.

Regnault, D., N. Shanbanel, and E. Thierry. 2010. On analysis of "simple" 2D stochastic cellular automata. *Discrete Mathematics and Theoretical Computer Science* 12:(2)263–94.

Reichle, D. E., and D. A. Crossley. 1965. Radiocesium dispersion in a cryptozoan food web. *Health Physics* 11:1375–84.

Reimchen, T. 2001. Salmon nutrients, nitrogen isotopes and coastal forests. *Ecoforestry* (Fall), 13–16. Quarterly journal of the Ecoforestry Institute, Victoria, British Columbia; http://www.ecoforestry.ca.

Reiners, W. A., and J. A. Lockwood. 2009. Philosophical foundations for the practices of ecology. Cambridge: Cambridge University Press.

Rejmánek, M., C. E. Sasser, and J. G. Gosselink. 1987. Modeling of vegetation dynamics in the Mississippi River deltaic plain. *Vegetatio* 69:133–40.

Resnik, D. B. 2011. A troubled tradition: It's time to rebuild trust among authors, editors and peer reviewers. *Am. Sci.* 99(1):24–27.

Reynolds, Richard T., Andrew J. Sanchez Meador, James A. Youtz, Tessa Nicolet, S. Matonis, Patrick L. Jackson, Donald G. DeLorenzo, and Andrew D. Graves. 2013. Restoring composition and structure in southwestern frequent fire forests: A science-based framework for improving ecosystem resiliency. *Gen. Tech. Rep. RMRS-GTR-310*, p. 76. Fort Collins, Colo.: USDA Forest Service, Rocky Mountain Research Station.

Ricotta, C. 2005. Through the jungle of biological diversity. *Acta Biotheoretica* 53:29–38.

Roberts, David W. 1984. Forest vegetation and site relations: Theory, methods, and application to the forests of Montana. PhD dissertation, University of Wisconsin, Madison.

———. 1986. Ordination on the basis of fuzzy set theory. *Vegetatio* 66:123–31.

———. 1987a. An anticommutative difference operator for fuzzy sets and relations. *Fuzzy Sets and Systems* 21:35–42.

———. 1987b. A dynamical systems perspective in vegetation theory. *Vegetatio* 69:27–33.

———. 1989. Analysis of forest succession with fuzzy graph theory. *Ecological Modelling* 45:261–74.

Rosen, Robert. 1962. The derivation of D'Arcy Thompson's theory of transformations from the theory of optimal design. *Bull. Math. Biophys.* 40:549–79.

———. 1979. Anticipatory systems in retrospect and prospect. *General Systems* 24:11–23.

———. 1981. The challenges of systems theory. *General Systems Bulletin* 11:2–5.

———. 1989. Similitude, similarity, and scaling. *Landscape Ecology* 3:207–16.

———. 1991. *Life Itself.* New York: Columbia University Press.

———. 2000. *Life Itself: Essays.* New York: Columbia University Press.

———. 2012. Anticipatory Systems: Philosophical, Mathematical and Methodological Foundations. 2d ed. New York: Springer.

Rosling, H. 2014. www.ted.com/speakers/hans_rosling.

Roughgarden, J., S. D. Gaines, and S. Pacala. 1987. Supply side ecology: The role of physical transport processes. In P. Giller and J. Gee, eds., *Organization of Communities: Past and Present.* London: Blackwell.

Roughgarden, J., and S. Pacala. 1989. Taxon cycling among Anolis lizards populations: Review of the evidence. In D. Otte and J. Endler, eds., *Speciation and Its Consequences.* Sunderland, Mass.: Sinauer.

Rowe, L., R. Repasky, and R. Palmer. 1997. Size-dependent asymmetry: fluctuating asymmetry versus antisymmetry and its relevance to condition-dependent signaling. *Evolution* 51: 1401–1408.

Ruckleshouse, M., C. Hartway, and P. Karieva. 1997. Assessing data requirements of spatially explicit dispersal models. *Conservation Biology* 11:1298–1306.

Saatchi, S., N. L. Harris, S. Brown, M. Lefsky, E. T. A. Mitchard, W. Salas, B. R. Zutta, W. Buermann, S. L. Lewis, S. Hagen, S. Petrova, L. White, M. Silman, and A. Morel. 2011. Benchmark map of forest carbon stocks in tropical regions across three continents. *PNAS* 108:9899–904. doi:10.1073/pnas. 1019576108.

Sala, O. E., W. J. Parton, L. A. Joyce, and W. K. Laurenroth. 1988. Primary production of the central grassland region of the United States. *Ecology* 69:40–45.

Salisbury, E. J. 1932. The East Anglian flora. *Trans. Norfolk Norwich Nat. Soc.* 13:191–263.

———. 1964. *Weeds and Aliens.* 2d ed. London: Collins.

Salthe, S. 1985. *Evolving Hierarchical Systems.* New York: Columbia University Press.

Schimper, A. F. W. 1898. *Pflanzengeographie auf physiologischer Grundlage.* Jena: G. Fischer. (*Plant Geography upon a Physiological Basis.* Translated by W. R. Fisher, edited and revised by P. Groom and I. B. Balfour. Oxford: Clarendon, 1903.)

Schmidt-Nielsen, K. 1984. Scaling: *Why Is Animal Size So Important?* Cambridge: Cambridge University Press.

Schneider, E., and J. Kay. 1994. Life as a manifestation of the second law of thermodynamics.. *Jour. of Adv. in Math. and Comp.* 19:25–48.

Schoff, Gretchen H. 1991. *Reflections, the Story of Cranes.* Baraboo, Wis.: International Crane Foundation.

Schooler, J. W. 2011. Unpublished results hide the decline effect. *Nature.* 2011 Feb 24;470(7335):437. doi: 10.1038/470437a.

Schooler, J. W., and Engstler-Schooler T. Y. 1990. Verbal overshadowing of visual memories: some things are better left unsaid. *Cogn Psychol.* 1990 22:36–7.

Schrodinger, Erwin. 1959. *Mind and Matter.* Cambridge: Cambridge University Press.

Schumpeter, J. A. 1950. *Capitalism, Socialism and Democracy.* New York: Harper and Rowe.

Scott, J. C. 1998. *Seeing Like a State: How Certain Schemes to Improve the Human Condition Have Failed.* New Haven, Conn.: Yale University Press.

Seneff, S. 2014. How statins really work explains why they don't really work; http://people .csail.mit.edu/seneff/why_statins_dont_really_work.html.

Seneff, S., A. Lauritzen, R. Davidson, and L. Lentz-Marino. 2012. Is endothelial nitric oxide synthase a moonlighting protein whose day job is cholesterol sulfate synthesis? Implications for cholesterol transport, diabetes and cardiovascular disease. *Entropy* 14: 2492–530. doi:10.3390/e14122492.

Seneff, S., G. Wainwright, and B. Hammarskjold. n.d. Atherosclerosis may play a pivotal role in protecting the myocardium in a vulnerable situation. *Hypotheses in the Life Sciences.*

———. n.d. Cholesterol sulfate supports glucose and oxygen transport into erythrocytes and myocytes: A novel evidence based theory. *Hypotheses in the Life Sciences.*

Seneff, S., G. Wainwright, and L. Mascitelli. 2011a. Is the metabolic syndrome caused by a high fructose, and relatively low fat, low cholesterol diet? *Archives of Medical Science* 7(1): 8–20. doi:10.5114/aoms.2011.20598.

———. 2011b. Nutrition and Alzheimer's disease: The detrimental role of a high carbohydrate diet. *European Journal of Internal Medicine.*22:134–40.

Shelford, V. E. 1918. Conditions of existence. In H. B. Ward and G. C. Whipple, eds., *Freshwater Biology,* 21–60.New York: Wiley.

Shepard, M. 2013. *Restoration Agriculture: Real World Permaculture for Farmers.* Austin, Tex.: Acres U.S.A.

Shugart, H. H. 2004. *How the Earthquake Bird Got Its Name, and Other Tales of an Unbalanced Nature.* New Haven, Conn.: Yale University Press.

Shugart, H. H., and D. C. West. 1977. Development of an Appalachian forest succession model and its application to assessment of the impact of the chestnut blight. *Journal of Environmental Management* 5:161–79.

Shugart, H. H., D. C. West, and W. R. Emanuel. 1981. Patterns and dynamics of forests: An application of simulation models. In D. C. West, H. H. Shugart, and D. B. Botkin, eds., *Forest Succession: Concepts and Application.* New York: Springer-Verlag.

Sibley, P. 1977. Predicting structural failures. PhD thesis, University of London.

Sibley, P., and A. C. Walker. 1977. Structural accidents and their causes. *Proceedings of the Institute of Civil Engineers* 62(1):191–208.

Simberloff, D. 1983. Competition theory, hypothesis-testing and other ecological buzzwords. *Amer. Nat.* 122:626–35.

Simberloff, D. A., J. A. Farr, J. Cox, and D. W. Mehlman. 1992. Movement corridors: Conservation bargains or poor investments? *Conservation Biology* 6:493–504.

Simberloff, D., and E. O. Wilson. 1969. Experimental zoography of islands: The colonization of empty islands. *Ecology* 40:23–47.

Simon, H. 1996. *The Sciences of the Artificial.* 3d ed. Cambridge, Mass.: MIT Press.

———. 2000. Can there be a science of complex systems? In *Proceedings from the International Conference on Complex Systems on Unifying Themes in Complex Systems,* 3–14. Cambridge, Mass.: Perseus Books.

Smil, V. 2005. Review of *Collapse* by Jared Diamond. *International Journal* 60:886–89.

———. 2008. *Global Catastrophes and Trends: The Next Fifty Years.* Cambridge, Mass.: MIT Press.

Smith, R. 1898. Plant associations in the Tay Basin. *Trans. Proc. Perthsh. Soc. Nat. Set.* 2:200–217.

———. 1899. On the study of plant associations. *Nat. Sci.* 14:109–20.

Snowden, D. J. 2003. Narrative Patterns: the perils and possibilities of using story in organisations. Edited for Oxford University Press. http://cognitive-edge.com/uploads/articles/41_narrative_patterns_-_perils_and_possibilities_final.pdf. Original version in *Knowledge Management* 4:10 2001. First edited for Oxford University Press in 2003. Edited 2004.

———. 2010. Naturalizing sensemaking. In K. L. Mosier and U. M. Fischer, eds., *Informed by Knowledge: Expert Performance in Complex Situations,* 223–34. New York: Psychology Press.

Snowden, D. J., and M. Boone. 2007. A leader's framework for decision making. *Harvard Business Review,* November, 69–76.

Southwood, T. R. E. 1976. Bionmomic strategies and population strategies. In R. May, ed., *Theoretical Ecology: Principals and Applications,* 26–48. Philadelphia: Saunders.

Spehn, E. M., et al. 2005. Ecosystem effects of biodiversity manipulations in European grasslands. *Ecological Monographs* 75(1):37–63.

Stahl, W. 1967. Scaling of respiratory variables in mammals. *J. Appl. Physiol.* 22: 435–60.

Sterner, R. W. 2012. Raymond Laurel Lindeman and the trophic dynamic viewpoint. *Limnology and Oceanography Bulletin* 21(2):38–51.

Stevens, P. 1974. *Patterns in Nature.* Boston, Mass.: Little Brown.

Stockner, J. G. 1981. Whole lake fertilization for the enhancement of sockeye salmon *Onchrhynchus nerka* in British Columbia, Canada. *Verh. Int. Ver. Limnol.* 21:293–99.

Strøm, K. W. 1928. Recent advances in limnology. *Proceedings of the Linnean Society of London* 2:96–110.

Sugihara, G., and R. M. May, 1990. Nonlinear forecasting as a way of distinguishing chaos from measurement error in time series. *Nature* 344:734–741.

Sukachev, V., and N. Dylis. 1964. *Fundamentals of Forest Biogeocoenology*. Translated by J. M. McLennan. London: Oliver and Boyd.

Summers, Peter W. 1989. The atmospheric region of influence for Kejimkujik, Nova Scotia. In *Proceedings of the International Air Quality Board of the International Joint Commission First Regional Workshop on Integrated Monitoring*, St. Andrews, New Brunswick, May 31–June 2, 1988. Washington, D.C.: International Joint Commission.

Sutherland. W. J. 2006. Predicting the ecological consequences of environmental change: A review of the methods. *Journal of Applied Ecology* 43:599–616.

Swetnam, T. W., and A. M. Lynch. 1993. Multi-century, regional-scale patterns of western spruce budworm history. *Ecological Monographs* 63(4):399–424.

Tainter, J. A. 1988. *The Collapse of Complex Societies*. Cambridge: Cambridge University Press.

———. 2008. Collapse, sustainability, and the environment: How authors choose to fail or succeed. *Reviews in Anthropology* 37:342–71.

Tainter, J. A., and T. F. H. Allen. 2014. Energy gain in historical anthropology: Ants, empires, the evolution of organization. In C. Isendahl and D. Stump, eds., *The Oxford Handbook of Historical Ecology and Applied Archeology*. Oxford: Oxford University Press.

Tainter J. A., and G. Lucas. 1983. The epistemology of the significance concept. *American Antiquity* 48:707–19.

Tansky, M. 1978. Stability of multispecies predator-prey systems. *Memoirs Coll. Sci. Univ. Kyoto, Ser. B*. 7(2):87–94.

Tansley, A. 1926. Succession: The concept and its value. In *Proceedings of the International Congress of Plant Sciences*, vol. 1, 677–86. Ithaca, New York, August 16–23. (Published in 1929, Menasha, Wis.: George Banta.)

———. 1935. The use and abuse of vegetational terms and concepts. *Ecology* 16(3):284–307.

Taylor, C. R. 1977. The energetics of terrestrial locomotion and body size in vertebrates. In T. J. Pedley. ed. *Scale Effects in Animal Locomotion* 127–41. New York: Academic Press.

Teal, J. M. 1962. Energy flow in the salt marsh ecosystems of Georgia. *Ecology* 43:614–24.

Therres, G. D. 1993. Integrating management of forest interior migratory birds with game in the Northeast. Pages 402–407 in D. Finch and P. W. Stangle (eds.). Status and management of neotropical migratory birds. *General Technical Report RM 229*. Fort Collins, Colo.: USDA Forest Service.

Thomas, L. 1975. *Lives of a Cell*. New York: Bantam.

Thomforde, S., and P. Allen. 2010. Bison, buffalo-fish and the canvas back duck: Understanding the prairie lakes from an ecosystem perspective. In *Restoring a National Treasure*, Cultural Prairie section of 22nd North American Prairie Conference, Northern Iowa University, August 1–5, p. 75; http://www.northamericanprairieconference.org/pdf/22nd_NAPC_Program+Abstracts.pdf.

Thompson, D'Arcy Wentworth. 1942. *On Growth and Form*. Vols. 1–2. Cambridge: Cambridge University Press.

Tilman, D. 1988. *Plant Strategies and the Structure and Dynamics of Plant Communities*. Princeton, N.J.: Princeton University Press.

Tilman, D., J. Knops, D. Wedin, P. Reich, M. Ritchie, and E. Siemann. 1997. The influence of functional diversity and composition on ecosystem processes. *Science* 277:1300–1302.

Tilman, D., R. M. May, C. L. Lehman, and M. A. Nowak. 1994. Habitat destruction and the extinction debt. *Nature* 371:65–66.

Transeau, E. N. 1926. The accumulation of energy by plants. *Ohio J. Sci.* 26:1–10.

Turner, M. G., and C. L. Ruscher. 1988. Changes in landscape patterns in Georgia, USA. *Landscape Ecology* 1(4) 241–251.

Turner, M. G., R. H. Gardner, V. H. Dale, and R. V. O'Neill. 1989. Predicting the spread of disturbance across heterogeneous landscapes. *Oikos* 55: 121–129.

Uexkull, J. von. 1920. *Umwelt und Imnenwelt der Tiere.* 2d ed. Berlin: Springer.

———. 1929. *Theoretische Biologic.* 2d ed. Berlin: Springer.

———. 1957. A stroll through the worlds of animals and men: A picture book of invisible worlds. In Clair H. Schiller, ed. and trans., *Instinctive Behavior.* New York: International Universities Press.

Uexkull, J. von, and G. Kriszat. 1934. *Streifuge durch die Umwelten von Tieren und Menschen.* Berlin: Springer.

Unger, F. 1836. *Ueber den Einfluss des Bodens auf die Vertheilung der Gewashse, nachgewiesen in der Vegetation des nordostlichen Tirols.* Wein: Rohrman und Schweigerd.

Urban, D. L., H. H. Shugart, D. L. DeAngelis, and R. V. O'Neill. 1988. Forest bird demography in a landscape mosaic. ORNL/TM-10332, ESD Publ. No. 2853. Environmental Sciences Division, Oak Ridge National Laboratory, Oak Ridge, Tennessee.

Van Voris, P., R. V. O'Neill, W. R. Emanuel, and H. H. Shugart. 1980. Functional complexity and functional stability. *Ecology.* 61:1352–60.

Vogel, S. 1988. *Life's Devices: The Physical World of Animals and Plants.* Princeton, N.J.: Princeton University Press.

Vogt, R. G., and L. M. Riddifo. 1981. Pheromone binding and inactivation by moth antenna. *Nature* 293:161–63. doi:10.1038/293161a0.

Volk, T. 2002. The humongous fungus: Ten years later. *Innoculum* 53(2):4–8 (supplement to *Mycologia*).

Volterra, V. 1926. Fluctuations in abundance of a species considered mathematically. *Nature* 100:603–9.

Volz, K.G., L. J. Schooler, R. I. Schubotz, M. Raab, G. Gigerenzer, and D. Y. von Cramon. 2006. Why you think Milan is larger than Modena: Neural correlates of the recognition heuristic. *J. Cogn. Neurosci.* 18(11):1924–36.

Wainwright, S. A., W. D. Biggs, J. D. Currey, and J. M. Gosline. 1982. Mechanical design in organisms. Princeton, N.J.: Princeton University Press.

Wald, A. 1943. A method of estimating plane vulnerability based on damage of survivors. Statistical Research Group, Columbia University. CRC 432, reprint from July 1980. Center for Naval Analyses; http://cna.org/sites/default/files/research/0204320000.pdf.

Walters, Carl J., and C. S. Holling. 1990. Large-scale management experiments and learning by doing. *Ecology* 71:2060–68.

Warming, Eugenius. 1896. *Lehrbuch der okologischen Pflanzengeogmphie: Eine Ein-fuhrung in die Kenntnis der Pflanzenvereine.* Translated by E. Knoblauch. Berlin: Borntraeger. (*Oecology of Plants: An Introduction to the Study of Plant-Communities.* Translated and edited by P. Groom and I. B. Balfour. Oxford: Clarendon, 1909.)

Watson, V. J., O. L. Loucks, and W. Wojner. 1981. The impact of urbanization on seasonal hydrologic and nutrient budgets of a small North American watershed. *Hydrobiologia* 77:87–96.

Watt, A. S. 1926. Yew communities of the South Downs. *Journal of Ecology* 14:282–316.

———. 1947. Pattern and process in the plant community. *Journal of Ecology* 35:1–22.

Way, D. 1973. *Terrain Analysis: A Guide to Site Selection Using Aerial Photographic Interpretation.* Stroudsberg, PA: Dowden, Hutchinson and Ross.

Webb, D. A. 1954. Pattern and process in the plant community. *Journal of Ecology* 51:362–70.

Webster, K. E., T. Kratz, C. Bowser, J. Magnusson, and W. Rose. 1986. The influence of landscape position on lake chemical responses to drought in Northern Wisconsin. *Limnology and Oceanography* 41: 977–84.

Wedin, D. 1990. Nitrogen cycling and competition among grass species. PhD dissertation, University of Minnesota, Minneapolis.

Wedin, D., and D. Tilman. 1990. Species effects on nitrogen cycling: A test with perennial grasses. *Oecologia* 84:433–41.

Wegwarth, O., L. M. Schwartz, S. Woloshin, W. Gaissmaier, and G. Gigerenzer. 2012. Do physicians understand cancer screening statistics? A national survey of primary care physicians in the United States. *Annals of Internal Medicine* 156:340–49, W-92–W-94.

Weinberg, G. M. 1975. An introduction to general systems thinking. Chichester, U.K.: Wiley.

Weins, J. A., and B. Milne. 1989 Scaling of 'landscapes' in landscape ecology from a beetle's perspective. *Landscape Ecology* 3:87–96.

Wessman, C. A., J. D. Aber, D. L. Peterson, and J. M. Melillo. 1988. Remote sensing of canopy chemistry and nitrogen cycling in temperate forest ecosystems. *Nature* 335:154–56. doi:10.1038/335154a0.

West, R. G. 1970. Pleistocene history of the flora of Britain. In D. Walker and R. West, eds., *Studies in the Vegetational History of the British Isles*, 1–11. Cambridge: Cambridge University Press.

White, L., Jr. 1967. The historical roots of our ecological crisis. *Science* 155(3767):1203–7.

Whitehead, A. N., and B. Russell. *Principia Mathematica* 1910–1913. (1927 2d ed.; reprinted 1962. London: Cambridge at the University Press.

Whittaker, R. H. 1956. Vegetation of the Great Smoky Mountains. *Ecol. Monogr.* 26:1–80.

Wiens, J. A., and B. T. Milne. 1989. Scaling of "landscapes" in landscape ecology, or, landscape ecology from a beetle's perspective. *Landscape Ecology* 3:87–96.

Wixon, D. 2011. Synthesizing microbial community change and soil carbon vulnerability in warmed soils. PhD dissertation, University of Wisconsin, Madison. Proquest/UMI Publication No. 3486760.

Wixon, D., and T. Balser. 2009. Complexity, global climate change and soil carbon: A systems approach to microbial temperature response. *Systems Research and Behavioral Science* 26:601–20.

Worster, D. 1977. *Nature's Economy: The Roots of Ecology.* San Francisco: Sierra Club.

Wright, P., G. Alward, J. L. Colby, T. W. Hoekstra, B. Tegler, and M. Turner. 2002. Monitoring for forest management unit scale sustainability. *USDA IMI Report 4.* Fort Collins, Colo.: USDA Forest Service.

Wright, P., and P. Greengrass. 1987. *Spycatcher.* New York: Penguin Viking.

Yoon, C. K. 2005. Ernst Mayr, pioneer in tracing geography's role in the origin of species, dies at 100. *New York Times*, February 5.

Zadeh, L. 1965a. Fuzzy sets. *Information and Control* 8:338–53.

————. 1965b. Fuzzy sets and systems. In J. Fox, ed., *System Theory*, 29–39. Brooklyn, N.Y.: Polytechnic.

Zellmer, A. J., T. F. H. Allen, Kirsten Koesseboehmer. 2006. The nature of ecological complexity: A protocol for building the narrative. *Ecological Complexity* 3:171–82.

Zhu, Y., H. Chen, J. Fan, Y. Wang, Y. Li, J. Chen, J. Fan, S. Yang, L. Hu, H. Leung, T. W. Mew, S. P. Teng, Z. Wang, and C. C. Mundt. 2000. Genetic diversity and disease control in rice. *Nature* 406:718–22.

Zimmerer, K. S. 1998. The ecogeography of Andean potatoes. *Bioscience* 48:445–54.

INDEX